T0297172

CAMBRIDGE TRACTS IN MATHEMATICS

General Editors

B. BOLLOBAS, P. SARNAK, C.T.C. WALL

118 Sets of Multiples

Richard R. Hall
York University

Sets of Multiples

CAMBRIDGE
UNIVERSITY PRESS

CAMBRIDGE UNIVERSITY PRESS
Cambridge, New York, Melbourne, Madrid, Cape Town, Singapore, São Paulo, Delhi

Cambridge University Press
The Edinburgh Building, Cambridge CB2 8RU, UK

Published in the United States of America by Cambridge University Press, New York

www.cambridge.org
Information on this title: www.cambridge.org/9780521109925

© Cambridge University Press 1996

This publication is in copyright. Subject to statutory exception
and to the provisions of relevant collective licensing agreements,
no reproduction of any part may take place without the written
permission of Cambridge University Press.

First published 1996
This digitally printed version 2009

A catalogue record for this publication is available from the British Library

Library of Congress Cataloguing in Publication data

Hall, R. R. (Richard Roxby)
Sets of multiples / Richard R. Hall,
p. cm. – (Cambridge tracts in mathematics : 118)
Includes bibliographical references (p. –) and - index.
ISBN 0 521 40424 X (hc)
1. Sequences. I. Title.
QA246.5.H33 1996
512′.72 – dc20 95-39233 CIP

ISBN 978-0-521-40424-2 hardback
ISBN 978-0-521-10992-5 paperback

To the memory of my mother and father

Contents

Preface

In this book I describe some of the developments which have taken place in the theory of sets of multiples since Halberstam and Roth's *Sequences* was published in 1966. My object is twofold: to give a coherent account of the general theory as it exists today, and to encourage others to study an elegant and, perhaps to some persons, surprisingly subtle subject in which I believe much progress is possible. There are still many unsolved problems, some of them arising from the most recent work, and I have attempted to fit these necessarily loose ends into the text in such a way that the reader can see at which point a new idea is required. One of the attractions of the subject is the great variety of techniques which can be employed: thus one chapter (not the easiest) consists entirely of elementary inequalities, another involves Dirichlet series, contour integration and exponential sums, while a third is probabilistic. Where probabilistic methods have been used, I have presented them in an accessible fashion as a non-probabilist writing for (perhaps mostly) non-probabilists.

This tract is a companion volume to Cambridge Tract No. 90, *Divisors*, written with Gérald Tenenbaum some years ago. Although there are references to *Divisors* (I refer to this book by its name throughout) at several points, *Sets of Multiples* is self-contained and can be read by persons unfamiliar with this area.

I have quoted freely from joint papers with two collaborators: Paul Erdös and Gérald Tenenbaum. It is almost automatic that in any long-standing collaboration some of the results will be due, on different occasions, entirely to one or the other author. Without embarking on any sort of catalogue at this juncture I should like to make clear that many of the ideas in the book are due to my collaborators, particularly in Chapter 1 where I quote from Hall and Tenenbaum (1992) and Erdös,

Hall and Tenenbaum (1994). For example the fine, short proof of Erdös' criterion for Besicovitch sequences is Tenenbaum's. Where results were published individually by any author this will of course be clear from the references.

I should like to acknowledge my great debt to Paul Erdös who introduced me to this subject and from whom I have learned, or had the opportunity to learn, so much. I shall always be grateful to Gérald Tenenbaum for his collaboration – his results in this area speak for themselves. I wish to record my thanks to Heini Halberstam for his encouragement during 1990 to write this book, which was essentially planned during the British Mathematical Colloquium at the University of East Anglia that year. Finally I am grateful to David Tranah for his patience and good spirits: I undertake not to write another book for Cambridge University Press in longhand.

York

Introduction

The study of sets of multiples began in the thirties as an abstraction from one special problem. A number of mathematicians, including Behrend, Chowla, Davenport, Erdös and Schur, had been interested in abundant numbers, (the positive integers not greater than the sum of their proper divisors), in particular whether the proportion of such integers $\leq n$ converged to a limit with increasing n. This was proved by Davenport (1933), using an analytic method involving the Stieltjes moment problem due to Schoenberg (1928), which Schoenberg had applied to a similar problem about the Euler ϕ-function. A few months later Erdös (1934) gave an elementary proof of this theorem; general ideas which developed into the subject now called sets of multiples can be discerned clearly in both these proofs. We shall not be concerned with abundant numbers in this book, nevertheless it may be helpful to use this historical example as an illustration. We note the property that any multiple of an abundant number is abundant. This leads to the idea of a *primitive* abundant number, which is minimal in the sense that its proper divisors are not abundant. The abundant numbers then comprise all the multiples of these primitives. This immediately raises general questions about the sequence, or set, of integers which are multiples of the elements of a given base sequence, for example (as above) whether the former, top, sequence possesses asymptotic density. In general the answer is no (Besicovitch (1934)). We call a sequence whose set of multiples does possess asymptotic density a Besicovitch sequence (this terminology is due to the present writer and Tenenbaum). Erdös observed that a sufficient condition for this conclusion would be the convergence of the series of reciprocals of the base sequence (which he could establish in the case of abundant numbers). This led to much work on general primitive sequences (in which no element divides another), and to Erdös' criterion (1948a) for a sequence to be Besicovitch. Meanwhile Davenport and Erdös (1937) proved that every set of multiples possesses logarithmic density, and Heilbronn (1937) and Rohrbach (1937) independently obtained the inequality which bears their names. This is a precursor of Behrend's inequality (1948).

The complement of a set of multiples comprises the integers not divisible by any elements of the base sequence, and so there is a connection with the theory of sieves. Usually the emphasis in the two subjects is rather different: in sieve problems the sieving set consists simply of primes, but the sifted sequence will be the values of a polynomial, possibly at prime arguments. In sets of multiples the sifted sequence is nearly always $\mathbf{Z}^+$: the complications arise from the fact that the elements of the base sequence need not be coprime. Also, if this sequence is infinite, there is no analogue of a sieving limit. We regard the base sequence as fixed and we have to say what we can about the distribution of the set of multiples. One question which has arisen repeatedly in Erdös' work is whether the set of multiples has density 1, or equivalently whether almost all integers have at least one divisor of a particular type. To formalize this I have called a sequence whose set of multiples has density 1 a Behrend sequence. Quite general necessary conditions for a sequence to be Behrend are now available; the sufficiency conditions are still rather specialized. Another leitmotif in Erdös' work has been to refine the base sequence to a threshold at which the set of multiples possesses positive density determined by a statistical law: examples occur in Chapters 4 and 7. More recent avenues of research include oscillation (we prove Ω-theorems, i.e. that the oscillations are large), and derived sequences, which relate to the following question. What can we say about the density of the integers which are multiples of two, or k, elements of the base sequence, given only the density of the set of multiples itself. This part of the subject is dominated by elementary but delicate inequalities and there are new problems which may be difficult.

Each chapter of the book has an introduction and so I shall not attempt a resumé of the contents here. However I think it might be useful at this point to suggest possible orders in which the chapters could be read. Chapter 0 is essential, and most readers would follow this with Chapter 1, which contains basic information about Besicovitch and Behrend sequences. Some choices are now possible. Chapters 2 and 3 go together in this order. They comprise the material on derived sequences and oscillations mentioned above and are elementary and combinatorial in nature, certainly presenting a new face to the subject. Chapters 5 and 6 also combine in order, dealing with divisor density (an alternative to asymptotic/logarithmic density which is particularly appropriate to this subject) and divisor uniform distribution. The techniques involved here are mathematically more sophisticated but also familiar: many analytic number theorists would find these chapters the easiest place to start after Chapter 1. A general problem about multiplicative functions, which seems to me fairly fundamental, arises and is explained at the end of §5.2. The discrepancy lower bounds and double variance lemma in Chapter 6 are the most recent work in this Tract. Chapter 4, on probabilistic group theory, concerns an idea of Erdös for constructing Behrend sequences (among many other applications), and is important for a full understanding of other parts of the book, but could be read at any time. I suggest that some of the technical details might be skimmed at first reading. Some of the hardest problems in the book, of importance outside number theory, occur here; a brief introduction to these is given in §4.5. Finally Chapter 7 presents the solution of a conjecture from *Divisors* about Tenenbaum's function $H(x, y, z)$ which counts the set of multiples of an interval. This is a central example, but no other chapter depends directly upon it.

Notation

$\mathscr{A}$ a positive integer sequence, or a family of sets.
$\mathscr{M}(\mathscr{A})$ the set of multples of $\mathscr{A}$.
$\mathbf{d}\mathscr{M}$ the asymptotic density of $\mathscr{M}$.
$\overline{\mathbf{d}}, \underline{\mathbf{d}}$ upper, lower asymptotic density.
$\boldsymbol{\delta}$ logarithmic density.
$\tau(n)$ the number of (positive) divisors of n.
$\tau(n, \mathscr{A})$ the number of divisors of n which belong to $\mathscr{A}$.
$\tau(n, y, z)$ as above, with $\mathscr{A} = (y, z]$.
$\omega(n)$ the number of distinct prime factors of n.
$\omega(n, t)$ the number of distinct prime factors p of n such that $p \le t$.
$\omega(n, s, t)$ the number of distinct prime factors p of n such that $s < p \le t$.
$\Omega(n)$ the number of prime factors of n counted according to multiplicity.
$\Omega(n, t), \Omega(n, s, t)$ similar to ω, but the prime factors are counted according to multiplicity.
$P^+(n), P^-(n)$ the greatest, least prime factor of n.
$a \mid b$ a divides b.
$a^r \parallel b$ a^r divides b but a^{r+1} does not.
$\Phi(x, y) = \text{card}\,\{n : n \le x,\ P^-(n) > y\}$.
$\Psi(x, y) = \text{card}\,\{n : n \le x,\ P^+(n) \le y\}$.
Exp expectation.

The following is an index, by section numbers, of notation introduced in the text.

$\mathscr{P}(\mathscr{A})$	0.2
$\Delta(\mathscr{A})$	0.3
$\delta = .086071\ldots$	0.3
$M(x)$	0.3
$\mathscr{T}(\mathscr{A})$	0.4

$\mathbf{t}(\mathscr{A})$	0.4
$\mathscr{A}'(b)$	0.4
$E(f;M)$	0.4
$\{d(S)\}$	0.5
$v_p(a)$	0.5
$\Theta(x,y,z)$	1.2
$H(x,y,z)$	1.2
$\mathscr{A}^*(t)$	1.3
$Q(\lambda) = \lambda \log \lambda - \lambda + 1$	1.3
$\mathscr{A}^{(k)}$	2.1
$\mathbf{t}_k(\mathscr{A})$	2.1
π_k	2.2
ρ_k	2.2
$\varphi_k(\sigma)$	2.2
$\varphi_k^*(\sigma)$	2.2
$t_{k,n}$	2.2
$E(x,\mathscr{A}), \overline{E}(x,\mathscr{A})$	3.1
$\langle \mathscr{A}, \mathscr{A} \rangle$	3.1
$\rho(x) = [x] - x + \frac{1}{2}$	3.1
$G_0(z)$	4.1
$\mathbf{D}\mathscr{A}$	5.1
$\Delta(n;f)$	6.1

0

First ideas

0.1 Introduction

In this chapter we state and prove some of the classical results about sets of multiples and we make some definitions which have arisen in the more recent theory. This is presented in the later chapters.

0.2 Sets of multiples and primitive sequences

Let $\mathscr{A} = \{a_1, a_2, a_3, \ldots\}$ be a non-decreasing sequence of positive integers. It may be finite or not, and we allow repetitions. We do not require that $a_1 > 1$ but this will usually be the case, when we shall refer to $\mathscr{A}$ as being *non-trivial*. The *set of multiples* of $\mathscr{A}$, which we denote by $\mathscr{M}(\mathscr{A})$, comprises all the distinct, positive multiples of elements of $\mathscr{A}$. Thus if we define

$$\tau(n, \mathscr{A}) = \text{card}\{a : a \mid n, a \in \mathscr{A}\} \tag{0.1}$$

then we have

$$\mathscr{M}(\mathscr{A}) = \{n : n \in \mathbf{Z}^+, \tau(n, \mathscr{A}) \geq 1\}. \tag{0.2}$$

It is traditional to refer to $\mathscr{M}(\mathscr{A})$ as a set of multiples, but for the most part it will be better to regard it as an ordered sequence.

We may have $\mathscr{A}_1 \neq \mathscr{A}_2$ but $\mathscr{M}(\mathscr{A}_1) = \mathscr{M}(\mathscr{A}_2)$. We begin with the observation that in this circumstance we also have

$$\mathscr{M}(\mathscr{A}_1 \cap \mathscr{A}_2) = \mathscr{M}(\mathscr{A}_1). \tag{0.3}$$

Let us prove this by contradiction. Suppose (0.3) is false. Plainly $\mathscr{M}(\mathscr{A}_1 \cap \mathscr{A}_2) \subseteq \mathscr{M}(\mathscr{A}_1)$ so there exists an integer $b \in \mathscr{M}(\mathscr{A}_1) \backslash \mathscr{M}(\mathscr{A}_1 \cap A_2)$: we may suppose it is the least such. Then $b = m_1 a_1 = m_2 a_2$ where $a_i \in \mathscr{A}_i$ for $i = 1, 2$ and $m_i \in \mathbf{Z}^+$. Since $b \notin \mathscr{M}(\mathscr{A}_1 \cap \mathscr{A}_2)$ we have $a_1 \neq a_2$. Suppose

$a_1 < a_2$. Then $m_1 > 1$ and $a_1 < b$. Since $a_1 \in \mathscr{M}(\mathscr{A}_1)$ this implies, by the minimal property of b, that $a_1 \in \mathscr{M}(\mathscr{A}_1 \cap \mathscr{A}_2)$, i.e. $a_1 = m_3 a_3$ where $a_3 \in \mathscr{A}_1 \cap \mathscr{A}_2$, $m_3 \in \mathbf{Z}^+$. We now have $b = (m_1 m_3) a_3 \in \mathscr{M}(\mathscr{A}_1 \cap \mathscr{A}_2)$, which is the required contradiction. This proves (0.3).

Let $\mathscr{P}(\mathscr{A})$ denote the intersection of all the sequences $\mathscr{A}'$ for which $\mathscr{M}(\mathscr{A}') = \mathscr{M}(\mathscr{A})$. Then

$$\mathscr{M}(\mathscr{P}(\mathscr{A})) = \mathscr{M}(\mathscr{A}), \tag{0.4}$$

moreover $\mathscr{P}(\mathscr{A})$ is a *primitive sequence,* that is a sequence of positive integers none of which divides any other. There is a substantial literature on primitive sequences, quite independent of the theory of sets of multiples, and we record some of the properties of these sequences here. Let $\mathscr{P}$ be primitive, with counting function

$$P(x) = \operatorname{card}\{\mathscr{P} \cap (0, x]\}. \tag{0.5}$$

We have

$$\sup_{\mathscr{P}} P(x) = \left[\frac{x+1}{2}\right] \tag{0.6}$$

on the one hand because $\mathscr{P} = \mathbf{Z} \cap (\frac{1}{2}x, x]$ is an example and on the other because for each odd integer $m \leq x$, $P(x)$ counts at most one integer n of the form $2^k m$, $(k = 0, 1, 2, \ldots)$. Erdös (1935b) showed that always

$$\sum_{a \in \mathscr{P}} \frac{1}{a \log a} < \infty, \qquad \mathscr{P} \neq \{1\}, \tag{0.7}$$

and in the same volume of J. London Math. Soc. Behrend proved that

$$\sum_{a \in \mathscr{P}} \left\{\frac{1}{a} : a \leq x\right\} \ll \frac{\log x}{\sqrt{(\log \log x)}}. \tag{0.8}$$

Neither of these results implies the other. Pillai (1939) gave an example to show that (0.8) is qualitatively best possible, and the question of the sharp constant on the right proved to be a resistant problem, still unsolved when Halberstam and Roth (1966) appeared. Erdös, Sarközy and Szemerédi (1967a,b) obtained essentially final results for this problem. We have

$$\sup_{\mathscr{P}} \sum_{a \in \mathscr{P}} \left\{\frac{1}{a} : a \leq x\right\} \leq (1 + o(1)) \frac{\log x}{\sqrt{(2\pi \log \log x)}} \tag{0.9}$$

but for each fixed $\mathscr{P}$, as $x \to \infty$,

$$\sum_{a \in \mathscr{P}} \left\{ \frac{1}{a} : a \leq x \right\} = o\left(\frac{\log x}{\sqrt{(\log \log x)}} \right). \tag{0.10}$$

Each of (0.9) and (0.10) is best possible.

In view of (0.4) we may, for many purposes assume that $\mathscr{A}$ is primitive, indeed this will be essential in Chapter 3, §3.4. However, in both Chapters 2 and 3 when we deal with *derived sequences,* (these are defined in §2.1) we shall find that such a reduction is inappropriate: to obtain the most general derived sequences we even allow $\mathscr{A}$ to contain repeated elements.

A review of the literature concerning sets of multiples reveals that primitive sequences have not played as large a part as may have been expected. The results above demonstrate that a primitive sequence need not be at all thin (certainly not enough to be useful), moreover the structural constraint $a_i \nmid a_j$ is quite awkward to handle.

0.3 Densities

Various definitions of the density of an integer sequence are possible. We shall be concerned with asymptotic, logarithmic and divisor densities, and we consider the first two of these here. divisor density is the subject of Chapter 5.

Let $\mathscr{K}$ be a subsequence of $\mathbf{Z}^+$ with characteristic function χ. We say that $\mathscr{K}$ possesses *asymptotic density* $\mathbf{d}\mathscr{K}$ if

$$\sum_{n \leq x} \chi(n) = (\mathbf{d}\mathscr{K} + o(1))\, x, \ (x \to \infty). \tag{0.11}$$

In any event there exist

$$\limsup_{x \to \infty} x^{-1} \sum_{n \leq x} \chi(n), \ \liminf_{x \to \infty} x^{-1} \sum_{n \leq x} \chi(n) \tag{0.12}$$

and we refer to these as the upper and lower asymptotic densities of $\mathscr{K}$, denoting them by $\overline{\mathbf{d}}\mathscr{K}$ and $\underline{\mathbf{d}}\mathscr{K}$.

We say that $\mathscr{K}$ possesses *logarithmic density* $\boldsymbol{\delta}\mathscr{K}$ if

$$\sum_{n \leq x} \frac{\chi(n)}{n} = (\boldsymbol{\delta}\mathscr{K} + o(1)) \log x, \ (x \to \infty) \tag{0.13}$$

and there are definitions of upper and lower logarithmic density analogous to (0.12). We sometimes say, simply, '$\mathbf{d}\mathscr{K}$ exists' instead of '$\mathscr{K}$

possesses asymptotic density $\mathbf{d}\mathscr{K}$'. If $\mathbf{d}\mathscr{K}$ exists then so does $\boldsymbol{\delta}\mathscr{K}$, moreover $\boldsymbol{\delta}\mathscr{K} = \mathbf{d}\mathscr{K}$. The converse is false: the most that we can say is that if $\boldsymbol{\delta}\mathscr{K}$ exists then

$$\underline{\mathbf{d}}\mathscr{K} \leq \boldsymbol{\delta}\mathscr{K} \leq \overline{\mathbf{d}}\mathscr{K}. \tag{0.14}$$

For example the sequence

$$\bigcup_{m=1}^{\infty}\{(2^{2m-1}, 2^{2m}] \cap \mathbf{Z}\} \tag{0.15}$$

has upper and lower asymptotic densities $\frac{2}{3}$ and $\frac{1}{3}$ respectively, and logarithmic density $\frac{1}{2}$. We prove the right-hand inequality in (0.14): if we apply this to the complement of $\mathscr{K}$ the left-hand inequality follows. Let the sum on the left of (0.11) be denoted by $K(x)$. We have

$$\sum_{n \leq x} \frac{\chi(n)}{n} = x^{-1}K(x) + \int_1^x K(t)\frac{dt}{t^2}. \tag{0.16}$$

By hypothesis $K(x) \leq (\overline{\mathbf{d}}\mathscr{K} + \varepsilon)x$ for $x > x_0(\varepsilon)$ and, trivially $K(x) \leq x$. Hence for $x > x_0(\varepsilon)$,

$$\sum_{n \leq x} \frac{\chi(n)}{n} \leq 1 + \log x_0(\varepsilon) + (\overline{\mathbf{d}}\mathscr{K} + \varepsilon) \log\left(\frac{x}{x_0(\varepsilon)}\right). \tag{0.17}$$

We divide through by $\log x$ and let $x \to \infty$ to obtain

$$\overline{\boldsymbol{\delta}}\mathscr{K} = \limsup_{x\to\infty} \frac{1}{\log x} \sum_{n \leq x} \frac{\chi(n)}{n} \leq \overline{\mathbf{d}}\mathscr{K} + \varepsilon. \tag{0.18}$$

This holds for every $\varepsilon > 0$ whence $\overline{\boldsymbol{\delta}}\mathscr{K} \leq \overline{\mathbf{d}}\mathscr{K}$, and this implies (0.14).

Our first two theorems which follow show that logarithmic density will play a central role in the theory of sets of multiples.

Theorem 0.1 *Let $\mathscr{A}$ be finite, or such that*

$$\sum_{i=1}^{\infty} \frac{1}{a_i} < \infty. \tag{0.19}$$

Then $\mathbf{d}\mathscr{M}(\mathscr{A})$ exists. This is best possible in the sense that if $\xi(x) \to \infty$ as $x \to \infty$, there exists a sequence $\mathscr{A}$ such that

$$\sum\left\{\frac{1}{a_i} : a_i \leq x\right\} < \xi(x) \tag{0.20}$$

but $\mathbf{d}\mathscr{M}(\mathscr{A})$ does not exist.

The essential part of this theorem, and the part which is not straightforward, is that there exist sequences $\mathscr{A}$ for which the set of multiples do not possess asymptotic density. This is due to Besicovitch (1934). The next theorem is due to Davenport and Erdös (1937), (1951).

Theorem 0.2 *For every $\mathscr{A}$, $\boldsymbol{\delta}\mathscr{M}(\mathscr{A})$ exists. Moreover the logarithmic density and lower asymptotic density are equal, that is $\boldsymbol{\delta}\mathscr{M}(\mathscr{A}) = \underline{\mathbf{d}}\mathscr{M}(\mathscr{A})$.*

Proof of Theorem 0.1 When $\mathscr{A}$ is finite $\mathscr{M}(\mathscr{A})$ is a finite union of arithmetic progressions and so possesses asymptotic density. Moreover we may derive a formula for the density from the inclusion–exclusion principle: if $\mathscr{A} = \{a_1, a_2, \dots, a_n\}$ then

$$\begin{aligned}\mathbf{d}\mathscr{M}(\mathscr{A}) &= \sum_{i \le n} \frac{1}{a_i} - \sum_{i<j\le n} \frac{1}{[a_i, a_j]} \\ &\quad + \sum_{i<j<k\le n} \frac{1}{[a_i, a_j, a_k]} - \cdots + \frac{(-1)^{n-1}}{[a_1, a_2, \dots, a_n]},\end{aligned} \tag{0.21}$$

in which we have used the standard square brackets notation for least common multiple.

Next, let $\mathscr{A}$ be infinite, and (0.19) hold. Let $\mathscr{M}_i(\mathscr{A})$ comprise the positive integers n for which $a_i | n$ but for $j < i$, $a_j \nmid n$. Let $\chi_i(n)$ be the characteristic function of $\mathscr{M}_i(\mathscr{A})$ and $\chi(n)$ be the characteristic function of $\mathscr{M}(\mathscr{A})$. Then

$$\chi(n) = \sum_{i=1}^{\infty} \chi_i(n). \tag{0.22}$$

Every $\mathscr{M}_i(\mathscr{A})$ is a finite union of arithmetic progressions and possesses asymptotic density $\mathbf{d}\mathscr{M}_i(\mathscr{A})$, for which there is a formula similar to (0.21). We deduce from (0.22) that for all N,

$$\sum_{i=1}^{N} \mathbf{d}\mathscr{M}_i(\mathscr{A}) \le \underline{\mathbf{d}}\mathscr{M}(\mathscr{A}) \tag{0.23}$$

whence

$$\Delta(\mathscr{A}) := \sum_{i=1}^{\infty} \mathbf{d}\mathscr{M}_i(\mathscr{A}) \le \underline{\mathbf{d}}\mathscr{M}(\mathscr{A}). \tag{0.24}$$

On the other hand we have for all N

$$\bar{\mathbf{d}}M(\mathscr{A}) \le \sum_{i=1}^{N} \mathbf{d}M_i(\mathscr{A}) + \sum_{i=N+1}^{\infty} \frac{1}{a_i} \tag{0.25}$$

and we deduce from (0.19) and (0.24) that for $\varepsilon > 0$,

$$\overline{\mathbf{d}}\mathcal{M}(\mathcal{A}) \le \Delta(\mathcal{A}) + \varepsilon. \tag{0.26}$$

Thus, subject to (0.19), $\mathbf{d}\mathcal{M}(\mathcal{A})$ exists and

$$\mathbf{d}\mathcal{M}(\mathcal{A}) = \Delta(\mathcal{A}). \tag{0.27}$$

Finally we drop condition (0.19). We require the following result of Erdös (1936): *as $T \to \infty$, we have*

$$\mathbf{d}(T) := \mathbf{d}\mathcal{M}(\mathbf{Z} \cap (T, 2T]) \to 0. \tag{0.28}$$

For completeness we shall prove this in Lemma 0.3 below; let us assume (0.28) for the moment. By the definition of asymptotic density, for each T there exists $x_0(T)$ such that for all $x > x_0(T)$ we have

$$\operatorname{card}\{n : n \le x,\ n \in \mathcal{M}(\mathbf{Z} \cap (T, 2T])\} \le 2\mathbf{d}(T)x. \tag{0.29}$$

Let $\{T_k : k \in \mathbf{Z}^+\}$ be a sequence of integers satisfying the three conditions

$$\sum_{k=1}^{\infty} \mathbf{d}(T_k) \le \varepsilon, \tag{0.30}$$

$$\operatorname{card}\{k : T_k \le x\} < \xi(x) \text{ for all } x, \tag{0.31}$$

$$T_{k+1} > x_0(T_k) \text{ for all } k. \tag{0.32}$$

These are consistent, all requiring that T_k should be large. Let

$$\mathcal{A} = \bigcup_{k=1}^{\infty} \{\mathbf{Z} \cap (T_k, 2T_k]\} \tag{0.33}$$

and consider the counting function $M(x)$ of $\mathcal{M}(\mathcal{A})$. For all k, $M(2T_k) \ge T_k$ whence $\overline{\mathbf{d}}\mathcal{M}(\mathcal{A}) \ge \frac{1}{2}$. Next,

$$M(T_k) \le \sum_{h<k} 2\mathbf{d}(T_h)T_k \le 2\varepsilon T_k \tag{0.34}$$

by (0.29) and (0.30). Therefore $\underline{\mathbf{d}}\mathcal{M}(\mathcal{A}) \le 2\varepsilon$, and we choose $\varepsilon < \frac{1}{4}$, whence $\mathbf{d}\mathcal{M}(\mathcal{A})$ does not exist. We notice that for positive integers T we have

$$\sum_{T<a\le 2T} \frac{1}{a} < \log 2 \tag{0.35}$$

so that (0.31) implies (0.20). This proves Theorem 0.1.

Notice that for this proof it is sufficient to have $\liminf \mathbf{d}(T) = 0$, the result proved and employed by Besicovitch (1934). The following quantitive version of (0.28) is Lemma 1 of Erdös and Hall (1980).

Lemma 0.3 *Let* $\mathbf{d}(T)$ *be defined by (0.28). Then*

$$\mathbf{d}(T) \ll (\log T)^{-\delta} \tag{0.36}$$

where

$$\delta = 1 - \frac{1}{\log 2}\left(1 - \log\left(\frac{1}{\log 2}\right)\right) = .086071\ldots \tag{0.37}$$

Tenenbaum (1984) obtained a more precise result in which the right-hand side of (0.36) is divided by $\sqrt{(\log\log T)}$, and gave a lower bound for $\mathbf{d}(T)$ which establishes that the exponent δ is sharp. We discuss the matter further in Chapter 7.

Proof Let $\Omega(n, T)$ denote the number of prime factors of n which do not exceed T, counted according to multiplicity. We begin by showing that for each fixed $y_0 < 2$ we have

$$\sum_{n\le x} y^{\Omega(n,T)} \ll x(\log T)^{y-1}, \tag{0.38}$$

uniformly for $0 < y \le y_0$, $2 \le T \le x$. This may be proved by the inequality of Halberstam and Richert (1979) (Theorem 00 of *Divisors*) which gives the upper bound

$$\ll \frac{x}{\log x}\sum_{n\le x}\frac{y^{\Omega(n,T)}}{n} \tag{0.39}$$

for the sum in (0.38). We have

$$\begin{aligned}\sum_{n\le x}\frac{y^{\Omega(n,T)}}{n} &\le \prod_{p\le T}\left(1-\frac{y}{p}\right)^{-1}\prod_{T<p\le x}\left(1-\frac{1}{p}\right)^{-1}\\ &\ll \prod_{p\le T}\left(1-\frac{1}{p}\right)^{-y}\frac{\log x}{\log T}\\ &\ll \log x(\log T)^{y-1}\end{aligned} \tag{0.40}$$

by Mertens' theorem. We combine (0.39) and (0.40) to obtain (0.38).

Now consider the integers $n \le x$ which have a divisor $d \in (T, 2T]$. We split these into two classes according to whether $\Omega(n, T) \le \kappa\log\log T$ or not, (with κ at our disposal). For arbitrary y, $(0 < y \le 1)$ the cardinality of the first class does not exceed

$$y^{-\kappa \log\log T} \sum_{n \le x} y^{\Omega(n,T)} \sum_{\substack{d|n \\ T<d\le 2T}} 1$$

$$\le (\log T)^{-\kappa \log y} \sum_{T<d\le 2T} y^{\Omega(d,T)} \sum_{m\le (x/d)} y^{\Omega(m,T)}, \tag{0.41}$$

on inverting summations and writing $n = md$. We estimate the inner sum by (0.38), replacing x/d by x/T, to obtain

$$\begin{aligned} &\ll \frac{x}{T}(\log T)^{y-1-\kappa \log y} \sum_{T<d\le 2T} y^{\Omega(d,T)} \\ &\ll x(\log T)^{2y-2-\kappa \log y}, \end{aligned} \tag{0.42}$$

using (0.38) again. We choose $y = \kappa/2$ (which is in order provided $\kappa \le 2$ as we need $y \le 1$) to minimize the exponent of $\log T$. This becomes $\kappa - 2 - \kappa \log(\kappa/2) = -\delta$ if $\kappa = 1/\log 2$.

We estimate the cardinality of the second class similarly, but more crudely since we ignore the divisibility condition. For arbitrary z, $(1 \le z \le y_0)$ this does not exceed

$$z^{-\kappa \log\log T} \sum_{n\le x} z^{\Omega(n,T)} \ll x(\log T)^{z-1-\kappa \log z} \tag{0.43}$$

by (0.38). We choose $z = \kappa = 1/\log 2$ which again makes the exponent of $\log T$ equal to $-\delta$. We assemble these estimates to complete the proof.

We have written out a detailed proof of Lemma 0.3 with two objectives, first that the reader should have a complete proof of Besicovitch's important theorem and second that he should understand the role of the variables y and z, since this method will appear repeatedly throughout the book.

We require a lemma for the proof of Theorem 0.2. This is the Tauberian theorem due to Hardy and Littlewood (see Hardy (1949)) and Karamata (1931). We state this with sufficient generality for applications later in the book.

Lemma 0.4 *Let $f(t)$ be non-decreasing and such that for some fixed $\alpha > 0$ we have*

$$\int_0^\infty e^{-\sigma t} df(t) \sim \frac{\Gamma(\alpha+1)}{\sigma^\alpha} \text{as } \sigma \to 0+. \tag{0.44}$$

Then as $t \to \infty$, we have $f(t) \sim t^{\alpha}$. In particular, if we set

$$f(t) = \sum_{n \le e^t} \frac{b_n}{n}, \ (b_n \ge 0 \text{for } n \in \mathbf{Z}^+), \tag{0.45}$$

then the hypothesis (0.44) becomes

$$\sum_{n=1}^{\infty} \frac{b_n}{n^{1+\sigma}} \sim \frac{\Gamma(\alpha+1)}{\sigma^{\alpha}} \text{as } \sigma \to 0+ \tag{0.46}$$

and the conclusion is that

$$\sum_{n \le x} \frac{b_n}{n} \sim (\log x)^{\alpha} \text{as } x \to \infty. \tag{0.47}$$

For a proof see for example Hardy (1949), Chap. VII.

Proof of Theorem 0.2 Let $\mathcal{M}_i(\mathcal{A})$ be as defined in the proof of Theorem 0.1, that is the sequence of positive integers such that $a_i \mid n$, $a_j \nmid n$ for $j < i$. $\mathcal{M}_i(\mathcal{A})$ has asymptotic density $\mathbf{d}\mathcal{M}_i(\mathcal{A})$ and characteristic function $\chi_i(n)$, and we have

$$\sum_{n=1}^{\infty} \frac{\chi_i(n)}{n^s} \sim \frac{\mathbf{d}\mathcal{M}_i(\mathcal{A})}{s-1} \text{ as } s \to 1+. \tag{0.48}$$

We denote the function of s on the left of (0.48) by $F_i(s)\zeta(s)$, where $\zeta(s)$ is the Riemann zeta function, and we have from (0.48),

$$0 \le F_i(s) \le 1, \ (s > 1); \ \lim_{s \to 1+} F_i(s) = \mathbf{d}\mathcal{M}_i(\mathcal{A}). \tag{0.49}$$

Let

$$G_N(s) = F_1(s) + F_2(s) + \cdots + F_N(s). \tag{0.50}$$

By (0.22), $\chi_1(n)+\chi_2(n)+\cdots+\chi_N(n) \le \chi(n) \le 1$ (where χ is the characteristic function of $\mathcal{M}(\mathcal{A})$), whence $G_N(s)\zeta(s) \le \zeta(s)$ and $G_N(s) \le 1$. Hence we may define

$$G(s) = \sum_{i=1}^{\infty} F_i(s) \tag{0.51}$$

and we notice that

$$\liminf_{s \to 1+} G(s) \ge \lim_{s \to 1+} G_N(s) = \sum_{i=1}^{N} \mathbf{d}\mathcal{M}_i(\mathcal{A}) \tag{0.52}$$

by (0.49), whence

$$\liminf_{s \to 1+} G(s) \ge \mathbf{\Delta}(\mathcal{A}) \tag{0.53}$$

where $\mathbf{\Delta}(\mathcal{A})$ is defined in (0.24).

For any large positive integer n_0 there exists N sufficiently large that

$$\chi(n) = \chi_1(n) + \chi_2(n) + \cdots + \chi_N(n), \quad 1 \le n \le n_0. \tag{0.54}$$

Hence for fixed $s > 1$ and $\varepsilon > 0$ there exists $N_0(s, \varepsilon)$ such that for $N > N_0(s, \varepsilon)$ we have

$$\sum_{n=1}^{\infty} \frac{\chi(n)}{n^s} - \varepsilon < G_N(s)\zeta(s) \le \sum_{n=1}^{\infty} \frac{\chi(n)}{n^s}. \tag{0.55}$$

Therefore

$$\sum_{n=1}^{\infty} \frac{\chi(n)}{n^s} = G(s)\zeta(s), \quad s > 1. \tag{0.56}$$

We claim that for each fixed N, we have

$$\frac{d}{ds} G_N(s) \le 0, \quad s > 1. \tag{0.57}$$

To see this, let us denote the sum on the right-hand side of (0.54) by $\chi^{(N)}(n)$ so that we have

$$\sum_{n=1}^{\infty} \frac{\chi^{(N)}(n)}{n^s} = G_N(s)\zeta(s). \tag{0.58}$$

We may differentiate the Dirichlet series term by term and (employing dashes in the usual way) we obtain

$$\sum_{n=1}^{\infty} \frac{\chi^{(N)}(n) \log n}{n^s} = -G_N'(s)\zeta(s) - G_N(s)\zeta'(s). \tag{0.59}$$

Since $\chi^{(N)}$ is the characteristic function of a set of multiples we have $\chi^{(N)}(n) \ge \chi^{(N)}(d)$ for every n and divisor d of n. Therefore

$$\begin{aligned} \chi^{(N)}(n) \log n &= \chi^{(N)}(n) \sum_{d|n} \Lambda\left(\frac{n}{d}\right) \\ &\ge \sum_{d|n} \chi^{(N)}(d) \Lambda\left(\frac{n}{d}\right) \end{aligned} \tag{0.60}$$

(where Λ is von Mangoldt's function), whence for $s > 1$ we have

$$\begin{aligned} \sum_{n=1}^{\infty} \frac{\chi^{(N)}(n) \log n}{n^s} &\ge \left(\sum_{d=1}^{\infty} \frac{\chi^{(N)}(d)}{d^s}\right)\left(\sum_{m=1}^{\infty} \frac{\Lambda(m)}{m^s}\right) \\ &\ge (G_N(s)\zeta(s))\left(-\frac{\zeta'(s)}{\zeta(s)}\right). \end{aligned} \tag{0.61}$$

It follows from this and (0.59) that $-G'_N(s) \geq 0$ as required. We deduce from (0.49), (0.50) and (0.57) that for all $s > 1$ we have

$$G_N(s) \leq \mathbf{d}\mathscr{M}_1(\mathscr{A}) + \mathbf{d}\mathscr{M}_2(\mathscr{A}) + \cdots + \mathbf{d}\mathscr{M}_N(\mathscr{A}) \tag{0.62}$$

whence

$$G(s) \leq \Delta(\mathscr{A}), \quad s > 1, \tag{0.63}$$

and so in view of (0.53) we have

$$\lim_{s \to 1+} G(s) = \Delta(\mathscr{A}). \tag{0.64}$$

We substitute $s = 1 + \sigma$, $(\sigma > 0)$ in (0.56) and we deduce from (0.64) that

$$\sum_{n=1}^{\infty} \frac{\chi(n)}{n^{1+\sigma}} \sim \frac{\Delta(\mathscr{A})}{\sigma}, \quad \sigma \to 0 \tag{0.65}$$

whence Lemma 0.3, with $\alpha = 1$, yields

$$\sum_{n \leq x} \frac{\chi(n)}{n} \sim \Delta(\mathscr{A}) \log x, \quad x \to \infty. \tag{0.66}$$

Therefore $\boldsymbol{\delta}\mathscr{M}(\mathscr{A})$ exists and is equal to $\Delta(\mathscr{A})$. We have $\underline{\mathbf{d}}\mathscr{M}(\mathscr{A}) \leq \boldsymbol{\delta}\mathscr{M}(\mathscr{A})$ by (0.14) and $\Delta(\mathscr{A}) \leq \underline{\mathbf{d}}\mathscr{M}(\mathscr{A})$ by (0.24) whence $\underline{\mathbf{d}}\mathscr{M}(\mathscr{A}) = \boldsymbol{\delta}\mathscr{M}(\mathscr{A})$. This proves Theorem 0.2.

Let $\mathscr{A}^{(N)}$ comprise the first N elements of $\mathscr{A}$. We notice that the proof also gives, by (0.24)

$$\boldsymbol{\delta}\mathscr{M}(\mathscr{A}) = \lim_{N \to \infty} \mathbf{d}\mathscr{M}(\mathscr{A}^{(N)}). \tag{0.67}$$

We have followed the original proof of Davenport and Erdös (1937). Perhaps influenced by the recent elementary proof of the Prime Number Theorem, in 1951 Davenport and Erdös gave a new proof of Theorem 0.2 which eliminated the Hardy–Littlewood–Karamata theorem and which is in this sense more elementary. The later proof is presented in Halberstam and Roth (1966). We believe the pre-war proof is the natural way to tackle this theorem; moreover we shall need the Hardy–Littlewood–Karamata theorem later in the book.

We conclude this section with two definitions. These are each quite recent, but they simply isolate and formalize two central problems in this subject which have been present for many years.

Definition 0.5 *We say that a sequence $\mathscr{A} \subseteq \mathbf{Z}^+$ is a Besicovitch sequence if its set of multiples $\mathscr{M}(\mathscr{A})$ possesses asymptotic density.*

For example, every finite sequence is a Besicovitch sequence. The formal definition appears in Erdös, Hall and Tenenbaum (1994). It is a little paradoxical that the sequence which Besicovitch constructed should be non-Besicovitch, nevertheless it seemed clear to these authors that his name should be attached to the sequences with a positive property.

Definition 0.6 *We say that a sequence $\mathscr{A} \subseteq \mathbf{Z}^+ \setminus \{1\}$ is a Behrend sequence if its set of multiples $\mathscr{M}(\mathscr{A})$ has logarithmic density 1.*

This definition appeared in Hall (1990b). By the Davenport–Erdös theorem (Theorem 0.2) we may deduce that if $\mathscr{A}$ is Behrend then $\mathbf{d}\mathscr{M}(\mathscr{A}) = 1$ and so an equivalent, but logically more complicated, definition would be to require that $\mathscr{M}(\mathscr{A})$ possesses asymptotic density 1. A further consequence of (0.67) is that $\mathscr{A}$ is Behrend if and only if it is non-trivial and

$$\lim_{N\to\infty} \mathbf{d}\mathscr{M}(\mathscr{A}^{(N)}) = 1, \quad \mathscr{A}^{(N)} = \{a_1, a_2, \ldots, a_N\}. \tag{0.68}$$

Let $\mathscr{A}$ be a (non-trivial) sequence of pairwise coprime elements. Then the familiar inclusion-exclusion formula gives

$$\mathbf{d}\mathscr{M}(\mathscr{A}^{(N)}) = 1 - \prod_{i=1}^{N} \left(1 - \frac{1}{a_i}\right)$$

whence $\mathscr{A}$ is Behrend, in this special case, if and only if

$$\sum \left\{ \frac{1}{a} : a \in \mathscr{A} \right\} = \infty. \tag{0.69}$$

It will emerge that in the general case this condition remains necessary, but can be rather weak. One thread of the book will be to find applicable, realistic criteria for a given sequence to be Besicovitch or Behrend and we begin this study in Chapter 1. We regard the notion of a Behrend sequence as the more fundamental: it would be possible to construct a coherent theory of sets of multiples without any reference to asymptotic density (an idea with which the author of this work, perhaps rather briefly, toyed).

Erdös and Tenenbaum have noticed recently the following interesting fact about Behrend sequences.

Theorem 0.7 *Let $\mathscr{A}$ be Behrend and $M(x)$ denote the counting function of $\mathscr{M}(\mathscr{A})$. Then we have, uniformly in x,*

$$M(x+y) - M(x) \sim y, \ (y \to \infty). \tag{0.70}$$

Proof Let $\varepsilon > 0$ be fixed. By (0.68) there exists $N = N(\varepsilon)$ such that $\mathbf{d}\mathcal{M}(\mathcal{A}^{(N)}) > 1 - (\varepsilon/2)$, where as before $\mathcal{A}^{(N)}$ comprises the first N elements of $\mathcal{A}$. Since $\mathcal{M}(\mathcal{A}^{(N)})$ is a finite union of arithmetic progressions its intersection with $(x, x+y]$ has cardinality $\mathbf{d}\mathcal{M}(\mathcal{A}^{(N)})y+O(1)$, uniformly in x. Hence there exists $y_0 = y_0(\varepsilon)$ such that for $y > y_0(\varepsilon)$ we have $M(x+y) - M(x) > (1-\varepsilon)y$, and (0.70) follows.

0.4 The Heilbronn–Rohrbach and Behrend inequalities

Clearly for all sequences $\mathcal{A}$ and $\mathcal{B}$ we have $\mathcal{M}(\mathcal{A}) \subseteq \mathcal{M}(\mathcal{A} \cup \mathcal{B})$. We consider the quantitative effect of adjoining $\mathcal{B}$ to $\mathcal{A}$ and we begin with the case where $\mathcal{B}$ has just one element b. Notice that the conditions

$$a \nmid n,\ b \mid n \tag{0.71}$$

are equivalent to

$$n = bm,\ a' \nmid m \text{ where } a' = \frac{a}{(a,b)}. \tag{0.72}$$

Let us write

$$\mathcal{T}(\mathcal{A}) = \mathbf{Z}^+ \setminus \mathcal{M}(\mathcal{A}) \tag{0.73}$$

and

$$\mathcal{A}'(b) = \left\{a' = \frac{a}{(a,b)} : a \in A\right\}. \tag{0.74}$$

We employ the letter $\mathcal{T}$ for the complement because of the now almost standard notation

$$\mathbf{t}(\mathcal{A}) = 1 - \boldsymbol{\delta}\mathcal{M}(\mathcal{A}). \tag{0.75}$$

We deduce from (0.72) that

$$\mathcal{M}(\mathcal{A} \cup \{b\}) = \mathcal{M}(\mathcal{A}) \cup b\mathcal{T}\left(\mathcal{A}'(b)\right). \tag{0.76}$$

If $b \notin \mathcal{M}(\mathcal{A})$ then $a' > 1$ always, that is $\mathcal{A}'(b)$ is non-trivial. For example if $\mathcal{A}$ is finite then $\mathcal{T}\left(\mathcal{A}'(b)\right)$ has positive asymptotic density and $\mathcal{M}(\mathcal{A} \cup \{b\})$ has strictly greater density than $\mathcal{M}(\mathcal{A})$.

Theorem 0.8 *Let $b \notin \mathcal{M}(\mathcal{A})$. A necessary and sufficient condition that*

$$\boldsymbol{\delta}\mathcal{M}(\mathcal{A} \cup \{b\}) > \boldsymbol{\delta}\mathcal{M}(\mathcal{A})$$

is that the sequence $\mathscr{A}'(b)$ defined in (0.74) should not be a Behrend sequence. In any case we have

$$\delta\mathscr{M}(\mathscr{A}\cup\{b\}) \le \left(1-\frac{1}{b}\right)\delta\mathscr{M}(\mathscr{A})+\frac{1}{b}. \tag{0.77}$$

Proof The union on the right-hand side of (0.76) is disjoint and so

$$\delta\mathscr{M}(\mathscr{A}\cup\{b\}) = \delta\mathscr{M}(\mathscr{A})+\frac{1}{b}\mathbf{t}\left(\mathscr{A}'(b)\right), \tag{0.78}$$

and $t\left(\mathscr{A}'(b)\right)=0$ if and only if $\mathscr{A}'(b)$ is Behrend. This proves the first part. For any b we have $\mathscr{T}(\mathscr{A}'(b))\subseteq\mathscr{T}(\mathscr{A})$ whence

$$\mathbf{t}\left(\mathscr{A}'(b)\right)\le \mathbf{t}(\mathscr{A}),$$

which we insert in (0.78) to obtain (0.77). This completes the proof.

A corollary is the next theorem, proved independently by Heilbronn (1937) and Rohrbach (1937).

Theorem 0.9 (Heilbronn–Rohrbach) *Let $\mathscr{A}=\{a_1,a_2,\dots,a_N\}$. Then*

$$\mathbf{t}(\mathscr{A})\ge\prod_{i=1}^{N}\left(1-\frac{1}{a_i}\right). \tag{0.79}$$

Proof We can rewrite (0.77) in the form

$$\mathbf{t}(\mathscr{A}\cup\{b\})\ge\left(1-\frac{1}{b}\right)\mathbf{t}(\mathscr{A}) \tag{0.80}$$

and (0.79) follows by induction.

In fact (0.80) is the most useful form of the inequality and is sometimes loosely referred to as the Heilbronn–Rohrbach inequality. Strictly this is (0.79). An immediate consequence is

Corollary 0.10 *A necessary condition for $\mathscr{A}$ to be a Behrend sequence is that (0.69) holds.*

The example $\mathscr{A}=2\mathbf{Z}^+$ shows that (0.69) is insufficient for $\mathscr{A}$ to be Behrend. We construct a rather more interesting example after the next theorem. Before proceeding to this we give another definition, suggested by Theorem 0.8.

Definition 0.11 *We say that the sequence $\mathscr{A}$ is taut if, for every $a\in\mathscr{A}$, we have*

$$\mathbf{t}(\mathscr{A}\setminus\{a\})>\mathbf{t}(\mathscr{A}). \tag{0.81}$$

A taut sequence is necessarily primitive, and every finite primitive sequence is taut. A Behrend sequence cannot be taut, by (0.80), and it may well be primitive: the sequence of primes is an example. We shall be able to classify the taut sequences later. (Corollary 0.19.)

The following result generalizes (0.80) and is due to Behrend (1948).

Theorem 0.12 (Behrend's Inequality) *For all $\mathscr{A}$ and $\mathscr{B}$ we have*

$$\mathbf{t}(\mathscr{A} \cup \mathscr{B}) \geq \mathbf{t}(\mathscr{A})\mathbf{t}(\mathscr{B}). \tag{0.82}$$

Equality holds if $(a, b) = 1$ for all $a \in \mathscr{A}$, $b \in \mathscr{B}$. If $\mathscr{A}$ and $\mathscr{B}$ are finite, primitive sequences this condition is necessary for equality to hold.

We give some examples of equality at the end of the proof.

Corollary 0.13 *Let $\mathscr{A} = \mathscr{A}_1 \cup \mathscr{A}_2 \cup \ldots$ be a Behrend sequence. Then*

$$\sum_{k=1}^{\infty} \delta\mathscr{M}(\mathscr{A}_k) = \infty. \tag{0.83}$$

We notice that the sequence $\mathscr{A}$ as above with

$$\mathscr{A}_k = \mathbf{Z} \cap \left(e^{k^{12}}, 2e^{k^{12}}\right] \tag{0.84}$$

is not Behrend, by Lemma 0.3 and (0.83). On the other hand it fulfils (0.69).

Corollary 0.14 *The sequence $\mathscr{A} \cup \mathscr{B}$ is a Behrend sequence if and only if at least one of $\mathscr{A}$ and $\mathscr{B}$ is Behrend. In particular any tail of a Behrend sequence is Behrend.*

The next corollary is due to Ruzsa and Tenenbaum (199x). I have not seen their proof of this interesting observation and give my own.

Corollary 0.15 (Ruzsa-Tenenbaum) *Any Behrend sequence may be split into an infinite disjoint union of Behrend sequences.*

Proof Let $\mathscr{A}$ be the given Behrend sequence. We construct $\mathscr{A}_1, \mathscr{A}_2, \mathscr{A}_3, \ldots$ by apportioning the successive elements of $\mathscr{A}$ to these subsequences in consecutive finite runs in the order $\mathscr{A}_1, \mathscr{A}_2, \mathscr{A}_1, \mathscr{A}_2, \mathscr{A}_3, \mathscr{A}_1, \mathscr{A}_2, \mathscr{A}_3, \mathscr{A}_4, \mathscr{A}_1, \ldots$ etc.; at each stage the remaining tail of $\mathscr{A}$ is Behrend by Corollary 0.14, moreover at the nth stage we ensure that the particular subsequence being augmented has $\mathbf{t} < 1/n$. We can do this by virtue of (0.68); this same result then establishes that every $\mathscr{A}_i$ is Behrend.

Corollary 0.16 *If $\mathscr{A}$ is Behrend then*

$$\tau(n, \mathscr{A}) \to \infty \ p.p. \tag{0.85}$$

This follows from the previous corollary. See also Hall and Tenenbaum (1992).

Theorem 0.12 is important and we give two proofs, Behrend's own and that of Ruzsa (1976). Another proof appears in Halberstam and Roth (1966). These proofs are quite distinct.

Behrend considered the cases of equality in his paper but Ruzsa did not. For the sake of completeness, and furthermore because we require his interesting method later in the book, we supply the extra ingredient needed for this question in Ruzsa's proof here. We begin with Behrend's proof: we note that originally Behrend restricted his result to finite sequences (the extension is easy) and both proofs begin with this case.

Proof (i) (Behrend)

Lemma 0.17 *Let $\mathscr{A}$ be any non-empty sequence and c be a positive integer. Let $\mathscr{B}$ be a (possibly empty) sequence all of whose elements are prime to c. Then*

$$\boldsymbol{\delta}\mathscr{M}(c\mathscr{A} \cup \mathscr{B}) = \frac{1}{c}\boldsymbol{\delta}\mathscr{M}(\mathscr{A} \cup \mathscr{B}) + \left(1 - \frac{1}{c}\right)\boldsymbol{\delta}\mathscr{M}(\mathscr{B}). \tag{0.86}$$

Proof Let $\mathscr{N} = \mathscr{M}(\mathscr{A}) \setminus \mathscr{M}(\mathscr{B})$. We consider the integers n belonging to $\mathscr{M}(c\mathscr{A} \cup \mathscr{B})$ but not to $\mathscr{M}(\mathscr{B})$. This requires that $n = mca$ for some $m \in \mathbf{Z}^+$, $a \in \mathscr{A}$ such that $b \nmid mca$ for any $b \in \mathscr{B}$. Since $(b, c) = 1$ by hypothesis, this last condition is equivalent to $b \nmid ma$ for any b, that is $n \in c\mathscr{N}$. Hence

$$\mathscr{M}(c\mathscr{A} \cup \mathscr{B}) = \mathscr{M}(\mathscr{B}) \cup c\mathscr{N} \tag{0.87}$$

is a disjoint union and so

$$\boldsymbol{\delta}\mathscr{M}(c\mathscr{A} \cup \mathscr{B}) = \boldsymbol{\delta}\mathscr{M}(\mathscr{B}) + \frac{1}{c}\boldsymbol{\delta}\mathscr{N}. \tag{0.88}$$

We may substitute $c = 1$ in (0.88) and eliminate $\boldsymbol{\delta}\mathscr{N}$ between the result obtained and (0.88) itself. This proves the lemma.

Now assume $\mathscr{A}$ and $\mathscr{B}$ finite, and write $\sigma(\mathscr{A})$ for the sum of the elements of $\mathscr{A}$. We proceed by induction on $N = \sigma(\mathscr{A}) + \sigma(\mathscr{B})$. If $\mathscr{A}$ is empty we put $\sigma(\mathscr{A}) = 0$ and $\mathbf{t}(\mathscr{A}) = 1$. We note that (0.82) holds if either $\mathscr{A}$ or $\mathscr{B}$ is empty or contains 1's (is trivial). It also holds if $(a, b) = 1$

for every $a \in \mathscr{A}$, $b \in \mathscr{B}$. This is a consequence of the inclusion–exclusion formula

$$\mathbf{t}(\mathscr{A}) = 1 - \sum_i \frac{1}{a_i} + \sum_{i<j} \frac{1}{[a_i, a_j]} - \sum_{i<j<k} \frac{1}{[a_i, a_j, a_k]} + \cdots \tag{0.89}$$

because we then have $[a_i, a_j, \ldots, b_r, b_s, \ldots] = [a_i, a_j, \ldots]\,[b_r, b_s, \ldots]$ throughout and the expression for $\mathbf{t}(\mathscr{A} \cup \mathscr{B})$ separates.

From the foregoing, (0.82) holds when $N = 0$. Let $N > 0$: the induction hypothesis is that (0.82) holds for every $\mathscr{A}$ and $\mathscr{B}$ such that $\sigma(\mathscr{A}) + \sigma(\mathscr{B}) < N$.

We may assume that $\mathscr{A}$ and $\mathscr{B}$ are non-trivial and that there exist a and b such that $(a, b) > 1$. Let p be a prime which divides at least one element of each sequence. We write

$$\mathscr{A} = p\mathscr{A}_1 \cup \mathscr{A}_2, \; \mathscr{B} = p\mathscr{B}_1 \cup \mathscr{B}_2 \tag{0.90}$$

where $\mathscr{A}_2$ and $\mathscr{B}_2$ but not $\mathscr{A}_1$ and $\mathscr{B}_1$ may be empty. By the lemma, we have

$$\begin{aligned}
&\mathbf{t}(\mathscr{A})\mathbf{t}(\mathscr{B}) \\
&= \left(\frac{1}{p}\mathbf{t}(\mathscr{A}_1 \cup \mathscr{A}_2) + \left(1 - \frac{1}{p}\right)\mathbf{t}(\mathscr{A}_2)\right)\left(\frac{1}{p}\mathbf{t}(\mathscr{B}_1 \cup \mathscr{B}_2) + \left(1 - \frac{1}{p}\right)\mathbf{t}(\mathscr{B}_2)\right) \\
&= \frac{1}{p}\mathbf{t}(\mathscr{A}_1 \cup \mathscr{A}_2)\mathbf{t}(\mathscr{B}_1 \cup \mathscr{B}_2) + \left(1 - \frac{1}{p}\right)\mathbf{t}(\mathscr{A}_2)\mathbf{t}(\mathscr{B}_2) \\
&\quad - \frac{1}{p}\left(1 - \frac{1}{p}\right)\{\mathbf{t}(\mathscr{A}_1 \cup \mathscr{A}_2) - \mathbf{t}(\mathscr{A}_2)\}\{\mathbf{t}(\mathscr{B}_1 \cup \mathscr{B}_2) - \mathbf{t}(\mathscr{B}_2)\}.
\end{aligned} \tag{0.91}$$

We have $\mathbf{t}(\mathscr{A}_1 \cup \mathscr{A}_2) \le \mathbf{t}(\mathscr{A}_2)$ etc. so that the last term on the right is ≤ 0. Also $\sigma(\mathscr{A}_1) + \sigma(\mathscr{A}_2) = \sigma(\mathscr{A}) - (p-1)\sigma(\mathscr{A}_1) < \sigma(\mathscr{A})$ whence

$$\sigma(\mathscr{A}_1 \cup \mathscr{A}_2) + \sigma(\mathscr{B}_1 \cup \mathscr{B}_2) < \sigma(\mathscr{A}) + \sigma(\mathscr{B}) = N. \tag{0.92}$$

By the induction hypothesis we have both

$$\mathbf{t}(\mathscr{A}_1 \cup \mathscr{A}_2)\mathbf{t}(\mathscr{B}_1 \cup \mathscr{B}_2) \le \mathbf{t}((\mathscr{A}_1 \cup \mathscr{B}_1) \cup (\mathscr{A}_2 \cup \mathscr{B}_2))$$

and

$$\mathbf{t}(\mathscr{A}_2)\mathbf{t}(\mathscr{B}_2) \le \mathbf{t}(\mathscr{A}_2 \cup \mathscr{B}_2).$$

We insert these inequalities into (0.91), striking out the non-positive final term. This yields

$$\begin{aligned}
\mathbf{t}(\mathscr{A})\mathbf{t}(\mathscr{B}) &\le \frac{1}{p}\mathbf{t}((\mathscr{A}_1 \cup \mathscr{B}_1) \cup (\mathscr{A}_2 \cup \mathscr{B}_2)) + \left(1 - \frac{1}{p}\right)\mathbf{t}(\mathscr{A}_2 \cup \mathscr{B}_2) \\
&= \mathbf{t}(p(\mathscr{A}_1 \cup \mathscr{B}_1) \cup (\mathscr{A}_2 \cup \mathscr{B}_2))
\end{aligned}$$

$$\begin{aligned} &= \mathbf{t}((p\mathscr{A}_1 \cup \mathscr{A}_2) \cup (p\mathscr{B}_1 \cup \mathscr{B}_2)) \\ &= \mathbf{t}(\mathscr{A} \cup \mathscr{B}) \end{aligned} \tag{0.93}$$

employing the lemma again. This completes the induction and establishes the theorem for finite $\mathscr{A}$ and $\mathscr{B}$.

The extension to the general case is common to the two proofs and is a straightforward application of (0.67).

To establish the last assertion of the theorem we show that if $\mathscr{A}$ and $\mathscr{B}$ are finite primitive sequences and for some a, b we have $(a, b) > 1$ then $\mathbf{t}(\mathscr{A} \cup \mathscr{B}) > \mathbf{t}(\mathscr{A})\mathbf{t}(\mathscr{B})$. Let p be a prime factor of (a, b) and write $\mathscr{A}$ and $\mathscr{B}$ as in (0.90). Our assertion will follow from (0.91) if we can show that

$$\mathbf{t}(\mathscr{A}_1 \cup \mathscr{A}_2) < \mathbf{t}(\mathscr{A}_2),\ \mathbf{t}(\mathscr{B}_1 \cup \mathscr{B}_2) < \mathbf{t}(\mathscr{B}_2). \tag{0.94}$$

We refer to Theorem 0.8. Let $a^{(1)} \in \mathscr{A}_1$ (which is non-empty). By hypothesis $\mathscr{A}$ is primitive, whence $a^{(2)} \nmid pa^{(1)}$ and so $a^{(2)} \nmid a^{(1)}$ for any $a^{(2)} \in \mathscr{A}_2$. Hence

$$\mathscr{A}_2'(a^{(1)}) = \left\{ \frac{a^{(2)}}{(a^{(1)}, a^{(2)})} : a^{(2)} \in \mathscr{A}_2 \right\} \tag{0.95}$$

is non-trivial and as it is finite, therefore not Behrend. We consider $\mathscr{B}$ similarly and apply Theorem 0.8 to obtain (0.94). This completes the proof.

Before proceeding to Ruzsa's proof we give some examples of equality in (0.82), in which $(a, b) > 1$ for some $a \in \mathscr{A}$, $b \in \mathscr{B}$. If $\mathscr{A}$ and $\mathscr{B}$ are finite but not primitive, Behrend's own example is $\mathscr{A} = \{2, 4\}$, $\mathscr{B} = \{3, 6\}$. The elements 4 and 6 are simply redundant.

Now let $\mathscr{A}$ and $\mathscr{B}$ be primitive but not finite. Let $\mathscr{P}_i$ denote the sequence of primes congruent to i (mod 8) and put

$$\mathscr{A} = 5\mathscr{P}_1 \cup \{35\}, \quad \mathscr{B} = 7\mathscr{P}_3 \cup \{35\}, \tag{0.96}$$

so that $\mathscr{A}$ and $\mathscr{B}$ are primitive and share an element. We have $\boldsymbol{\delta}\mathscr{M}(\mathscr{A}) = \frac{1}{5}$, $\boldsymbol{\delta}\mathscr{M}(\mathscr{B}) = \frac{1}{7}$ because $\mathscr{P}_1$ and $\mathscr{P}_3$ are Behrend. The element 35 is not redundant but it does not affect the densities: we hardly need Theorem 0.8 in this case but it is useful to check that the sequences

$$(5\mathscr{P}_1)'(35) = \mathscr{P}_1,\ (7\mathscr{P}_3)'(35) = \mathscr{P}_3 \tag{0.97}$$

are each Behrend. Clearly $\boldsymbol{\delta}\mathscr{M}(\mathscr{A} \cup \mathscr{B}) = 11/35$. Finally we remark that if either $\mathscr{A}$ or $\mathscr{B}$ is Behrend then (0.82) holds with equality.

Proof (ii) (Ruzsa*) We say that the arithmetic function f is *multiplicatively non-increasing* if $f(md) \le f(d)$ for all $m, d \in \mathbf{Z}^+$. The proof involves

* The purely number theoretic version of this is due to Ruzsa and Tenenbaum (199x), except for the equality condition, which is new.

sums of the form

$$E(f;M) = \frac{1}{M}\sum_{d|M}\varphi\left(\frac{M}{d}\right)f(d) \tag{0.98}$$

where φ is Euler's function and $M \in \mathbf{Z}^+$.

Lemma 0.18 *Let f and g be multiplicatively non-increasing arithmetic functions. Then for all $M \geq 1$ we have*

$$E(fg;M) \geq E(f;M)E(g;M). \tag{0.99}$$

There is equality in (0.99) if and only if f and g split M, that is there exist F and G such that $M = FG$, $(F,G) = 1$, and for every divisor d of M we have

$$f(d) = f\left((d,F)\right),\ g(d) = g\left((d,G)\right). \tag{0.100}$$

Proof of lemma We proceed by induction on $k = \omega(M)$. If $k = 1$ then $M = p^\nu$ for some prime p, and the property that f and g are non-increasing implies

$$\left(f(p^j) - f(p^h)\right)\left(g(p^j) - g(p^h)\right) \geq 0,\ (0 \leq h,\ j \leq \nu). \tag{0.101}$$

We multiply by $\varphi(p^{\nu-h})\varphi(p^{\nu-j})$ and sum over all h and j to obtain

$$2E(fg;M) - 2E(f;M)E(g;M) \geq 0. \tag{0.102}$$

Notice that f and g split p^ν if and only if at least one of them is constant on the divisors of p^ν. Therefore if f and g do not split p^ν we have both $f(1) > f(p^\nu)$, $g(1) > g(p^\nu)$ by the monotonicity, so that the left-hand side of (0.101) is positive when $h = 0$, $j = \nu$. We therefore have strict inequality in (0.102) and (0.99) as required. Hence the lemma holds when $k = 1$.

Let $\omega(M) = k > 1$, and $p^\nu \parallel M$. We put $M = M_1 p^\nu$ and define, for each $j = 0, 1, \ldots, \nu$, the functions f_j, g_j by setting $f_j(d) = f(p^j d)$, $g_j(d) = g(p^j d)$. These are obviously multiplicatively non-increasing. Now

$$\begin{aligned} E(fg;M) &= \frac{1}{M_1 p^\nu}\sum_{j=0}^{\nu}\sum_{d|M_1}\varphi\left(M_1(p^{\nu-j}/d)\right)f_j(d)g_j(d) \\ &= \frac{1}{p^\nu}\sum_{j=0}^{\nu}\varphi(p^{\nu-j})E(f_j g_j;M_1) \\ &\geq \frac{1}{p^\nu}\sum_{j=0}^{\nu}\varphi(p^{\nu-j})E(f_j;M_1)e(g_j;M_1) \end{aligned} \tag{0.103}$$

where we have used, in the second step, the fact that Euler's φ-function is multiplicative, and in the third step the induction hypothesis that (0.99) holds for $\omega(M_1) = k - 1$. We define $\tilde{f}(p^j) = E(f_j; M_1)$, $\tilde{g}(p^j) = E(g_j; M_1)$, $(0 \le j \le \nu)$, and we have shown that

$$E(fg; M) \ge E(\tilde{f}\tilde{g}; p^\nu). \tag{0.104}$$

The functions $\tilde{f}, \tilde{g}$ are multiplicatively non-increasing and we may apply (0.99) with $k = 1$ to obtain

$$E(\tilde{f}\tilde{g}; p^\nu) \ge E(\tilde{f}; p^\nu)E(\tilde{g}; p^\nu). \tag{0.105}$$

Since

$$E(\tilde{f}; p^\nu) = \frac{1}{p^\nu}\sum_{j=0}^{\nu} \varphi(p^{\nu-j})E(f_j; M_1) = E(f; M)$$

with a similar equation involving g, we may assemble (0.104) and (0.105) to obtain (0.99). It remains to consider the cases of equality, and we leave it to the reader to check that the condition that f and g split M is sufficient for this. Suppose, conversely, that there is equality in (0.99). Then we must have equality in (0.105), moreover in (0.103) we must have $E(f_j g_j; M_1) = E(f_j; M_1)E(g_j; M_1)$ for every j, $(0 \le j \le \nu)$. We apply the induction hypothesis in each case. Firstly, either $\tilde{f}$ or $\tilde{g}$, say $\tilde{f}$, is constant. In view of the monotonicity of f, this implies that f_j is independent of j, that is $f(p^j d) = f(d)$ for every divisor d of M_1. Secondly we require that for every j, f_j and g_j split M_1, that is f and g split M_1. This means that for every j there is a factorization $M_1 = F_j G_j$ in which $(F_j, G_j) = 1$ and $f(d) = f\left((d, F_j)\right)$, $g_j(d) = g_j\left((d, G_j)\right)$, $d | M_1$. Let $F = \text{h.c.f.}(F_0, F_1, \ldots, F_\nu)$. We deduce that $f(d) = f\left((d, F)\right)$ for every d dividing M_1, and hence for every d dividing M. Also $g_j(d) = g_j\left((d, M_1/F)\right)$ for every divisor of M_1 and hence $g(d) = g\left((d, G)\right)$ for every divisor of M, where $G = p^\nu M_1/F = M/F$. Thus f and g split M as required, and (0.99) holds when $\omega(M) = k$. This completes the induction and the proof of the lemma.

Now let $\chi_{\mathscr{A}}$ be the characteristic function of the set of *non-multiples* of $\mathscr{A}$. We notice first that $\chi_{\mathscr{A}}$ is multiplicatively non-increasing, and second that, provided $M\,(\in \mathbf{Z}^+)$ is a common multiple of the elements of $\mathscr{A}$, we have

$$\chi_{\mathscr{A}}(n) = \chi_{\mathscr{A}}\left((n, M)\right), \ (n \in \mathbf{Z}^+). \tag{0.106}$$

For each divisor d of M there are $\varphi(M/d)$ congruence classes (mod M) in which $(n, M) = d$ whence

$$\mathbf{t}(\mathscr{A}) = \frac{1}{M}\sum_{d|M}\varphi\left(\frac{M}{d}\right)\chi_{\mathscr{A}}(d) = E(\chi_{\mathscr{A}}; M) \tag{0.107}$$

and we deduce (0.82) from (0.99), taking M to be the least common multiple of the elements of both $\mathscr{A}$ and $\mathscr{B}$.

It remains to establish the last part of the theorem concerning the cases of equality. Let $\mathscr{A}$ and $\mathscr{B}$ be finite and $(a, b) = 1$ for all $a \in \mathscr{A}$, $b \in \mathscr{B}$. Then $\chi_{\mathscr{A}}$ and $\chi_{\mathscr{B}}$ split M (as above): we take F to be the least common multiple of the a's and G that of the b's, whence (0.99) and (0.82) hold with equality. The extension to the infinite case with $(a, b) = 1$ is clear. Next let $\mathscr{A}$ and $\mathscr{B}$ be finite primitive sequences and for some a and b let $(a, b) > 1$. Let $M = FG$ where $(F, G) = 1$. Then either $(a, F) < a$ or $(b, G) < b$, whence $\chi_{\mathscr{A}}((a, F)) \neq \chi_{\mathscr{A}}(a)$ or $\chi_{\mathscr{B}}((b, G)) \neq \chi_{\mathscr{B}}(b)$, because $\mathscr{A}$ and $\mathscr{B}$ are primitive. Hence M is not split and (0.99) and (0.82) are strict. This completes proof (ii) of Theorem 0.12.

We mention one further consequence of Behrend's inequality at this point, concerning *taut* sequences. We recall that $\mathscr{A}$ is taut if and only if $\mathbf{t}(\mathscr{A} \setminus \{a\}) > \mathbf{t}(\mathscr{A})$ for every $a \in \mathscr{A}$. We count $\mathscr{A} = \{1\}$ among the taut sequences.

Corollary 0.19 *The non-trivial sequence $\mathscr{A}$ is taut if and only if it is primitive and does not contain a sequence $c\mathscr{B}$ in which $\mathscr{B}$ is Behrend.*

Proof Let $c\mathscr{B} \subseteq \mathscr{A}$ with $\mathscr{B}$ Behrend. Then $\boldsymbol{\delta}(\mathscr{M}(\mathscr{A}) \cap c\mathbf{Z}) = 1/c$, and this is unchanged if we remove an element of $c\mathscr{B}$. This cannot affect $\mathscr{M}(\mathscr{A}) \cap (\mathbf{Z} \setminus c\mathbf{Z})$ whence $\boldsymbol{\delta}\mathscr{M}(\mathscr{A})$ is unchanged and $\mathscr{A}$ is not taut. Conversely if $\mathscr{A}$ is not taut, there exists $a \in \mathscr{A}$ such that $t(\mathscr{A}_1) = t(\mathscr{A})$ where $\mathscr{A}_1 = \mathscr{A} \setminus \{a\}$. By Theorem 0.8 the sequence

$$\mathscr{A}_1'(a) = \left\{\frac{a^{(1)}}{(a^{(1)}, a)} : a^{(1)} \in \mathscr{A}_1\right\} \tag{0.108}$$

is Behrend. By corollary 0.14 (extended to a finite union of sequences), at least one of the sequences

$$\left\{\frac{a^{(1)}}{c} : a^{(1)} \in A_1, (a^{(1)}, a) = c\right\}, c \mid a \tag{0.109}$$

is Behrend, and we denote it by $\mathscr{B}$. Hence $\mathscr{A} \supseteq c\mathscr{B}$ as required. This proves the corollary.

0.5 Total decomposition sets

Some branches of number theory inevitably involve calculations with highest common factors and least common multiples. When just two integers occur we can write $a = (a,b)a'$, $b = (a,b)b'$ in the obvious way, but even with three integers the business can be somewhat tiresome. Total decomposition sets provide a mechanism which sometimes copes with these difficulties. They occur in Chapters 2 and 3 of the present book. Part of Theorem 0.20 below was proved independently by Ruzsa (1988).

We begin with the example of three (positive) integers a, b and c. By analogy with the factorization of a and b above, we can write

$$\begin{aligned} a &= a'vwd, \\ b &= b'uwd, \\ c &= c'uvd, \end{aligned} \tag{0.110}$$

where $d = (a,b,c)$ and, for example, $(a,b) = wd$. That is, the highest common factors are the products of the *visibly common elements*. It is not quite obvious that we can always do this but at any rate if a, b and c are square-free the reader will see that (0.110) works by arranging their prime factors in a Venn diagram. In this scenario the h.c.f.'s become intersections and the l.c.m.'s unions. The general case of (0.110) is covered by the theorems which follow. The rule for lowest common multiples is to be that we multiply together *all visibly different elements*. For example $[a,b] = a'b'uvwd$, $[a,b,c] = a'b'c'uvwd$.

The decomposition into factors is not the end of the matter since in applications we often require to know whether the various elements $a', b', \ldots, d$ are relatively prime. We emphasize that in the general case some, but not all, pairs of these elements are necessarily coprime.

The decomposition (0.110) involves in all seven factors and for n integers $a, b, c, \ldots$ there will be $2^n - 1$ factors. Our notation needs some care and we make this quite formal. In applications some shorthand is usually appropriate.

Theorem 0.20 *Let $\mathscr{A} = \{a_1, a_2, \ldots, a_n\}$ be a set of positive integers which may contain repetitions and 1's. Then $\mathscr{A}$ possesses a unique total decomposition set $\{d(S)\}$, comprising $2^n - 1$ positive integers $d(S)$ labelled by the non-empty subsets S of $\{1, 2, 3, \ldots, n\}$, with the following properties:*

(i) *For every non-empty subset $T \subseteq \{1, 2, \ldots, n\}$ we have*

$$\text{h.c.f.}(a_i : i \in T) = \prod \{d(S) : S \supseteq T\}, \tag{0.111}$$

(ii) *For every non-empty subset* $T \subseteq \{1,2,\ldots,n\}$ *we have*

$$\text{l.c.m.}[a_i : i \in T] = \prod\{d(S) : S \cap T \neq \emptyset\}. \qquad (0.112)$$

The proof in Hall (1989) is a bit clumsy and we give a new one. We use the notation

$$\nu_p(a) = \max\{\alpha : p^\alpha \mid a\} \qquad (0.113)$$

for primes p and non-zero integers a.

Proof of Theorem 0.20 We write down a formula for $d(S)$ and verify that these numbers satisfy (0.111) and (0.112).

Let p be a fixed prime and let the distinct values of $\nu_p(a_i)$, for $1 \leq i \leq n$, arranged in decreasing order be $v_1, v_2, \ldots, v_k$. Thus $k \leq n$ and

$$v_1 > v_2 > \cdots > v_k \geq 0. \qquad (0.114)$$

Put

$$Z_j(p) = \{i : p^{v_j} | a_i\}, \quad 1 \leq j \leq k \qquad (0.115)$$

so that the sets $Z_j(p)$ satisfy (with strict inclusion)

$$Z_1 \subset Z_2 \subset Z_3 \subset \cdots \subset Z_k. \qquad (0.116)$$

Let $u_j = v_j - v_{j+1}$ if $j < k$, $u_k = v_k$. We put

$$d(S,p) = \begin{cases} p^{u_j} & \text{if } S = Z_j(p) \text{ for some } j, \\ 1 & \text{else.} \end{cases} \qquad (0.117)$$

This defines $d(S,p)$ for every non-empty $S \subseteq \{1,2,\ldots,n\}$ and every prime p, and we put

$$d(S) = \prod_p d(S,p)$$

where the product is over all primes (or if we prefer, all prime factors of $a_1 a_2 \ldots a_n$).

We have to show that the set of numbers $\{d(S)\}$ so defined has the properties (i) and (ii) claimed in the statement of the theorem. In each case we consider, for each prime p and subset T the exact power of p which divides the left- and right-hand side of (0.111) or (0.112). We begin with (i) and we have

$$\nu_p\,(\text{h.c.f.}(a_i : i \in T)) = \min\{\nu_p(a_i) : i \in T\} = v_{j(T)} = \sum_{j(T) \leq l \leq k} u_l, \qquad (0.118)$$

where

$$j(T) = \min\{j : Z_j(p) \supseteq T\}.$$

The power of p dividing the right-hand side of (0.111) is

$$\prod\{d(S,p) : S \supseteq T\} = \prod\{p^{u_l} : Z_l(p) \supseteq T\} = \prod\{p^{u_l} : j(T) \le l \le k\}$$

and this agrees with (0.118). This establishes (i). We proceed to (ii) and we have

$$v_p(\text{l.c.m.}[a_i : i \in T]) = \max\{v_p(a_i) : i \in T\} = v_{h(T)} = \sum_{h(T)\le l\le k} u_l, \quad (0.119)$$

where

$$h(T) = \min\{j : Z_j(p) \cap T \neq \emptyset\}, \quad (0.120)$$

(notice the *minimum* is required on the right). The power of p dividing the right-hand side of (0.112) is

$$\begin{aligned}\prod\{d(S,p) : S \cap T \neq \emptyset\} &= \prod\{p^{u_l} : Z_l(p) \cap T \neq \emptyset\} \\ &= \prod\{p^{u_l} : h(T) \le l \le k\} \quad (0.121)\end{aligned}$$

which agrees with (0.119). This completes the proof of Theorem 0.20.

The next result classifies the total decomposition sets.

Theorem 0.21 *Let $\{e(S) : S \subseteq \{1,2,\ldots,n\}\}$ be a set of $2^n - 1$ labelled positive integers. This is a total decomposition set (associated with a suitable $\mathscr{A}$) if and only if, for every pair of sets R and S such that $R \not\subseteq S$ and $S \not\subseteq R$ we have $(e(R), e(S)) = 1$.*

Proof We begin by showing that the condition stated is necessary. Let $\{e(S)\}$ be a total decomposition set and $p \mid (e(R), e(S))$ for some pair of sets R, S. By (0.116) this requires $R = Z_l(p)$, $S = Z_m(p)$ for some l and m. By (0.116), one set lies inside the other. This establishes necessity.

For $1 \le i \le n$ let

$$a_i = \prod\{e(S) : i \in S\}; \quad (0.122)$$

we claim that if $\{e(S) : S \subseteq \{1,2,\ldots,n\}\}$ satisfies the condition stated in the theorem then this set of numbers is the total decomposition set of $\mathscr{A} = \{a_1, a_2, \ldots, a_n\}$, defined by (0.122). By Theorem 0.20 $\mathscr{A}$ possesses a total decomposition set $\{d(S)\}$ and we show that

$$d(S) = e(S) \text{ for all } S. \quad (0.123)$$

Let p be a prime. If $p \mid e(R)$, $p \mid e(S)$ then either $R \subseteq S$ or $S \subseteq R$ whence the sets S for which $p \mid e(S)$ are ordered by inclusion; suppose they are

$$S_1 \subset S_2 \subset S_3 \subset \cdots \subset S_h. \quad (0.124)$$

We distinguish two cases. If $S_h = \{1, 2, \ldots, n\}$ we put $h = k$. If not we put $h = k - 1$ and set $S_k = \{1, 2, \ldots, n\}$. For $1 \leq j \leq k$ let

$$u_j = v_p\left(e(S_j)\right), \tag{0.125}$$

so that in the second case above, $u_k = 0$. For $1 \leq i \leq n$ let

$$m(i) = \min\{j : i \in S_j\}. \tag{0.126}$$

Then we have, from (0.122), (0.125–0.126),

$$v_p(a_i) = \sum_{m(i) \leq j \leq k} u_j. \tag{0.127}$$

Thus $v_p(a_i)$ takes exactly k distinct values

$$v_j = \sum_{j \leq l \leq k} u_l, \quad 1 \leq j \leq k, \tag{0.128}$$

as in (0.114). We find that the sets $Z_j(p)$ defined in (0.115) are the present S_j whence by (0.117),

$$d(S, p) = \begin{cases} p^{u_j} & \text{if } S = S_j \text{ for some } j, \\ 1 & \text{else.} \end{cases} \tag{0.129}$$

By (0.125) we now have

$$v_p\left(d(S)\right) = v_p\left(e(S)\right) \text{ for all } S, p \tag{0.130}$$

and (0.123) holds. Therefore $\{e(S)\}$ is a total decomposition set. This completes the proof of Theorem 0.21.

This theorem appeared in Hall (1990a) with a different proof.

1

Besicovitch and Behrend sequences

1.1 Introduction

We recall two definitions from §0.3: an integer sequence $\mathscr{A}$ is a Besicovitch sequence if its set of multiples $\mathscr{M}(\mathscr{A})$ possesses asymptotic density. $\mathscr{A}$ is a Behrend sequence if $1 \notin \mathscr{A}$ and $\mathscr{M}(\mathscr{A})$ has logarithmic density 1. It is then a consequence of Theorem 0.2 that $\mathbf{d}\mathscr{M}(\mathscr{A}) = 1$, so that any Behrend sequence is Besicovitch.

In this chapter we are concerned with general criteria to decide whether or not a sequence $\mathscr{A}$ has one or other of these properties. We may restrict our attention to non-trivial infinite $\mathscr{A}$, indeed we may assume that

$$\sum \left\{ \frac{1}{a} : a \in \mathscr{A} \right\} = \infty, \tag{1.1}$$

since, as in (0.81), this is necessary for $\mathscr{A}$ to be Behrend, moreover if (1.1) is false, $\mathscr{A}$ is automatically Besicovitch by Theorem 0.1.

1.2 Erdös' criterion

Erdös (1948a) (this paper and Behrend (1948) are adjacent in *Bull. AMS*) gave a necessary and sufficient condition for a sequence to be Besicovitch. We suppose that $\mathscr{A} = \{a_1, a_2, \dots\}$ is strictly increasing, and note that $a_1 > 1$.

Theorem 1.1 *Let $\mathscr{M}_i(\mathscr{A})$ denote the set of positive integers n such that $a_i | n$, $a_j \nmid n$ if $j < i$. Let*

$$M_i(x) = \text{card}\{n : n \in \mathscr{M}_i(\mathscr{A}), n \le x\}. \tag{1.2}$$

Then $\mathscr{A}$ is a Besicovitch sequence if and only if

$$\lim_{\varepsilon\to 0}\limsup_{x\to\infty} x^{-1} \sum_{x^{1-\varepsilon}<a_i\le x} M_i(x) = 0. \tag{1.3}$$

The proof which follows appeared in Erdös, Hall and Tenenbaum (1994).

Proof of Theorem 1.1 We recall from the proof of Theorem 0.1 that $\mathscr{M}_i(\mathscr{A})$ is Besicovitch for every $\mathscr{A}$ and i. From (0.24) and (0.67) we have

$$\sum_{i=1}^{\infty} \mathbf{d}\mathscr{M}_i(\mathscr{A}) = \boldsymbol{\delta}\mathscr{M}(\mathscr{A}). \tag{1.4}$$

For each fixed k, we have

$$\sum_{i=1}^{k} M_i(x) \sim \left(\sum_{i=1}^{k} \mathbf{d}\mathscr{M}_i(\mathscr{A})\right)x \tag{1.5}$$

and by the selection principle and (1.4) we can allow $k = k(x)$ to tend to infinity as $x \to \infty$ so slowly that

$$\sum_{i\le k(x)} M_i(x) \sim \boldsymbol{\delta}\mathscr{M}_i(\mathscr{A})x. \tag{1.6}$$

So far we have not used any hypothesis. Suppose now that $\mathscr{A}$ is Besicovitch. We have $\mathbf{d}\mathscr{M}(\mathscr{A}) = \boldsymbol{\delta}\mathscr{M}(\mathscr{A})$, whence from (1.6),

$$\sum_{i>k(x)} M_i(x) = o(x) \tag{1.7}$$

which implies (1.3). Hence (1.3) is necessary.

Suppose that (1.3) holds. To show that $\mathscr{A}$ is Besicovitch, it will be sufficient to prove that for every $\varepsilon > 0$ there exists $T = T(\varepsilon)$ such that, uniformly

$$\sum_{T<a_i\le x^{1-\varepsilon}} M_i(x) \ll \varepsilon x. \tag{1.8}$$

We write $n = lm$ where l is the product of the prime factors of n which do not exceed y. We recall from *Divisors* §0.3 the function

$$\Theta(x,y,z) = \operatorname{card}\left\{n \le x : \prod_{p^\nu \| n, p\le y} p^\nu > z\right\} \tag{1.9}$$

and the uniform bound, for $2 \le y \le z \le x$,

$$\Theta(x,y,z) \ll x\exp\left(-c\frac{\log z}{\log y}\right), \tag{1.10}$$

(with a suitable $c > 0$). We put $z = x^{\varepsilon/2}$, $y = x^{\varepsilon^2}$ so that

$$\Theta(x, y, z) \ll x \exp\left(-\frac{c}{2\varepsilon}\right) \ll \varepsilon x, \tag{1.11}$$

and it follows that to establish (1.8) we may restrict our attention to integers $n \le x$ for which $l \ll x^{\varepsilon/2}$. We employ the following result.

Lemma 1.2 *Let $M_i(x, \varepsilon)$ denote the number of integers $n = lm \le x$ such that $a_i | n$ but $a_j \nmid n$ if $j < i$, and in addition such that*

$$l = \prod_{p^\nu \| n, p \le y} p^\nu \le x^{\varepsilon/2}, \quad y = x^{\varepsilon^2}. \tag{1.12}$$

Then uniformly for $a_i \le x^{1-\varepsilon}$ we have

$$M_i(x, \varepsilon) \ll_\varepsilon \mathbf{d}\mathscr{M}_i(\mathscr{A})x. \tag{1.13}$$

Proof of lemma Let us write

$$n = lm = a_i n' = a_i l' m' \tag{1.14}$$

so that (whatever the prime factors of a_i may be) we have $l' | l$, $m' | m$. Thus $l' \le x^{\varepsilon/2}$ and $x/a_i l' \ge x^{\varepsilon/2}$. Since $n \in \mathscr{M}_i(\mathscr{A})$, so does $a_i l'$. Hence

$$M_i(x, \varepsilon) \le \sum_{l'} \text{card}\left\{m' : m' \le \frac{x}{a_i l'}\right\} \tag{1.15}$$

where the only constraint on m' is that its prime factors exceed y. Since $y < x^{\varepsilon/2} \le x/a_i l'$ this yields (see for example Halberstam and Richert (1974), Theorem 3.5)

$$M_i(x, \varepsilon) \ll \sum_{l'} \frac{x}{a_i l'} \prod_{p \le y} \left(1 - \frac{1}{p}\right) \tag{1.16}$$

where we require on the right that $a_i l' \in \mathscr{M}_i(\mathscr{A})$ and that the prime factors of l' do not exceed y. For y as in (1.12), Mertens' formula implies

$$M_i(x, \varepsilon) \ll x\varepsilon^{-2} \sum_{l'} \frac{1}{a_i l'} \prod_{p \le x} \left(1 - \frac{1}{p}\right). \tag{1.17}$$

For $n \in \mathscr{M}_i(\mathscr{A})$ we may write $n = a_i r s$ where

$$s = \prod_{p^\nu \| n, p > x} p^\nu. \tag{1.18}$$

Clearly $(s, a_j) = 1$ for every $j < i$, whence $n \in \mathcal{M}_i(\mathcal{A})$ if and only if $a_i r \in \mathcal{M}_i(\mathcal{A})$. Therefore

$$\mathbf{d}\mathcal{M}_i(\mathcal{A}) = \sum_r \frac{1}{a_i r} \prod_{p \le x} \left(1 - \frac{1}{p}\right) \tag{1.19}$$

where we require on the right that $a_i r \in \mathcal{M}_i(\mathcal{A})$ and that the prime factors of r do not exceed x. Every l' in (1.17) is an r, whence

$$M_i(x, \varepsilon) \ll \varepsilon^{-2} \mathbf{d}\mathcal{M}_i(\mathcal{A}) x. \tag{1.20}$$

This proves the lemma.

We deduce from the lemma that

$$\sum_{T < a_i \le x^{1-\varepsilon}} M_i(x, \varepsilon) \ll_\varepsilon \left(\sum_{a_i > T} \mathbf{d}\mathcal{M}_i(\mathcal{A}) \right) x \ll \varepsilon x, \tag{1.21}$$

if $T = T(\varepsilon)$ is sufficiently large. As shown above, this is all we need for (1.8), and $\mathcal{A}$ is Besicovitch as required. This proves Theorem 1.1.

Corollary 1.3 (Erdös (1948a)) *Let $\mathcal{A}$ be such that*

$$card\{\mathcal{A} \cap [1, x]\} \ll \frac{x}{\log x}. \tag{1.22}$$

Then $\mathcal{A}$ is a Besicovitch sequence. Moreover this is best possible in the sense that the right-hand side of (1.22) cannot be multiplied by $\xi(x)$ for any function $\xi(x) \to \infty$.

The first part of the corollary is a straightforward exercise in partial summation employing no more than the upper bound $M_i(x) \le x/a_i$. We shall not prove the second part here and refer the reader to Erdös' original paper or to Halberstam and Roth (1966). We give another corollary of Theorem 1.1 as part of Theorem 1.6 below.

A sufficient condition for $\mathcal{A}$ to be a Besicovitch sequence is that its elements should be pairwise coprime, because in this case (1.1) implies that $\mathcal{A}$ is Behrend. The next theorem provides a similar but considerably weaker condition for $\mathcal{A}$ to be Besicovitch.

Theorem 1.4 *Let k and s be fixed positive integers and $\mathcal{A}$ be such that for all $i_0 < i_1 < \cdots < i_s$ we have*

$$\omega\left((a_{i_0}, a_{i_1}, \ldots, a_{i_s})\right) \le k. \tag{1.23}$$

Then $\mathcal{A}$ is Besicovitch. This is best possible in the following sense: if

$\xi(x) \to \infty$ as $i \to \infty$ then we can find a non-Besicovitch sequence such that for all i,

$$\Omega(a_i) \le \xi(i). \tag{1.24}$$

This result is in all essentials due to Erdös, Hall and Tenenbaum (1994) – in the paper $s = 1$ – I am grateful to Tenenbaum for correcting my proof of the generalization. We require the following lemma.

Lemma 1.5 *Let $k \in \mathbf{Z}^+$ and $\eta \in (0, \frac{1}{2})$ be fixed, and $V_k(z, \eta)$ denote the number of integers $n \le z$ free of prime factors $\le z^\eta$ and such that $\omega(n) \le k$. Then for $z > z_0(\eta)$ we have*

$$V_k(z,\eta) \ll_k \frac{z}{\log z}\left(\log\frac{1}{\eta}\right)^{k-1}. \tag{1.25}$$

Proof of lemma Let $V_k'(z, \eta)$ count the integers as above which, in addition, exceed $\sqrt{z}$ and are square-free. We have

$$V_k(z,\eta) - V_k'(z,\eta) \le \sqrt{z} + \sum_{p>z^\eta} p^{-2} \ll z^{1-\eta} \tag{1.26}$$

and suppose that $z > z_0(\eta)$ implies $z^\eta > \log z$ so that this is $\ll z/\log z$. Let $P^+(n)$ denote the largest prime factor of n. If n is counted by $V_k'(z, \eta)$ then $P^+(n)^k > \sqrt{z}$ whence

$$\frac{\log z}{2k} V_k'(z,n) \le \sum_{n\le z}{}^{*} \mu^2(n)\log P^+(n) \tag{1.27}$$

where the asterisk denotes that n satisfies the conditions of the lemma. The primes contribute $\ll z$ on the right. If n is composite put $n = mp$ where $\omega(m) = j < k$ and $p = P^+(n)$. The contribution of these integers is

$$\begin{aligned}
\sum_{\substack{m\le z\\ \omega(m)<k}}{}^{*} \sum_{p\le z/m} \log p \quad &\ll \quad \sum_{\substack{m\le z\\ \omega(m)<k}}{}^{*} \frac{z}{m} \ll z\sum_{j<k}\ \sum_{z^\eta<p_i\le z} (p_1p_2\dots p_j)^{-1}\\
&\ll \quad z\sum_{j<k}\left(\sum_{z^\eta<p\le z}\frac{1}{p}\right)^j\\
&\ll_k \quad z\left(\log\frac{1}{\eta}\right)^{k-1}.
\end{aligned} \tag{1.28}$$

Thus

$$V_k'(z,\eta) \ll_k \frac{z}{\log z}\left(\log\frac{1}{\eta}\right)^{k-1} \tag{1.29}$$

and together with (1.26) this gives the result stated.

Proof of Theorem 1.4 Let (1.23) hold. We employ Theorem 1.1, and we require an upper bound for the sum

$$\sum_{x^{1-\varepsilon}<a_i\le x} M_i(x). \tag{1.30}$$

Let n be counted by $M_i(x)$ so that $n = a_i b$, say, moreover $a_j \nmid n$ for $j < i$. Put $y = x^{\varepsilon^2}$ and $a_i = u_i v_i$ where the prime factors of u_i do not exceed y, and the prime factors of v_i do exceed y. We first count the integers n for which $u_i > x^\varepsilon$: by (1.10) we have

$$\operatorname{card}\Big\{n \le x : \prod_{\substack{p^\nu \| n \\ p\le y}} p^\nu > x^\varepsilon\Big\} = \Theta(x, y, x^\varepsilon) \ll x \exp\left(-\frac{c}{\varepsilon}\right), \tag{1.31}$$

which contributes zero on the left of (1.3). Next we have $a_i > x^{1-\varepsilon}$ and $u_i \le x^\varepsilon$ whence $v_i > x^{1-2\varepsilon}$. Let us split these a_i into two classes according as $\omega(v_i) \le k$ or not, and let S_1 and S_2 be the contributions of these classes to (1.30). Then

$$S_1 \le \sum_{\substack{x^{1-2\varepsilon}<v\le x \\ P^-(v)>y,\ \omega(v)\le k}} \operatorname{card}\{n \le x : v|n\}, \tag{1.32}$$

where $P^-(v)$ denotes the smallest prime factor of v, whence by the definition of $V_k(z, \eta)$ in Lemma 1.5 we have

$$S_1 \le x \int_{x^{1-2\varepsilon}}^{x} z^{-1} dV_k(z, \eta) \tag{1.33}$$

in which $z^\eta \le y$ for $x^{1-2\varepsilon} \le z \le x$: we put $\eta = \varepsilon^2$. We apply the lemma which yields

$$\begin{aligned} S_1 &\ll_k x\left\{\frac{1}{\log x} + \int_{x^{1-2\varepsilon}}^{x} \frac{dz}{z\log z}\right\}\left(\log\frac{1}{\varepsilon}\right)^{k-1} \\ &\ll_k x\left(\frac{1}{\log x} + \varepsilon\right)\left(\log\frac{1}{\varepsilon}\right)^{k-1} \end{aligned} \tag{1.34}$$

which also contributes zero on the left of (1.3). We turn to S_2, in which $\omega(v_i) > k$.

For integers v such that $\omega(v) > k$ let $\overline{v} = g(v)$ be the product of the $k+1$ largest distinct prime factors of v. If $a_i = u_i v_i$ where $\omega(v_i) > k$ then $g(v_i) = g(a_i)$, and if $i_0 < i_1 < \cdots < i_s$ then $g(a_{i_0}) = g(a_{i_1}) = \cdots = g(a_{i_s})$ is impossible, by (1.23). Thus the function $v \to g(v)$, restricted to these v_i, is at most s onto 1. Let V denote the set of v_i such that $\omega(v_i) > k$

arising from $a_i \in (x^{1-\varepsilon}, x]$ and $V(z)$ be the counting function of V; from the above argument we have $V(z) \leq sV_{k+1}(z, \varepsilon^2)$ whence

$$\begin{aligned} S_2 &\leq x\sum_{v\in V}\frac{1}{v} = x\int_{x^{1-2\varepsilon}}^{x} z^{-1}dV(z) \\ &\leq V(x) + x\int_{x^{1-2\varepsilon}}^{x} V(z)\frac{dz}{z^2} \\ &\ll_k \frac{sx}{\log x}\left(\log\frac{1}{\varepsilon}\right)^k + sx\int_{x^{1-2\varepsilon}}^{x}\frac{(\log\frac{1}{\varepsilon})^k dz}{z\log z} \\ &\ll_{k,s} x\left(\frac{1}{\log x}+\varepsilon\right)\left(\log\frac{1}{\varepsilon}\right)^k \end{aligned} \tag{1.35}$$

which again contributes zero on the left-hand side of (1.3). Hence by Theorem 1.1 $\mathscr{A}$ is a Besicovitch sequence as required.

In order to prove the second part of the theorem we modify the construction of a non-Besicovitch sequence given in the proof of Theorem 0.1, and we need a generalization of Lemma 0.3 due to Tenenbaum (1984) and contained in Theorem 21(iii) of *Divisors*. Let $H(x,y,z)$ denote the number of integers n not exceeding x possessing at least one divisor in the interval $(y,z]$. Let $2y < z \leq y^{\frac{3}{2}}$ and put $z = y^{1+u}$. Then we have, uniformly for $y \geq 2$ and $x \geq z^2$,

$$H(x,y,z) \ll u^{\delta}x, \tag{1.36}$$

where $\delta = .086071\ldots$ is defined in (0.37). (We have given a slightly weakened form of this result for convenience. Some new results on the function $H(x,y,z)$ are contained in Chapter 7, where we are interested in 'short' intervals $(y,z]$ for which $z \leq 2y$.)

Let $T_1 \geq 2$ and $T_{j+1} > T_j^3$ for $j = 1,2,\ldots$; let $\varepsilon_j = j^{-2/\delta}$ for all j and

$$\mathscr{A}_j = \left\{a : T_j < a \leq T_j^{1+\varepsilon_j},\ P^-(a) > T_j^{\varepsilon_j^2}\right\}, \tag{1.37}$$

finally (with J at our disposal),

$$\mathscr{A} = \bigcup_{j>J}\mathscr{A}_j. \tag{1.38}$$

By (1.36), ignoring the condition on $P^-(a)$, we have

$$\underline{\mathbf{d}}\mathscr{M}(\mathscr{A}) \ll \sum_{j<J}\varepsilon_j^{\delta} \ll J^{-1}. \tag{1.39}$$

Now let $n \in (T_j, T_j^{1+\varepsilon_j}]$ and put $n = uv$ where $P^+(u) \leq T_j^{\varepsilon_j^2} < P^-(v)$. We allow $v = 1$, (we may formalize this if we wish by defining $P^-(1) = \infty$ as in *Divisors*), but our object is to show that usually $v \neq 1$, indeed $v > T_j$, so that $v \in \mathscr{A}_j$ and $n \in \mathscr{M}(\mathscr{A})$. If not, either $n \leq T_j^{1+(\varepsilon_j/2)}$ or $u > T_j^{\varepsilon_j/2}$, and the number of such integers $n \in (T_j, T_j^{1+\varepsilon_j})$ is, employing (1.10),

$$\begin{aligned} &\leq \quad T_j^{1+(\varepsilon_j/2)} + \Theta(T_j^{1+\varepsilon_j}, T_j^{\varepsilon_j^2}, T_j^{\varepsilon_j/2}) \\ &\ll \quad T_j^{1+\varepsilon_j}\left(T_j^{-\varepsilon_j/2} + \exp\left(-\frac{c}{2\varepsilon_j}\right)\right) = o(T_j^{1+\varepsilon_j}), \qquad (1.40) \end{aligned}$$

as $j \to \infty$. Thus $\overline{\mathbf{d}}\mathscr{M}(\mathscr{A}) = 1$, and by choosing J large enough in (1.39) we may ensure that $\mathscr{A}$ is not Besicovitch. Now let $a \in \mathscr{A}$ so that for some j, $a \in \mathscr{A}_j$. By (1.37) we have

$$\Omega(a) < \frac{(1+\varepsilon_j)}{\varepsilon_j^2}. \qquad (1.41)$$

On the other hand if $a = a_i$ then (for large j)

$$i > \sum_{J<h<j} |\mathscr{A}_h| > T_{j-1}, \qquad (1.42)$$

since every prime in $(T_h, T_h^{1+\varepsilon_h}]$ lies in $\mathscr{A}_h$, and we can arrange that T_j increases so fast that (1.41) and (1.42) imply (1.24). This completes the proof.

The next theorem also appeared in Erdös, Hall and Tenenbaum (1994); it answers a question which might well have been asked fifty years ago but apparently was not. The result is a little surprising too.

Theorem 1.6 *Let $\mathscr{A}_1$ and $\mathscr{A}_2$ be Besicovitch sequences. Then $\mathscr{A}_1 \cup \mathscr{A}_2$ is Besicovitch. There exist primitive Behrend sequences $\mathscr{A}_1$ and $\mathscr{A}_2$ such that $\mathscr{A}_1 \cap \mathscr{A}_2$ is not Besicovitch.*

The first part is a straightforward deduction from Erdös' criterion, Theorem 1.1, and we deal with this before discussing the second part.

Proof of part (i) Let us write $\mathscr{A}_1 \cup \mathscr{A}_2 = \{a_i\}$ where $a_1 < a_2 < \cdots$ and $M_i(x)$ denote the number of integers $n \leq x$ such that $a_i|n$, but for all $j < i$, $a_j \nmid n$. By Theorem 1.1 it will be sufficient to show that

$$\lim_{\varepsilon\to 0} \limsup_{x\to\infty} x^{-1} \sum_{x^{1-\varepsilon}<a_i\leq x} M_i(x) = 0. \qquad (1.43)$$

We consider the a_i belonging to $\mathscr{A}_1$. For such i, let $M_i^{(1)}(x)$ denote the number if integers $n \leq x$ such that $a_i|n$ but n has no smaller divisor in

$\mathscr{A}_1$. It may have such a divisor in $\mathscr{A}_2 \setminus \mathscr{A}_1$ and we see that

$$M_i(x) \le M_i^{(1)}(x). \tag{1.44}$$

If $a_i \in \mathscr{A}_2 \setminus \mathscr{A}_1$ we define $M_i^{(2)}(x)$ similarly, and $M_i(x) \le M_i^{(2)}(x)$. Since $\mathscr{A}_1$ and $\mathscr{A}_2$ are Besicovitch sequences (1.43) applies to $M_i^{(1)}(x)$ or $M_i^{(2)}(x)$ and the conclusion follows.

It is not difficult to construct an example to show that $\mathscr{A}_1$ and $\mathscr{A}_2$ may be Besicovitch sequences while $\mathscr{A}_1 \cap \mathscr{A}_2$ is not. Let $\mathscr{A}$ be any non-Besicovitch sequence and put $\mathscr{A}_1 = \{2\} \cup 2\mathscr{A}$. Then $\mathscr{M}(\mathscr{A}_1) = 2\mathbf{Z}^+$ and $\mathscr{A}_1$ is Besicovitch. Also $\mathscr{A}_2 = \{a : a \ge 3\}$ is Besicovitch, indeed it is a Behrend sequence. We have $\mathscr{A}_1 \cap \mathscr{A}_2 = 2\mathscr{A}$, and since $\mathscr{M}(2\mathscr{A}) = 2\mathscr{M}(\mathscr{A})$ this is not Besicovitch.

Consideration of this easy example might suggest that the extra condition that $\mathscr{A}_1$ and $\mathscr{A}_2$ should be primitive would be enough to make $\mathscr{A}_1 \cap \mathscr{A}_2$ Besicovitch, but this is not the case.

Proof of part (ii) The idea of the proof is as follows. We begin with a primitive, non-Besicovitch sequence $\mathscr{B}$ which is actually going to be $\mathscr{A}_1 \cap \mathscr{A}_2$. To construct $\mathscr{A}_1$ and $\mathscr{A}_2$ we extend $\mathscr{B}$ in two ways, disjointly, into a Behrend sequence. Of course the awkward part is to arrange that $\mathscr{A}_1$ and $\mathscr{A}_2$ are primitive, since it is an easy matter to extend any sequence into a Behrend sequence for example by adding primes.

The construction of $\mathscr{B}$ involves a slight modification of the sequence $\mathscr{A}$ used in the proof of Theorem 1.4 and defined by (1.38), after which we set $\mathscr{B} = \mathscr{P}(\mathscr{A})$. The modification occurs in (1.37) which becomes

$$\mathscr{A}'_j = \left\{a : T_j < a \le T_j^{1+\varepsilon_j},\ P^-(a) > T_j^{\varepsilon_j^2},\ P^+(a) \le T_j^{\frac{1}{2}}\right\}. \tag{1.45}$$

The extra condition on the greatest prime factor of a makes $\mathscr{A}$ thinner so that (1.39) still holds: we may choose J so large that $\underline{\mathbf{d}}\mathscr{M}(\mathscr{A}) < \frac{1}{9}$. We claim that $\overline{\mathbf{d}}\mathscr{M}(\mathscr{A}) \ge \frac{1}{9}$, whence $\mathscr{A}$ is non-Besicovitch. To see this, we consider the integers $n \le T_j^{1+\varepsilon_j}$, writing $n = uv$ where, as in the argument following (1.39), we have $P^+(u) \le T_j^{\varepsilon_j^2} < P^-(v)$. Suppose that n has no divisor in $\mathscr{A}'_j$. Then there are three possibilities, namely

(i) $n \le T_j^{1+(\varepsilon_j/2)}$

(ii) $u > T_j^{\varepsilon_j/2}$

(iii) $P^+(n) > T_j^{1/2}$.

The number of integers in cases (i) or (ii) may be estimated as in (1.40), and for j sufficiently large does not exceed $\frac{1}{8}T_j^{1+\varepsilon_j}$. Otherwise $v = \frac{n}{u} > T_j$

whence $v \in \mathscr{A}_j$, unless $P^+(v) > T_j^{\frac{1}{2}}$. But then n is in case (iii). The number of such integers n does not exceed

$$\sum_{n \le T_j^{1+\varepsilon_j}} \sum_{\substack{p|n \\ p > \sqrt{T_j}}} 1 \le \left(\log 2 + \varepsilon_j + o(1)\right) T_j^{1+\varepsilon_j}, \tag{1.46}$$

(inverting summations on the left) and this is less than $\frac{3}{4} T_j^{1+\varepsilon_j}$ (assuming j is large). This leaves at least $[T_j^{1+\varepsilon_j}] - \frac{7}{8} T_j^{1+\varepsilon_j} > \frac{1}{9} T_j^{1+\varepsilon_j}$ integers n in $\mathscr{M}(\mathscr{A}'_j)$, whence $\bar{\mathbf{d}}\mathscr{M}(\mathscr{A}) \ge \frac{1}{9}$. Our claim is valid and $\mathscr{A}$ is non-Besicovitch. As stated above, we put $\mathscr{B} = \mathscr{P}(\mathscr{A})$. Next we construct $\mathscr{A}_1$ and $\mathscr{A}_2$. For $l = 1, 2$ (and p, q primes), put

$$\mathscr{C}_l = \{pq : q > \exp\left((\log p)^2\right), q \equiv (-1)^l \pmod 4\} \tag{1.47}$$

and set

$$\mathscr{A}_l = \mathscr{B} \cup \mathscr{C}_l. \tag{1.48}$$

Clearly $\mathscr{C}_1$ and $\mathscr{C}_2$ are disjoint whence $\mathscr{A}_1 \cap \mathscr{A}_2 = \mathscr{B}$. We have to show that $\mathscr{A}_1$ and $\mathscr{A}_2$ are primitive, and since $\mathscr{B}$ is primitive and each $\mathscr{C}_l$ is primitive (its members all have the same number of prime factors) this reduces to showing that $b \nmid c_l$, $c_l \nmid b$ (with obvious notation). The former assertion is clear because from (1.45), $\Omega(a) \ge 3$ for every $a \in \mathscr{A}'_j$ whence $\Omega(b) \ge 3 > \Omega(c_l)$. Also (1.45) implies, for $a \in \mathscr{A}'_j$,

$$\log P^+(a) < \frac{1}{2\varepsilon_j^2} \log P^-(a) < \left(\log P^-(a)\right)^2 \tag{1.49}$$

if, as we may assume, $\log T_j > \frac{1}{2} j^{8/\delta} = \frac{1}{2}\varepsilon_j^{-4}$. This justifies the latter assertion that $c_l \nmid b$. It remains to show that $\mathscr{A}_1$ and $\mathscr{A}_2$ are Behrend sequences: in fact $\mathscr{C}_1$ and $\mathscr{C}_2$ are already Behrend. Consider $\mathscr{C}_1$ for example. Let $Q^+(n)$ denote the greatest prime factor $\equiv 1 \pmod 4$ of n; if n has no such prime factor then $Q^+(n) = 0$. We have to show that

$$Q^+(n) > \exp\left(\left(\log P^-(n)\right)^2\right) \ p.p. \tag{1.50}$$

i.e. that the number of integers $n \le x$ for which the above inequality is false is $o(x)$. Let $\xi(x)$ be at our disposal. For such an integer n, either $P^-(n) > \exp \xi(x)$ or $Q^+(n) \le \exp \xi(x)^2$, whence the number of these integers is

$$\ll x \prod_{p \le e^{\xi}} \left(1 - \frac{1}{p}\right) + x \prod_{\substack{\exp \xi^2 < q \le x \\ q \equiv 1 \pmod 4}} \left(1 - \frac{1}{q}\right). \tag{1.51}$$

Let χ be the non-principal Dirichlet character (mod 4). The product on the right of (1.51) is

$$\ll \prod_{\exp\xi^2<q\le x}\left(1-\frac{1}{q}\right)^{\frac{1}{2}}\left(1-\frac{\chi(q)}{q}\right)^{\frac{1}{2}}$$

$$\ll \prod_{\exp\xi^2<q\le x}\left(1-\frac{1}{q}\right)^{\frac{1}{2}}$$

$$\ll \frac{\xi}{\sqrt{\log x}} \tag{1.52}$$

because partial products of the convergent product

$$\prod_q\left(1-\frac{\chi(q)}{q}\right)^{1/2} = L(1,\chi)^{-1/2} = \frac{2}{\sqrt{\pi}} \tag{1.53}$$

are bounded. By (1.51) (and Mertens' theorem) the number of exceptional integers n not exceeding x is

$$\ll x\xi^{-1} + \frac{x\xi}{\sqrt{\log x}} \ll x(\log x)^{-1/4} \tag{1.54}$$

if we choose $\xi(x)$ optimally. Therefore $\mathscr{C}_1$, and similarly $\mathscr{C}_2$, is Behrend. This proves part (ii) of Theorem 1.6.

1.3 Behrend sequences

Erdös (1979) wrote "It seems very difficult to obtain a necessary and sufficient condition that if $a_1 < \cdots$ is a sequence of integers then almost all integers n should be a multiple of one of the a's." This problem had concerned Erdös for many years and is merely formalized by our definition at the end of §0.3 of a Behrend sequence. Erdös himself, and other authors, have given many examples of (what we now call) Behrend sequences, some of them far from easy, but of course special cases.

In this section we make a start on this problem, and it is taken up again at other points of the book. We begin with a general, necessary condition for a sequence to be Behrend, and we show by examples that it is in some respects best possible.

Theorem 1.7 (Hall (1990b)) *Let $\mathscr{A}$ be a Behrend sequence. Then for each y, $(0 < y \le 1)$ and β, $\beta > y - 1 - \log y$, we have*

$$\sum_{i=1}^{\infty} a_i^{-1}(\log a_i)^{\beta} y^{\Omega(a_i)} = \infty. \tag{1.55}$$

For every y and $\beta < y-1-\log y$ there exists a Behrend sequence $\mathscr{A}(y,\beta)$ for which the series in (1.55) is convergent.

It is likely that the condition on β can be weakened to $\beta \geq y-1-\log y$, but we are unable to settle this point. The theorem would then include, as the special case $y = 1$, the only other general condition of this sort which we have, Corollary 0.10. This is

$$\sum_{i=1}^{\infty} a_i^{-1} = \infty. \tag{1.56}$$

We make two remarks at this point. The first is that $y-1-\log y > 0$ for $0 < y < 1$, whence we must have $\beta > 0$ and (1.55) is weaker than (1.56) if we have no information about $\Omega(a_i)$. We should expect this to be the case because (1.56) is both necessary and sufficient for $\mathscr{A}$ to be Behrend when (for example) all the a_i are prime. Our second remark concerns the status of the variable y. It will emerge that in an important special case we may choose $y = \frac{1}{2}$ optimally. However this could be misleading: we show by example that if any open interval of the range $(0,1)$ be omitted then there exists a non-Behrend sequence which the theorem consequently fails to detect.

Proof of Theorem 1.7 We prove the first part here, and then we describe the special sequences $\mathscr{A}(y,\beta)$ which establish that the condition on β is, apart from the moot point concerning equality, best possible. Then we justify our second remark above about y. We require a lemma which is a useful variant of Erdös' law of the iterated logarithm discussed in detail in *Divisors*, Chapter 1. It is a similar result to that indicated in Exercise 10 of *Divisors*.

Lemma 1.8 *Let $\Omega(n,t)$ denote the number of prime factors not exceeding t of n, counted according to multiplicity. For each $\kappa \in (0,1)$ and $t_0 \geq 3$ let $\mathscr{B}(\kappa,t_0)$ denote the sequence of integers n for which*

$$\sup_{t_0 \leq t \leq n} \left| \frac{\Omega(n,t)}{\log\log t} - 1 \right| \geq \kappa. \tag{1.57}$$

Then we have

$$\overline{\mathbf{d}}\mathscr{B}(\kappa,t_0) \ll_\kappa (\log t_0)^{-K} \tag{1.58}$$

where

$$K = (1+\kappa)\log(1+\kappa) - \kappa. \tag{1.59}$$

Proof of lemma We use a result from the proof of Lemma 0.3, that is (0.38). For fixed $y_0 < 2$ we have uniformly for $0 < y \le y_0$ and $2 \le t \le x$ that

$$\sum_{n \le x} y^{\Omega(n,t)} \ll_{y_0} x(\log t)^{y-1}. \tag{1.60}$$

We fix $y_0 = 1 + \kappa$. Let $t \in [t_0, x]$ be fixed and $0 < \lambda \le \kappa$. By (1.60), the number of integers $n \le x$ such that

$$\Omega(n,t) \ge (1+\lambda)\log\log t \tag{1.61}$$

does not exceed

$$\sum_{n \le x} (1+\lambda)^{\Omega(n,t)-(1+\lambda)\log\log t} \ll_\kappa x(\log t)^{\lambda-(1+\lambda)\log(1+\lambda)}. \tag{1.62}$$

Similarly, the number of integers $n \le x$ such that

$$\Omega(n,t) \le (1-\lambda)\log\log t \tag{1.63}$$

does not exceed

$$\sum_{n \le x} (1-\lambda)^{\Omega(n,t)-(1-\lambda)\log\log t} \ll_\kappa x(\log t)^{-\lambda-(1-\lambda)\log(1-\lambda)}. \tag{1.64}$$

Put $\lambda - (1+\lambda)\log(1+\lambda) = -L$. The exponent of $\log t$ in (1.64) is $< -L$, and we have shown that for $0 < \lambda \le \kappa$ and each fixed t we have

$$\operatorname{card}\left\{ n \le x : \left| \frac{\Omega(n,t)}{\log\log t} - 1 \right| \ge \lambda \right\} \ll_\kappa x(\log t)^{-L}. \tag{1.65}$$

This inequality applies for fixed t. To obtain a uniform result such as (1.57) we introduce a sequence of discrete *checkpoints* $t_k \in [t_0, x]$ and employ the monotonicity of $\Omega(n,t)$ and the slow variation of the function $\log\log t$ to interpolate between these checkpoints.

Let $\theta > 1$ be such that

$$\log\theta < \frac{\kappa}{1+\kappa}\log\log 3 \tag{1.66}$$

and let t_k satisfy

$$\log t_k = \theta^k \log t_0,\ k = 1, 2, 3, \dots; \tag{1.67}$$

then if $t \in (t_k, t_{k+1})$ and

$$|\Omega(n,t) - \log\log t| \ge \kappa \log\log t \tag{1.68}$$

we have either $\Omega(n, t_{k+1}) \geq (1+\kappa)\log\log t_k$ or $\Omega(n, t_k) \leq (1-\kappa)\log\log t_{k+1}$, whence

$$\left|\frac{\Omega(n,t_j)}{\log\log t_j} - 1\right| \geq \kappa - \frac{(1+\kappa)\log\theta}{\log\log t_j} \tag{1.69}$$

for at least one of $j = k$ or $j = k+1$. Since $t_0 \geq 3$, the right-hand side of (1.69) is positive, by (1.66), and we denote it by λ_j. We put $L_j = (1+\lambda_j)\log(1+\lambda_j) - \lambda_j$, and we note that $0 < \lambda_j \leq \kappa$ so that (1.65) is applicable. Let K be as in (1.59). By the mean value theorem, we have $K - L_j < (\kappa - \lambda_j)\log(1+\kappa)$ whence $(K - L_j)\log\log t_j \ll 1$ (uniformly). We apply inequality (1.65) with $t = t_k$, $\lambda = \lambda_k$, $L = L_k$ for every k, and we deduce that the number of integers $n \leq x$ for which (1.57) holds is

$$\begin{aligned} &\ll \sum_{k=0}^{\infty} x(\log t_k)^{-L_k} \ll_\kappa x\sum_{k=0}^{\infty}(\log t_k)^{-K} \\ &\ll x(\log t_0)^{-K}\sum_{k=0}^{\infty}\theta^{-kK} \ll_\kappa x(\log t_0)^{-K}, \end{aligned} \tag{1.70}$$

after choosing θ suitably in (1.66). This proves the lemma.

(In *Divisors*, Exercise 10 the specified checkpoints are $t_k = t_0^{2^k}$, that is $\theta = 2$; this works because we have $K < \frac{1}{2}\kappa^2 < \kappa$ and so we may assume that $\kappa\log\log t_0$ is large, else (1.58) is nugatory. We replace 3 by t_0 in (1.66).)

We proceed to the proof of the theorem. We may assume $y < 1$, and we have $\beta > y - 1 - \log y$ so we may fix $\kappa > 0$ so that $\beta = y - 1 - (1+\kappa)\log y$ (making β smaller if this would involve $\kappa \geq 1$). We next fix t_0 in Lemma 1.8 so that $\overline{\mathbf{d}}\mathscr{B}(\kappa, t_0) \leq \frac{1}{3}$, and we let l be such that $a_l \geq t_0$. We consider the tail $\mathscr{A}_l = \{a_l, a_{l+1}, \ldots\}$ of $\mathscr{A}$. We argue by contradiction, showing that if the series (1.55) is convergent and l is large enough then $\mathscr{A}_l$ is not Behrend. Let

$$\chi(n) = \min(1, \tau(n, \mathscr{A}_l)) \tag{1.71}$$

so that if $\mathscr{A}_l$ were Behrend we should have $\chi(n) = 1$ *p.p.* whence

$$\lim_{\sigma\to 1+}(\sigma - 1)\sum_{n=1}^{\infty} n^{-\sigma}\chi(n) = 1. \tag{1.72}$$

By the construction of $\mathscr{B}(\kappa, t_0)$ we have

$$\limsup_{\sigma\to 1+}(\sigma - 1)\sum_{n\in\mathscr{B}(\kappa,t_0)} n^{-\sigma}\chi(n) \leq \frac{1}{3} \tag{1.73}$$

since $\chi(n) \le 1$. For $n \notin \mathscr{B}(\kappa, t_0)$ we have

$$\chi(n) \le \tau(n, \mathscr{A}_l) < \sum_{\substack{a_i | n \\ i \ge l}} y^{\Omega(n,a_i)-(1+\kappa)\log\log a_i} \tag{1.74}$$

because $y < 1$ and the exponent is negative by (1.57). Therefore

$$\sum_{n \notin \mathscr{B}} n^{-\sigma}\chi(n) < \sum_{i \ge l} (\log a_i)^{-(1+\kappa)\log y} \sum_{n \equiv 0 (\mathrm{mod}\ a_i)} n^{-\sigma} y^{\Omega(n,a_i)} \tag{1.75}$$

and we put $n = ma_i$ in the inner sum, to obtain

$$< \sum_{i \ge l} a_i^{-\sigma} (\log a_i)^{-(1+\kappa)\log y} y^{\Omega(a_i)} \sum_{m=1}^{\infty} m^{-\sigma} y^{\Omega(m,a_i)}. \tag{1.76}$$

The inner sum in (1.76) is

$$\begin{aligned} \prod_{p \le a_i} \left(1 - \frac{y}{p^\sigma}\right)^{-1} \prod_{p > a_i} \left(1 - \frac{1}{p^\sigma}\right)^{-1} &= \zeta(\sigma) \prod_{p \le a_i} \left(\frac{1 - p^{-\sigma}}{1 - yp^{-\sigma}}\right) \\ &\ll \zeta(\sigma) \exp\left((y-1)\sum_{p \le a_i} p^{-\sigma}\right). \end{aligned}$$

We have

$$\begin{aligned} \sum_{p \le a_i} p^{-\sigma} &\ge \sum_{p \le a_i} \frac{1}{p} - (\sigma - 1) \sum_{p \le a_i} \frac{\log p}{p} \\ &\ge \log\log a_i - (\sigma - 1)\log a_i + O(1) \end{aligned} \tag{1.77}$$

and (1.75)–(1.77) imply that

$$\sum_{n \notin \mathscr{B}} n^{-\sigma}\chi(n) \ll (\sigma - 1)^{-1} \sum_{i \ge l} a_i^{-1} (\log a_i)^\beta y^{\Omega(a_i)}$$

(recalling that $y - 1 - (1+\kappa)\log y = \beta$). By hypothesis the series (1.55) is convergent whence if l is sufficiently large we have

$$\limsup_{\sigma \to 1+} (\sigma - 1) \sum_{n \notin \mathscr{B}} n^{-\sigma}\chi(n) \le \frac{1}{3}. \tag{1.78}$$

This, together with (1.73), contradicts (1.72) whence $\mathscr{A}$ is not Behrend. This completes the proof of the first part of the theorem.

It remains to describe the sequences $A(y, \beta)$. For this purpose we introduce a family of special Behrend sequences $\mathscr{A}^*(t)$ depending on the parameter $t \in (0, 1)$. Let

$$\|u\| = \min\{|u - m| : m \in \mathbf{Z}\}, \ (u \in \mathbf{R}), \tag{1.79}$$

and

$$\mu(t) = -t\log t - (1-t)\log(1-t), \quad t \in (0,1); \tag{1.80}$$

furthermore let $\xi(a) \to \infty$ as $a \to \infty$. We define

$$\mathscr{A}^*\left(\frac{1}{2}\right) = \{a > 1 : \| \log a\| < (\log a)^{-\log 2}\exp(\xi(a)\sqrt{\log\log a})\} \tag{1.81}$$

and for $t < \frac{1}{2}$, $t > \frac{1}{2}$ respectively

$$\begin{aligned} \mathscr{A}^*(t) &= \{a > 1 : \Omega(a) \le t\log\log a + \xi(a)\sqrt{\log\log a}, \\ &\qquad \| \log a\| < T^{-\log\log a + \xi(a)\sqrt{\log\log a}}\}, \end{aligned} \tag{1.82}$$

$$\begin{aligned} \mathscr{A}^*(t) &= \{a > 1 : \Omega(a) \ge t\log\log a - \xi(a)\sqrt{\log\log a}, \\ &\qquad \| \log a\| < T^{-\log\log a + \xi(a)\sqrt{\log\log a}}\}, \end{aligned} \tag{1.83}$$

where $T = \exp\mu(t)$. We prove in Theorem 4.23 that these are Behrend sequences. Some remarks about their structure may be helpful here. First, $\mu(t)$ is symmetric about $\frac{1}{2}$ where it takes its maximum value $\log 2$; therefore as t increases from 0 to 1, T first increases from 1 to 2 and then decreases to 1, so that the condition on $\| \log a\|$ in (1.81)–(1.83) is most stringent when $t = \frac{1}{2}$. This condition is relaxed progressively as $| t - \frac{1}{2} |$ increases, but there is a complementary condition on $\Omega(a)$ in (1.82) and (1.83) which we observe may be written in the form

$$\left|\Omega(a) - \frac{1}{2}\log\log a\right| \ge \left|t - \frac{1}{2}\right| \log\log a - \xi(a)\sqrt{\log\log a}. \tag{1.84}$$

Normally, a random divisor of a large random integer will have about $\frac{1}{2}\log\log a$ prime factors. We explain this statement heuristically here: it will be made precise in Theorem 5.4. If $a|n$ then $\Omega(a)$ is (essentially) binomially distributed with mean $\Omega(n)/2$, (because we can assume that n has few repeated prime factors). Also $\Omega(n)$ will have about $\log\log n$ prime factors by the Hardy–Ramanujan theorem, moreover we may expect that $\log\log a$ and $\log\log n$ will be almost indistinguishable. In view of this, we see that (1.84) becomes more taxing as $\left|t - \frac{1}{2}\right|$ increases.

We wish to apply the first part of Theorem 1.7 to the sequences $\mathscr{A}^*(t)$ to show that both the theorem and the specification of these sequences are in some respects best possible. We require the following result.

Lemma 1.9 *Let* $0 < y \le 1$. *Then we have*

$$\sum_{x-w<n\le x} y^{\omega(n)} \ll w(logw)^{y-1} \tag{1.85}$$

uniformly in y, *and for* $x > w \ge 2$.

Proof We may assume $x > 2w$ else (1.85) is an easy application of the Halberstam–Richert inequality (*Divisors,* Theorem 00). We follow Shiu (1980) very closely. Put

$$z = w^{1/3}, \quad v = \log z . \log\log z, \tag{1.86}$$

assuming as we may that $z > 16$: we need that $\log\log z > 1$ later. For each $n \in (x-w, x]$ we write $n = b_n d_n$ where b_n is maximal, subject to the constraints

$$b_n \le z, \quad P^+(b_n) < P^-(d_n). \tag{1.87}$$

Notice that $n > x - w > w = z^3$, whence $d_n > z^2 > 1$. We split the integers $n \in (x-w, x]$ into four classes, and we write $\sum^{(i)}$ for summation over the ith class. These classes are:

(i) $P^-(d_n) > \sqrt{z}$,
(ii) $P^-(d_n) \le \sqrt{z}$, $b_n \le \sqrt{z}$,
(iii) $P^-(d_n) \le v$, $b_n > \sqrt{z}$,
(iv) $v < P^-(d_n) \le \sqrt{z}$, $b_n > \sqrt{z}$.

We note that $y \le 1$. We have

$$\sum\nolimits^{(1)} y^{\omega(n)} \le \sum_{b \le z} y^{\omega(b)} \sum_{\substack{(x-w)/b<d\le x/b \\ P^-(d)>\sqrt{z}}} 1, \tag{1.88}$$

and we may apply Selberg's upper bound sieve to the inner sum because $(w/b) \ge z^2$. The above is

$$\begin{aligned} &\ll \sum_{b\le z} y^{\omega(b)} . \frac{w}{b \log z} \ll \frac{w}{\log w} \prod_{p \le z} \left(1 + \frac{y}{p-1}\right) \\ &\ll w(\log w)^{y-1} \end{aligned} \tag{1.89}$$

as required. Next, let $n \in$ class (ii). Let $P^-(d_n) = q$ and $q^s \| d_n$. By the maximal property of b_n we have $b_n q^s > z$, whence $q^s > \sqrt{z}$. For each prime $q \le \sqrt{z}$ let $s(q)$ denote the smallest integer such that $q^{s(q)} > \sqrt{z}$. Then $s \ge s(q) \ge 2$, moreover $q^{s(q)} \le z$. Hence

$$\begin{aligned} \sum\nolimits^{(2)} y^{\omega(n)} &\le \sum_{q \le \sqrt{z}} \sum_{\substack{x-w<n\le x \\ n \equiv 0 (\mathrm{mod}\ q^{s(q)})}} 1 \\ &\ll \sum_{q\le\sqrt{z}} \frac{w}{q^{s(q)}} \\ &\ll w\Bigg(\sum_{q \le z^{1/4}} z^{-1/2} + \sum_{q > z^{1/4}} q^{-2}\Bigg) \ll wz^{-1/4}, \end{aligned} \tag{1.90}$$

well within (1.85). Next

$$\sum\nolimits^{(3)} y^{\omega(n)} \le \sum\nolimits^{(3)} 1 \ll \sum_{\substack{\sqrt{z}<b\le z\\ P^+(b)<v}} \frac{w}{b} \le \frac{w}{\sqrt{z}}\Psi(z,v). \tag{1.91}$$

Shiu's Lemma 1, which is an application of Rankin's method, gives $\Psi(z,v) \le \exp\{3\log z/\sqrt{\log\log z}\}$, again more than we require.

In order to cope with class (iv) we require the following inequality (valid by the way for all $y > 0$), which is a special case of Shiu's lemma 4. As Shiu points out, this derives from Wolke (1971). We have, uniformly for $z \ge 6$ and $1 \le r \le \log z/\log\log z$, that

$$\sum_{\substack{b>\sqrt{z}\\ P^+(b)\le z^{1/r}}} \frac{y^{\omega(b)}}{b} \ll (\log z)^y \exp\left\{-\frac{1}{10} r\log r\right\}. \tag{1.92}$$

This again depends on Rankin's method. Let

$$r_0 = \left[\frac{\log z}{\log v}\right] < \frac{\log z}{\log\log z}, \tag{1.93}$$

(because $z \ge 16$). Then $v^{r_0+1} > z$, i.e. $z^{1/(r_0+1)} < v$. We split the range for $P^-(d_n)$ in class (iv) into the sub-ranges

$$z^{1/(r+1)} < P^-(d_n) \le z^{1/r}, \qquad 2 \le r \le r_0, \tag{1.94}$$

and we note from (1.87) that $P^-(b_n) < z^{1/r}$ in this sub-range, whence

$$\begin{aligned}
\sum\nolimits^{(4)} y^{\omega(n)} &\le \sum_{r=2}^{r_0} \sum_{\substack{\sqrt{z}<b\le z\\ P^+(b)\le z^{1/r}}} y^{\omega(b)} \sum_{\substack{(x-w)/b<d\le x/b\\ P^-(d)>z^{1/(r+1)}}} 1\\
&\ll \sum_{r=2}^{r_0} \sum_{\substack{\sqrt{z}<b\le z\\ P^+(b)\le z^{1/r}}} y^{\omega(b)} \frac{(r+1)w}{b\log z}\\
&\ll \frac{w}{\log w}\sum_{r=2}^{r_0}(r+1)(\log z)^y \exp\left\{-\frac{1}{10}r\log r\right\}
\end{aligned} \tag{1.95}$$

by the sieve, in line 2, and (1.92) in line 3. The sum over r is $O(1)$, and (1.95) together with (1.89), (1.90) and (1.91) yields the desired bound. This proves Lemma 1.9.

We consider the infinite series (1.55) when $\mathscr{A}$ is one of the sequences $\mathscr{A}^*(t)$. We note that we may replace ω by Ω in (1.85) because $y \le 1$.

Let us set

$$X_k = \sum \left\{ a^{-1}(\log a)^\beta y^{\Omega(a)} : \left|a - e^k\right| \le e^k k^{-\mu(t)} \exp(\eta(k)\sqrt{\log k}), \right.$$
$$\left. \Omega(a)t \log k + \eta(k)\sqrt{\log k} \right\}, \quad (1.96)$$

in which $\eta(k) \to \infty$ as $k \to \infty$, and $t < \frac{1}{2}$. We consider whether or not we have

$$\sum_{k=2}^{\infty} X_k < \infty. \quad (1.97)$$

For $0 < z \le 1$ at our disposal, we have $X_k \ll X_k(z)$ where

$$X_k(z) = e^{-k} k^{\beta - t\log z} z^{-\eta(k)\sqrt{\log k}} \sum \{(yz)^{\Omega(a)} :$$
$$\left|a - e^k\right| \le e^k k^{-\mu(t)} \exp(\eta(k)\sqrt{\log k})\} \quad (1.98)$$

and we estimate the inner sum by means of Lemma 1.9, which yields

$$X_k(z) \ll k^{\beta - t\log z - \mu(t) + yz - 1} \exp\left\{(1 - \log z)\eta(k)\sqrt{\log k}\right\}. \quad (1.99)$$

We choose $z = t/y$ to minimize the exponent of k: we require $t \le y$ at this point. The exponent is

$$\beta + t\log y + (1-t)\log(1-t) + t - 1 \quad (1.100)$$

which, as a function of t, has a minimum $\beta + \log y - y$, at $t = 1 - y$. If $y > \frac{1}{2}$ then $t < \frac{1}{2}$, $z < 1$. If $\beta < y - 1 - \log y$ the exponent of k is < -1 and (1.97) holds. The series (1.55) converges although $A(y, \beta)$ is Behrend. Next, let X_k' be as in (1.96) except that now $t > \frac{1}{2}$ and the condition on $\Omega(a)$ is $\Omega(a) \ge t\log k - \eta(k)\sqrt{\log k}$. We have $X_k' \ll X_k'(z)$ provided $z \ge 1$, where $X_k'(z)$ is as in (1.98), with the exponent of z (outside the sum) changed to $\eta(k)\sqrt{\log k}$. We apply Lemma 1.9 – we need $yz \le 1$ – and we obtain

$$X_k'(z) \ll k^{\beta - t\log z - \mu(t) + yz - 1} \exp\{(1 + \log z)\eta(k)\sqrt{\log k}\}. \quad (1.101)$$

As before we choose $z = t/y$ to minimize the exponent of k, which is in order provided $1 \ge t \ge y$. If $y < \frac{1}{2}$ we choose $t = 1 - y > \frac{1}{2}$ and the exponent, given by (1.100), is < -1. The conclusion is the same as before. We leave the case $y = \frac{1}{2}$ as an exercise. This completes the proof of Theorem 1.7.

We draw two further conclusions from the argument given above. The first is that $T = \exp \mu(t)$ with $\mu(t)$ defined by (1.80) is the best possible T in (1.81)–(1.83). For if $\mu(t)$ were larger we could make the exponent of k in (1.99) and (1.101) < -1, with $\beta > y - 1 - \log y$, $y = 1 - t$. By the theorem, the corresponding sequences $\mathscr{A}^*(t)$ would not be Behrend. In this sense the specification of $\mathscr{A}^*(t)$ is best possible. Finally we may now justify the second of our remarks following the statement of Theorem 1.7. Let $Y \in (0,1)$ be an open interval. Choose t such that $1-t \in Y$ and put $\mu_1(t) = \mu(t) + \delta$, so that the sequence $\mathscr{A}_1$ obtained by replacing $\mu(t)$ by $\mu_1(t)$ in the definition of $\mathscr{A}^*(t)$ is non-Behrend. Let us consider the exponent of k in (1.99) or (1.101), with $z = t/y$ optimized, $y \notin Y$. This is

$$\begin{aligned}\beta - t\log t + t\log y - \mu_1(t) + t - 1 \\ > y - 1 - \log y - t\log t + t\log y - \mu_1(t) + t - 1 \\ > -1 + \varepsilon - \delta, \qquad y \notin Y,\ \beta > y - 1 - \log y,\end{aligned}$$

where $\varepsilon > 0$, and depends on Y only. We put $\delta = \varepsilon$ and we see that we cannot detect that $\mathscr{A}_1$ is non-Behrend. (The loss in the inequalities $X_k \le X_k(z)$, $X_k' \le X_k'(z)$ with z optimized is not significant here.)

We give two corollaries of Theorem 1.7. The first of these was the original motivation behind this theorem.

Corollary 1.10 *Let $\mathscr{A}$ have the form*

$$\mathscr{A} = \bigcup_{k=1}^{\infty} (\mathbf{Z}^+ \cap (T_k, T_k + V_k]) \tag{1.102}$$

where $T_1 > 1$, $T_{k+1} \ge T_k + V_k$ for every k, and for some $c = c(\mathscr{A}) > 0$ we have

$$V_k \ge T_k^c. \tag{1.103}$$

Then a necessary condition for $\mathscr{A}$ to be Behrend is that for every $\gamma >$ $\log 2 - 1$ we have

$$\sum_{a \in \mathscr{A}} a^{-1}(\log a)^{\gamma} = \infty. \tag{1.104}$$

Proof We may assume that $V_k \le T_k$ always; if this condition is not met we may insert extra T's as necessary. Then by Lemma 1.9, we have (for any β, and $0 < y \le 1$),

$$\begin{aligned}\sum_{T_k < a \le T_k + V_k} a^{-1}(\log a)^{\beta} y^{\Omega(a)} &\ll T_k^{-1}(\log T_k)^{\beta} \sum_{T_k < a \le T_k + V_k} y^{\Omega(a)} \\ &\ll T_k^{-1}(\log T_k)^{\beta} V_k (\log V_k)^{y-1} \\ &\ll_{y,c} T_k^{-1} V_k (\log T_k)^{\beta+y-1} \\ &\ll_{y,c} \sum_{T_k < a \le T_k + V_k} a^{-1}(\log a)^{\beta+y-1}.\end{aligned} \tag{1.105}$$

We apply Theorem 1.7 and deduce that for $y \in (0,1]$ and $\beta > y-1-\log y$, we have

$$\sum_{a\in\mathscr{A}} a^{-1}(\log a)^{\beta+y-1} = \infty. \tag{1.106}$$

Put $\gamma = \beta + y - 1$. We require $\gamma > 2y - 2 - \log y$ and we choose $y = \frac{1}{2}$ optimally. This completes the proof.

There is a heuristic explanation for the condition $\gamma > \log 2 - 1$. By the Hardy–Ramanujan theorem we have

$$(\log n)^{\log 2-\varepsilon} < \tau(n) < (\log n)^{\log 2+\varepsilon} p.p. \tag{1.107}$$

and we may expect that if $\mathscr{A}$ is Behrend, and in some sense random, then an integer $d \le n$ should have a probability at least $1/\tau(n)$ of belonging to $\mathscr{A}$, say

$$\operatorname{card}\{\mathscr{A} \cap (1,n]\} \ge \frac{n}{(\log n)^{\log 2-\varepsilon}}. \tag{1.108}$$

This leads to (1.104). The *'randomness'* of $\mathscr{A}$ may be interpreted as meaning *'multiplicatively structureless'*, and the corollary shows that sufficiently long blocks of consecutive integers fulfil this albeit somewhat vague criterion.

These blocks of integers occur frequently in the theory for the reason given. Their significance was first understood by Erdös, and they made their appearance in the conjecture due to Erdös which begins §5.3.

We can cope with much shorter blocks than envisaged in (1.103), with a weaker outcome.

Corollary 1.11 *Let $\mathscr{A}$ be as in (1.102) and suppose that for some $\alpha = \alpha(\mathscr{A}) > 0$ we have*

$$\log V_k \ge (\log T_k)^{\alpha}. \tag{1.109}$$

Then a necessary condition for $\mathscr{A}$ to be Behrend is that (1.104) should hold for every $\gamma > \log(1+\alpha) - \alpha$.

The proof goes through as before, except that we use (1.109) in (1.105). We arrive at the condition $\gamma > (1+\alpha)(y-1) - \log y$, and we choose $y = 1/(1+\alpha)$.

Notice that in general (1.104), with $\gamma < 0$, is *not* a necessary condition for $\mathscr{A}$ to be Behrend. If the a_i are pairwise coprime then (1.56) is both necessary and sufficient.

On the other hand Corollary 1.10 is best possible in the sense that the constant $\log 2 - 1$ cannot be reduced, even if $c(\mathscr{A}) > 1 - \varepsilon$, by reference to the sequence $\mathscr{A}^*(\frac{1}{2})$ in (1.81). Another example is provided by the sequence

$$\mathscr{A}_\lambda = \bigcup_{k=1}^{\infty} \left(\mathbf{Z}^+ \cap (e^{k^\lambda}, 2e^{k^\lambda}] \right). \tag{1.110}$$

Erdös conjectured that this is Behrend for some $\lambda > 1$ and this was proved by Hall and Tenenbaum (1986), (1992) for $\lambda < 1.31457\ldots$, $\lambda \le 1/(1 - \log 2)$ respectively. The second of these results, due to Tenenbaum, depends on the technique of Maier and Tenenbaum (1984). It is best possible, by reference to Corollary 1.10, moreover it lends some support to the hypothesis that (1.104) holds for $\gamma = \log 2 - 1$, and that we may take $\beta = y - 1 - \log y$ in Theorem 1.7.

Definition 1.12 *A block sequence is a sequence of the form (1.102) for which (1.103) holds, for some* $c = c(\mathscr{A}) > 0$. *A weak block sequence satisfies the condition (1.109) only, for some* $\alpha = \alpha(\mathscr{A}) > 0$. *We refer to* $\mathbf{Z}^+ \cap (T_k, T_k + V_k]$ *as a short block if* $V_k \le T_k$, *else it is a long block, and we write* $T_k + V_k = H_k T_k$ *in this case, so that* $H_k > 2$. *We require that* $T_{k+1} \ge T_k + V_k$ *for all* k.

Sometimes we shall assume (after splitting some blocks if necessary), that $H_k \le T_k$.

Let $\mathscr{A}$ be Behrend and comprise long blocks. We can apply Corollary 1.10 and we have

$$\sum_{k=1}^{\infty} \frac{\log H_k}{(\log T_k)^{1-\log 2-\varepsilon}} = \infty. \tag{1.111}$$

This is useful only if H_k is much smaller than T_k and we want a better result. This depends on the following lemma, which has some independent interest.

Lemma 1.13 *Let* $\kappa > 0$, $z_0 > 16$ *and* $B > B(\kappa)$ *where*

$$B(\kappa) = 2\frac{1+\kappa}{Q(1+\kappa)}, \qquad Q(\lambda) = \lambda \log \lambda - \lambda + 1. \tag{1.112}$$

Then we have, except for $n \in \mathscr{B}(\kappa, B, z_0)$, *that*

$$\left| \Omega(n; w, z) - \log\left(\frac{\log z}{\log w}\right) \right| < \kappa \log\left(\frac{\log z}{\log w}\right) + B \log\log\log z, \tag{1.113}$$

for all w, z *such that* $2 \le w \le z \le n$, $z \ge z_0$, *where the sequence* $\mathscr{B}(\kappa, B, z_0)$

of exceptional n satisfies

$$\bar{\mathbf{d}}\mathscr{B}(\kappa, B, z_0) \to 0, \qquad z_0 \to \infty, \kappa, B \text{ fixed.} \tag{1.114}$$

This is a further development of results of Erdös (1969), (Theorem 1) and Hall and Tenenbaum (1992), (Lemma 2.1). Albeit (1.112) may not be best possible the second term on the right of (1.113) (or one very like it) really is necessary: this is best understood from Erdös' paper, in which he states the following theorem. There exist functions $f^+ : (0,\infty) \to (1,\infty)$ (decreasing continuously from ∞ to 1) and $f^- : (0,\infty) \to (0,1)$ $(= 0$ on $(0,1]$ and increasing continuously to 1 on $(1,\infty))$ such that for almost all n,

$$\begin{aligned} \max \Omega(n; w, z) &\sim f^+(x) \log\left(\frac{\log z}{\log w}\right) \\ \min \Omega(n; w, z) &\sim f^-(x) \log\left(\frac{\log z}{\log w}\right) \end{aligned} \tag{1.115}$$

each extremal being subject to the constraint

$$\log\left(\frac{\log z}{\log w}\right) > x \log\log\log n. \tag{1.116}$$

Erdös does not specify $f^{\pm}$, and we leave it as an exercise for the reader to show from Lemma 1.13 that

$$f^{\pm}(x) = 1 + O(x^{-1/3}),\ (x \to \infty), \quad f^+(x) \ll x^{-1}\ (x \to 0). \tag{1.117}$$

Proof of lemma Let $\eta > 1/(Q(1+\kappa))$, $2(1+\kappa)\eta \le B$. We set up checkpoints at the points $t_k = \exp\exp(\eta k \log k)$. Let (1.113) be false, say

$$\Omega(n; w, z) \ge (1+\kappa)\log\left(\frac{\log z}{\log w}\right) + B \log\log\log z \tag{1.118}$$

where $w \in (t_j, t_{j+1}]$, $z \in (t_k, t_{k+1}]$. (We may assume $w \ge 3$.) Thus $k \ge \max(j, k_0(z_0))$ moreover (1.118) implies that

$$\begin{aligned} \Omega(n; t_j, t_{k+1}) &\ge (1+\kappa)\log\left(\frac{\log t_k}{\log t_{j+1}}\right) + B \log\log\log t_k \\ &\ge (1+\kappa)\log\left(\frac{\log t_{k+1}}{\log t_j}\right), \end{aligned} \tag{1.119}$$

provided

$$B \log\log\log t_k \ge 2(1+\kappa)\log\left(\frac{\log t_{k+1}}{\log t_k}\right). \tag{1.120}$$

We have $B \ge 2(1+\kappa)\eta$ and (1.120) then follows if $k \ge 17$, as we may assume. Let $y = 1 + \kappa$. The number of integers $n \le x$ such that (1.119) holds is

$$\begin{aligned} &\le \sum_{n\le x} y^{\Omega(n;t_j,t_{k+1})} \left(\frac{\log t_{k+1}}{\log t_j}\right)^{-(1+\kappa)\log y} \\ &\ll x\left(\frac{\log t_{k+1}}{\log t_j}\right)^{y-1-(1+\kappa)\log y} \end{aligned} \tag{1.121}$$

by the Halberstam–Richert inequality (*Divisors,* Theorem 01). The exponent is $-Q(1+\kappa)$, moreover we have $\log\log t_{k+1} - \log\log t_j \ge \eta(\kappa + 1 - j)(1 + \log j)$. Hence the number of exceptional integers for which (1.118) holds for some w, z is

$$\ll x \sum_{j\ge 1} \sum_{k\ge \max(j,k_0)} (ej)^{-Q(1+\kappa)\eta(k+1-j)}. \tag{1.122}$$

Suppose next that, instead of (1.118), we have

$$\Omega(n; w, z) \le (1-\kappa)\log\left(\frac{\log z}{\log w}\right) - B\log\log\log z. \tag{1.123}$$

This is impossible if the right-hand side is negative, and we deduce that $j + 1 < k$, and so

$$\Omega(n; t_{j+1}, t_k) \le (1-\kappa)\log\left(\frac{\log t_{k+1}}{\log t_j}\right) - B\log\log\log t_{k+1}, \tag{1.124}$$

(we may assume that z_0 is so large that the right-hand side of (1.123) increases for $z \ge z_0$), whence by (1.120),

$$\Omega(n; t_{j+1}, t_k) \le (1-\kappa)\log\left(\frac{\log t_k}{\log t_{j+1}}\right). \tag{1.125}$$

We proceed as before, but now set $y = 1 - \kappa$, and arrive at a sum like (1.122) with $k \ge \max(j+2, k_0)$ and $-Q(1-\kappa)\eta(k-1-j)$ in the exponent. We note that $Q(1-\kappa) > Q(1+\kappa)$, moreover $Q(1+\kappa)\eta > 1$, whence each of these sums is $o(x)$ as z_0 and therefore k_0 tends to infinity. Hence (1.114) holds as required.

Theorem 1.14 *Let $\mathscr{A}$ be a weak block sequence, and Behrend. Let $\varepsilon > 0$, $0 < \varepsilon' < 1 - \log 2$. For the short blocks, $V_k \le T_k$, set*

$$W_k = \frac{V_k}{T_k}(\log T_k)^{\log(1+\alpha_k)-\alpha_k+\varepsilon} \tag{1.126}$$

where

$$\log V_k = (\log T_k)^{\alpha_k}. \tag{1.127}$$

For the long blocks, $V_k > T_k$, $T_k + V_k = H_k T_k$, *set*

$$W_k = \min\left\{(\log\log T_k)^A\left(\frac{\log T_k}{\log H_k}\right)^{\log 2-1+\varepsilon'}, \left(\frac{\log T_k}{\log H_k}\right)^{-\delta}\right\} \tag{1.128}$$

where $A > A(\varepsilon')$, *with* $A(\varepsilon')$ *as in* (1.156) *below, and*

$$\delta = 1 - \frac{1}{\log 2} + \frac{1}{\log 2}\log\left(\frac{1}{\log 2}\right) = .086071\ldots. \tag{1.129}$$

Then

$$\sum_{k=1}^{\infty} W_k = \infty. \tag{1.130}$$

We assume in the above that $H_k \leq T_k$, *splitting blocks if required.*

This is a development of Theorem 1 of Hall and Tenenbaum (1992), which was restricted to block sequences. It has a rather technical appearance and we begin with some explanatory remarks. First, notice that this result is much better than what we should obtain from a direct application of Corollary 0.13, that is

$$\sum_{k=1}^{\infty} \boldsymbol{\delta}\mathscr{M}\left(\mathbf{Z}^+ \cap (T_k, T_k + V_k]\right) = \infty. \tag{1.131}$$

For example, Theorem 21(iii) of *Divisors* implies that for the long blocks $\mathscr{A}_k = \mathbf{Z}^+ \cap (T_k, H_k T_k]$, we have

$$\left(\frac{\log H_k}{\log T_k}\right)^{\delta+\varepsilon} \ll \boldsymbol{\delta}\mathscr{M}(\mathscr{A}_k) \ll \left(\frac{\log H_k}{\log T_k}\right)^{\delta}, \tag{1.132}$$

and W_k is smaller than this whenever $\log H_k < \log T_k/(\log\log T_k)^{C(\varepsilon')}$. For the short blocks $\mathscr{A}_k = \mathbf{Z}^+ \cap (T_k, T_k + V_k]$ we have, for $V_k < T_k(\log T_k)^{1-\log 4-\varepsilon}$,

$$\boldsymbol{\delta}\mathscr{M}(\mathscr{A}_k) \sim \frac{V_k}{T_k} \tag{1.133}$$

by Theorem 21(i) of *Divisors*. Of course the point is that the 'probabilities' $\boldsymbol{\delta}\mathscr{M}(\mathscr{A}_k)$ are not independent. Technically, the improvement comes because we apply Lemmas 1.8, 1.13 *once only* to each integer $n \in \mathscr{M}(\mathscr{A})$.

We want the exponent of $\log T_k$ in (1.126) to be negative, i.e. $\varepsilon < \alpha(\mathscr{A}) - \log(1 + \alpha(\mathscr{A}))$. A small ε' may not be optimal in (1.128) because $A(\varepsilon')$ is a decreasing function of ε'. The minimum value of $C(\varepsilon')$ occurs when $\varepsilon' \approx \frac{1}{6}$, and is ≈ 504.

Proof of Theorem 1.14 Let $\mathscr{A} = \mathscr{A}^\flat \cup \mathscr{A}^\sharp$ where $\mathscr{A}^\flat$ comprises the short, and $\mathscr{A}^\sharp$ the long blocks. At least one of these sequences is Behrend by Corollary 0.14, and so it will be sufficient to show that the corresponding sub-series in (1.130) diverges.

First we suppose that $\mathscr{A}^\flat$ is Behrend, following the proof of Corollary 1.11 except that we are going to make y depend on the individual blocks. We go back to the proof of Theorem 1.7 to see how a variable y may be introduced. Let $\kappa > 0$ be at our disposal, and fix $t_0 = t_0(\kappa)$ in Lemma 1.8 so that $\overline{\mathbf{d}}\mathscr{B}(\kappa, t_0(\kappa)) \le \frac{1}{3}$. Let k_0 be such that $T_{k_0} \ge t_0$, and consider the tail $\mathscr{A}_0^\flat$ of $\mathscr{A}^\flat$ of blocks for which $k \ge k_0$. $\mathscr{A}_0^\flat$ is of course Behrend. Put $W_k^\flat = W_k$ if $\mathscr{A}_k$ is a short block, $W_k^\flat = 0$ else. We suppose that

$$\sum_{k=1}^{\infty} W_k^\flat < \infty \tag{1.134}$$

and seek a contradiction. Instead of (1.74) we write

$$\chi(n) \le \tau(n, \mathscr{A}_0^\flat) \le \sum_{\substack{a|n \\ a\in\mathscr{A}_0^\flat}} y(k)^{\Omega(n,a)-(1+\kappa)\log\log a} \tag{1.135}$$

where $y(k)$ is to depend on the block $\mathscr{A}_k$. Here χ is the characteristic function of $\mathscr{M}(\mathscr{A}_0^\flat)$. After (1.75)–(1.77) we arrive at

$$\begin{aligned}\sum_{n\in\mathscr{B}(\kappa,t_0)} n^{-\sigma}\chi(n) &\ll (\sigma-1)^{-1} \sum_{a\in\mathscr{A}_0^\flat} a^{-1}(\log a)^{y(k)-1-(1+\kappa)\log y(k)} y(k)^{\Omega(a)} \\ &\ll (\sigma-1)^{-1} {\sum_{k\ge k_0}}^{\flat} \frac{V_k}{T_k} (\log T_k)^{y(k)-1-(1+\kappa)\log y(k)} (\log V_k)^{y(k)-1}\end{aligned} \tag{1.136}$$

employing Lemma 1.9. $\sum^\flat$ is restricted to short blocks. We substitute $\log V_k = (\log T_k)^{\alpha_k}$ from (1.127) and set $y(k) = 1/(1+\alpha_k)$. Provided $\kappa \log 2 \le \varepsilon$ we may deduce that

$$\sum_{n\notin\mathscr{B}(\kappa,t_0)} n^{-\sigma}\chi(n) \ll (\sigma-1)^{-1} \sum_{k\ge k_0} W_k^\flat. \tag{1.137}$$

If (1.134) holds then we may take k_0 so large that

$$\limsup_{\sigma\to 1+} (\sigma-1) \sum_{n\notin\mathscr{B}(\kappa,t_0)} n^{-\sigma}\chi(n) \le \frac{1}{3} \tag{1.138}$$

as in (1.78). We deduce that $\mathscr{A}^\flat$ is not Behrend by the argument used in the proof of Theorem 1.7; this is the required contradiction and we conclude that the series (1.134) is divergent.

Next, we suppose that $\mathscr{A}^\#$ is Behrend. We denote by $\mathscr{A}^\#_L$ the union of the blocks for which the first term on the right of (1.128) is the minimum, and by $\mathscr{A}^\#_R$ the union of the remaining blocks. If $\mathscr{A}^\#_R$ is Behrend then (1.130) follows from (1.132) and there is nothing more to prove; so we suppose that $\mathscr{A}^\#_L$ is Behrend. Let $\mathscr{K}$ denote the sequence of integers k for which the block $\mathscr{A}_k$ is contained in $\mathscr{A}^\#_L$, and write $\mathscr{N}_k = \mathscr{M}(\mathscr{A}_k)$. Since any tail of $\mathscr{A}^\#_L$ is Behrend we have for every k_1,

$$\mathbf{d}\left(\bigcup_{\substack{k\in\mathscr{K}\\ k\geq k_1}} \mathscr{N}_k\right) = 1. \tag{1.139}$$

We apply Lemma 1.13. Let $\kappa \in (0,1)$, $B > B(\kappa)$. Let z_0 be so large that $\bar{\mathbf{d}}\mathscr{B}(\kappa, B, z_0) \leq \frac{1}{4}$; and let us take k_1 in (1.139) so large that $T_{k_1} \geq z_0$. We write $\mathscr{N}'_k = \mathscr{N}_k \setminus \mathscr{B}(\kappa, B, z_0)$ and we deduce from (1.139) that

$$\sum_{\substack{k\in\mathscr{K}\\ k\geq k_1}} \bar{\mathbf{d}}\mathscr{N}'_k \geq \frac{3}{4}. \tag{1.140}$$

Next, let $\mathscr{N}''_k$ denote the subsequence of $\mathscr{N}'_k$ for which

$$\prod_{\substack{p^\nu \| n\\ p\leq H_k}} p^\nu \leq H_k^{v_k}, \qquad n \in \mathscr{N}''_k, \tag{1.141}$$

with v_k at our disposal. By Theorem 0.7 of *Divisors*, we have

$$\bar{\mathbf{d}}(\mathbf{Z}^+ \setminus \mathscr{N}''_k) \ll \exp(-c_0 v_k) \tag{1.142}$$

where c_0 is an absolute positive constant. We put

$$v_k = \frac{1}{c_0}\log\left(\frac{\log T_k}{\log H_k}\right) \tag{1.143}$$

whence we have

$$\sum_{\substack{k\in\mathscr{K}\\ k\geq k_1}} \bar{\mathbf{d}}(\mathscr{N}'_k \setminus \mathscr{N}''_k) \ll \sum_{\substack{k\in\mathscr{K}\\ k\geq k_1}} \frac{\log H_k}{\log T_k}. \tag{1.144}$$

Either (1.130) holds or we may take k_1 so large that the sum on the left of (1.144) is $\leq \frac{1}{4}$, so that we have, from (1.140),

$$\sum_{\substack{k\in\mathscr{K}\\ k\geq k_1}} \bar{\mathbf{d}}\mathscr{N}''_k \geq \frac{1}{2}. \tag{1.145}$$

Let $n \in \mathcal{N}''_k$, and $a|n$, $a \in \mathcal{A}_k$. We write $a = st$ where $P^+(s) \le H_k < P^-(t)$, whence $s \le H_k^{v_k}$ in view of (1.141). Since $T_k < st \le H_k T_k$ we have

$$T_k H_k^{-v_k} < t \le H_k T_k. \tag{1.146}$$

Since $n \notin \mathcal{B}(\kappa, B, z_0)$ we have from (1.113)

$$\Omega(n; H_k, T_k) < (1+\kappa)\log\left(\frac{\log T_k}{\log H_k}\right) + B\log\log\log T_k \tag{1.147}$$

whence for $y \le 1$, and large x,

$$\sum_{\substack{n\le x\\ n\in\mathcal{N}''_k}} 1 \le \sum_{n\le x} y^{\Omega(n;H_k,T_k)}\left(\frac{\log T_k}{\log H_k}\right)^{-(1+\kappa)\log y}(\log\log T_k)^{-B\log y}\sum_{t|n}{}^{*}1 \tag{1.148}$$

where the star denotes that t satisfies (1.146), moreover $P^-(t) > H_k$. We put $n = mt$ and invert summations, employing the inequality

$$\sum_{m\le x/t} y^{\Omega(m;H_k,T_k)} \ll \frac{x}{t}\left(\frac{\log T_k}{\log H_k}\right)^{y-1} \tag{1.149}$$

(see *Divisors,* Theorem 01). Hence

$$\bar{\mathbf{d}}\mathcal{N}''_k \ll \left(\frac{\log T_k}{\log H_k}\right)^{y-1-(1+\kappa)\log y}(\log\log T_k)^{-B\log y}Z_k \tag{1.150}$$

where

$$Z_k = \sum{}^{*} t^{-1} y^{\Omega(t;H_k,t_k)}. \tag{1.151}$$

We estimate Z_k by partial summation. Provided $u \ge T_k$, we have

$$\sum_{\substack{t\le u\\ P^-(t)>H_k}} y^{\Omega(t;H_k,T_k)} \ll u(\log H_k)^{-y}(\log T_k)^{y-1}, \tag{1.152}$$

(see *Divisors,* Theorem 01) whence by (1.146)

$$Z_k \ll (1+v_k)\left(\frac{\log T_k}{\log H_k}\right)^{y-1}. \tag{1.153}$$

We insert this into (1.150), and choose $y = (1+\kappa)/2$ optimally, which yields

$$\bar{\mathbf{d}}\mathcal{N}''_k \ll (1+v_k)\left(\frac{\log T_k}{\log H_k}\right)^{\log 2-1-Q(1+\kappa)+\kappa\log 2}(\log\log T_k)^{B\log(\frac{2}{1+\kappa})}. \tag{1.154}$$

The function $\kappa \log 2 - Q(1+\kappa)$ increases from 0 to $1-\log 2$ as κ increases from 0 to 1; we choose κ to be the unique root of the equation

$$\kappa \log 2 - Q(1+\kappa) = \varepsilon' \tag{1.155}$$

and then we define

$$A(\varepsilon') = B(\kappa) \log\left(\frac{2}{1+\kappa}\right). \tag{1.156}$$

By (1.143), (1.154)–(1.156), provided $A > A(\varepsilon')$ in (1.128) we can choose $B > B(\kappa)$ so that we have

$$\bar{\mathbf{d}}\mathscr{N}_k'' \ll W_k, \ (k \in \mathscr{K}). \tag{1.157}$$

Since k_1, in (1.145), is unbounded, (1.130) follows. This deals with the long blocks and thereby completes the proof.

The next result is a refinement of Theorem 3 of Hall and Tenenbaum (1992).

Theorem 1.15 *Let $\mathscr{A}$ be a union of long blocks*

$$\mathscr{A}_k = \mathbf{Z}^+ \cap (T_k, H_k T_k], \ (2 < H_k \le T_k). \tag{1.158}$$

Then $\mathscr{A}$ is Behrend if **either**

$$\sum_{k=1}^{\infty} \frac{\log H_k}{\log T_k} = \infty \tag{1.159}$$

or, *for some $\varepsilon > 0$ we have*

$$\sum_{k=1}^{\infty} \left(\frac{\log H_k}{\log T_k}\right)^{\delta+\varepsilon} = \infty \tag{1.160}$$

where $\delta = .086071\ldots$ is defined in (1.129) and in addition, for all $k \ge 2$,

$$\log(H_{k-1}T_{k-1}) \le \frac{\log H_k}{C(\varepsilon) + (2\delta + 3\varepsilon)\log(\frac{\log T_k}{\log H_k})}. \tag{1.161}$$

Here $C(\varepsilon)$ is a sufficiently large constant.

The idea of the proof which follows is that we 'thin down' the $\mathscr{A}_k$ to subsequences $\mathscr{A}_k'$ in such a way that the events $n \in \mathscr{M}(\mathscr{A}_k')$ are independent, when it will be sufficient to have

$$\sum_{k=1}^{\infty} \boldsymbol{\delta}\mathscr{M}(\mathscr{A}_k') = \infty. \tag{1.162}$$

To do this we associate with each $\mathscr{A}_k$ a set of primes $\mathscr{P}_k$ in such a way that the sets $\mathscr{P}_k$ are disjoint. Then for every k,

$$\mathscr{A}_k' = \{a \in \mathscr{A}_k : p|a \Rightarrow p \in \mathscr{P}_k\}. \tag{1.163}$$

This is simple but effective: if we compare (1.160) with (1.128) we see that δ, as an exponent, is best possible in both instances. It will emerge that (1.159) is best possible too, as it stands.

Proof of Theorem 1.15 We begin with (1.159). We put

$$\mathscr{A}'_k = \mathscr{P}_k = \{p \in \mathscr{A}_k,\ p \text{ prime}\} \tag{1.164}$$

and we have

$$\boldsymbol{\delta}\mathscr{M}(\mathscr{A}'_k) = 1 - \prod_{t_k < p \leq H_k T_k} \left(1 - \frac{1}{p}\right) \gg \frac{\log H_k}{\log T_k} \tag{1.165}$$

whence (1.162) holds and $\mathscr{A}$ is Behrend.

We turn to the alternative condition comprising (1.160) and (1.161). In this case we let

$$\mathscr{P}_k = \{p : H_k^{\eta_k} < p \leq H_k T_k\} \tag{1.166}$$

where η_k is a small number at our disposal. The $\mathscr{P}_k$ have to be disjoint so we require that

$$H_{k-1} T_{k-1} \leq H_k^{\eta_k}. \tag{1.167}$$

Let $\mathscr{A}'_k$ be as in (1.163) and

$$\mathscr{A}''_k = (H_k^{1-\varepsilon} T_k, H_k T_k] \cap \mathbf{Z}^+, \tag{1.168}$$

where we have assumed that k is so large that this is non-empty. Let n have a divisor $a'' \in \mathscr{A}''_k$ and put $a'' = a'b$ where $P^-(a') > H_k^{\eta_k} \geq P^+(b)$. Then either $a' \in \mathscr{A}'_k$ and $n \in \mathscr{M}(\mathscr{A}'_k)$, or $b > H_k^{1-\varepsilon}$. Therefore

$$\boldsymbol{\delta}\mathscr{M}(\mathscr{A}'_k) \geq \boldsymbol{\delta}\mathscr{M}(\mathscr{A}''_k) - \boldsymbol{\delta}\Bigg\{n : \prod_{\substack{p^\nu \| n \\ p \leq H_k^{\eta_k}}} p^\nu > H_k^{1-\varepsilon}\Bigg\}. \tag{1.169}$$

By Theorem 21(iii) of *Divisors*, there exists $c_1(\varepsilon) > 0$ such that

$$\boldsymbol{\delta}\mathscr{M}(\mathscr{A}''_k) \geq c_1(\varepsilon)\left(\frac{\log H_k T_k}{\log H_k^{1-\varepsilon} T_k} - 1\right)^{\delta+\varepsilon} \tag{1.170}$$

(provided, as we may assume from (1.161), $H_k^\varepsilon > 2$), whence

$$\boldsymbol{\delta}\mathscr{M}(\mathscr{A}''_k) \geq c_2(\varepsilon)\left(\frac{\log H_k}{\log T_k}\right)^{\delta+\varepsilon} \tag{1.171}$$

because $H_k \leq T_k$. To estimate the second density on the right of (1.169) we employ Theorem 07 of *Divisors* with an explicit value of the constant, $c_0 = \frac{1}{2}$, provided by Tenenbaum (1990), (p.437 Ex.5). This density is

$$\ll \exp\left(-\frac{1}{2}.\frac{1-\varepsilon}{\eta_k}\right) \tag{1.172}$$

and we put

$$\eta_k = \left\{C(\varepsilon) + (2\delta + 3\varepsilon)\log\left(\frac{\log T_k}{\log H_k}\right)\right\}^{-1} \tag{1.173}$$

and deduce from (1.169), (1.171)–(1.173) that if $C(\varepsilon)$ is sufficiently large then

$$\delta\mathscr{M}(\mathscr{A}'_k) \geq \frac{1}{2}c_2(\varepsilon)\left(\frac{\log H_k}{\log T_k}\right)^{\delta+\varepsilon} \tag{1.174}$$

Together with (1.160) this gives (1.162); moreover (1.167) follows from (1.161) and the $\mathscr{P}_k$ are disjoint. Hence $\mathscr{A}$ is Behrend as required. An example of a Behrend sequence of this sort is furnished by

$$T_k = \exp\{(k!)^\Lambda\},\ H_k = \exp\left\{\frac{(k!)^\Lambda}{k^\lambda}\right\}, \quad \lambda < \Lambda < \frac{1}{\delta}. \tag{1.175}$$

The proof that (1.159) is sufficient for the long block sequence $\mathscr{A}$ to be Behrend is easy and this might suggest that with more work, a weaker sufficient condition could be found. We show next that, given any function $F(\xi) \to \infty$ as $\xi \to \infty$, we can construct a sequence $\mathscr{A}$ of long blocks, as in (1.158), such that

$$\sum_{k=1}^{\infty} F\left(\frac{\log T_k}{\log H_k}\right)\frac{\log H_k}{\log T_k} = \infty \tag{1.176}$$

but $\mathscr{A}$ is not Behrend. Our construction follows Hall and Tenenbaum (1992), (eq. (1.17) *et seq.*).

We assume as we may that F is non-decreasing, $F \geq 1$. We put $f(t) = F(\log t)$ and we consider blocks

$$(S_h, G_hS_h], \quad G_h = e^{g(h)}, \quad g(h) = \left[\frac{\log S_h}{f(S_h)}\right], \tag{1.177}$$

in which we let $S_h \to \infty$ so fast that both

$$\sum_{h=1}^{\infty} f(S_h)^{-\delta} < \infty,\ S_{h+1} > S_h^2, \tag{1.178}$$

where δ is defined in (1.129). By (1.177)–(1.178) we have

$$\sum_{h=1}^{\infty}\left(\frac{\log G_h}{\log S_h}\right)^{\delta} < \infty \tag{1.179}$$

so that

$$\mathscr{A} = \bigcup_{h=1}^{\infty} \{\mathbf{Z}^+ \cap (S_h, G_h S_h]\} \tag{1.180}$$

is not Behrend, by Theorem 1.14. Let $k(h)$ be the sequence such that $k(h+1) = k(h) + g(h)$, $h \geq 1$, $k(1) = 1$, and put

$$\begin{aligned} T_{k(h)} &= S_h, \quad h = 1, 2, 3, \ldots \\ T_{k(h)+j} &= T_{k(h)} e^j, \ j < g(h), \quad H_k = e \ \forall k \end{aligned} \tag{1.181}$$

so that we also have

$$\mathscr{A} = \bigcup_{k=1}^{\infty} \{\mathbf{Z}^+ \cap (T_k, H_k T_k]\}. \tag{1.182}$$

For $k(h) \leq k < k(h+1)$ we have $T_k \leq S_h^2$, and

$$F\left(\frac{\log T_k}{\log H_k}\right) = f(T_k) \geq f(S_h), \tag{1.183}$$

whence

$$\sum_{k(h) \leq k < k(h+1)} F\left(\frac{\log T_k}{\log H_k}\right) \frac{\log H_k}{\log T_k} \geq \frac{g(h) f(S_h)}{2 \log S_h}. \tag{1.184}$$

Then (1.177) and (1.184) imply (1.176), and the required construction is complete.

It is not clear whether such a sequence can be constructed if both the function F and a sequence $\{H_k : H_k > 2\}$ are given in advance. This problem is left open.

In a sense the above construction is a trick, in that most of the blocks are contiguous or nearly so (obviously we could space them out a little, e.g. by striking out alternate blocks). We may draw the conclusion that, to make progress with block sequences, we should for the most part concentrate on the case of 'well-spaced' blocks. Condition (1.161) achieves this but is rather strong. Notice that we may assume $\log H_k = o(\log T_k)$ else (1.159) and (1.160) are indistinguishable; but then (1.161) requires $\log T_{k-1} = o(\log T_k)$, indeed if (1.159) does not apply, rather more. An innocuous looking problem, again left open*, is to determine what extra spacing condition, together with

$$\sum_{k=1}^{\infty} \frac{\log H_k}{\log T_k} \log^m \left(\frac{\log T_k}{\log H_k}\right) = \infty \tag{1.185}$$

implies that $\mathscr{A}$ is Behrend. Here $m \in \mathbf{Z}^+$ is fixed. We have in mind a

* See Tenenbaum (199x) (note added in proof).

possible analogy with (1.159), i.e. the case $m = 0$, in which we restrict to a subsequece $\mathscr{A}'$ of $\mathscr{A}$ comprising numbers with $m+1$ prime factors. In this circle of ideas there is the following (straightforward) generalization of a result of Erdös (1959).

Theorem 1.16 *Let k be a positive integer and for $1 \le i \le k$ let $\varepsilon_i(p)$ be a positive valued function of the prime p. Let*

$$\mathscr{A} = \{pp_1p_2\dots p_k : p < p_i \le p^{1+\varepsilon_i(p)},\ (i \le k)\}. \tag{1.186}$$

Then $\mathscr{A}$ is Behrend if and only if we have

$$\sum_p \frac{1}{p}\prod_{i=1}^{k} \min(1, \varepsilon_i(p)) = \infty. \tag{1.187}$$

Erdös proved this in the case $k = 1$ and stated that $\mathscr{A}$ is in any event Besicovitch. The second part of his assertion also holds in the general case and is a consequence of Theorem 1.4.

Proof We begin with the proof of necessity, which is by contradiction. Let us suppose that the series (1.187) is convergent. We begin by splitting the primes into 2^k disjoint sequences labelled by the subsets of $\{1, 2, 3, \dots, k\}$. The prime p belongs to a particular sequence if and only if $\varepsilon_i(p) \le 1$ for precisely those i belonging to the label subset. We denote these subsets by I, and associate with I the sequence

$$\mathscr{A}(I) = \left\{p\prod_{i\in I} p_i\ :\ p < p_i \le p^{1+\varepsilon_i(p)},\ i \in I\right\}. \tag{1.188}$$

Since the series (1.187) converges by hypothesis, we have

$$\sum\left\{\frac{1}{a}\ :\ a \in \mathscr{A}(I)\right\} < \infty \tag{1.189}$$

(it clearly does not matter whether or not we count repetitions at this point), because

$$\sum_{w<p\le w^{1+\varepsilon}} \frac{1}{p} = \log(1+\varepsilon) + O\left(\frac{1}{\log w}\right). \tag{1.190}$$

Hence none of the $\mathscr{A}(I)$ is Behrend, and neither is $\mathscr{A}$.

We turn to sufficiency. We may assume $\varepsilon_i(p) \le 1$ always, simply by disallowing any $p_i > p^2$. We have to show that $x + o(x)$ integers $n \le x$ belong to $\mathscr{M}(\mathscr{A})$, and we restrict $\mathscr{A}$ by the condition $p < x^\alpha$ where $\alpha = 1/(4k+2)$, so that we have $pp_1p_2\dots p_k < \sqrt{x}$ always. Let $f(n)$ denote the number of divisors of n of this form, counting repetitions multiply. We employ Turán's method, that is we consider the sum

$$\sum_{n\le x}(f(n)-A(x))^2 = S_2(x) - 2A(x)S_1(x) + [x]A(x)^2 \qquad (1.191)$$

say, where we define

$$A(x) = \sum\left\{\frac{1}{pp_1p_2\cdots p_k} : p < x^\alpha,\, p < p_i \le p^{1+\varepsilon_i(p)}\right\}. \qquad (1.192)$$

Then we have

$$S_1(x) = \sum_{n\le x} f(n) = \sum\left[\frac{x}{pp_1p_2\cdots p_k}\right] = xA(x) + o(x) \qquad (1.193)$$

the range of summation on the right being as in (1.192). By similar reasoning we have

$$S_2(x) = xB(x) + o(x) \qquad (1.194)$$

where

$$B(x) = \sum\sum\frac{1}{[pp_1p_2\cdots p_k, qq_1q_2\cdots q_k]} \qquad (1.195)$$

and the primes q, q_i satisfy the same conditions as the p, p_i; we have $q < x^\alpha$, $q < q_i \le q^{1+\varepsilon_i(q)}$. (The error term in (1.194) is $O\left(\prod_{2k+2}(x)\right)$ where $\prod_l(x)$ is the number of integers $m \le x$ such that $\Omega(m) \le l$: for fixed l this is $\ll x(\log x)^{-1}(\log\log x)^{l-1}$.) We split $B(x)$ into $k+2$ subsums $B_m(x)$, $0 \le m \le k+1$; the mth subsum contains the terms from (1.195) for which

$$\Omega([pp_1p_2\cdots p_k, qq_1q_2\cdots q_k]) = k+1+m \qquad (1.196)$$

and we introduce the auxiliary sum, over these same primes

$$B_m^*(x) = \sum\sum\frac{1}{pp_1p_2\cdots p_k qq_1q_2\cdots q_k} \qquad (1.197)$$

so that $B_m^*(x) \le B_m(x)$, with equality if $m = k+1$. Also

$$\sum_{m=0}^{k+1} B_m^*(x) = A(x)^2, \qquad (1.198)$$

whence we have

$$A(x)^2 \le B(x) \le A(x)^2 + \sum_{m=0}^{k} B_m(x). \qquad (1.199)$$

We claim that for $0 \le m \le k$, we have

$$B_m(x) \ll_k A(x). \tag{1.200}$$

This is clear in the case $m = 0$ when the q's are a permutation of the p's. For $1 \le m \le k$ let us write

$$B_m(x) \ll \sum \sum{}' \frac{1}{pp_1p_2 \dots p_k rr_1r_2 \dots r_m} \tag{1.201}$$

where the dash means that the r_j are primes such that

$$[pp_1p_2 \dots p_k, qq_1q_2 \dots q_k] = pp_1p_2 \dots p_k rr_1r_2 \dots r_m. \tag{1.202}$$

If we demand that $r_1 \le r_2 \le \cdots \le r_m$ then these primes are uniquely determined, moreover any particular product $r_1r_2 \dots r_m$ only occurs $\ll_k 1$ times. Next, because $m < k+1$ in (1.196), at least one of $q, q_1, q_2, \dots, q_k$ is equal to one of $p, p_1, \dots, p_k$ and, by checking the various possibilities, we find that this involves

$$\sqrt{p} \le q \le p^2 \tag{1.203}$$

because every $\varepsilon_i \le 1$. Hence

$$\sqrt{p} \le r_j \le p^4, \quad 1 \le j \le m \tag{1.204}$$

and

$$\sum \frac{1}{r_j} \le \log 8 + O\left(\frac{1}{\log p}\right) = O(1). \tag{1.205}$$

We insert this into (1.201) to yield (1.200). Then we assemble (1.191), (1.193)–(1.194), (1.199)–(1.200) to obtain

$$\sum_{n \le x} (f(n) - A(x))^2 \ll_k xA(x) + o(x). \tag{1.206}$$

Let $n \notin \mathcal{M}(\mathcal{A})$. Then $f(n) = 0$, whence from (1.206),

$$\operatorname{card}\{n : n \le x, n \notin \mathcal{M}(\mathcal{A})\} \ll \frac{x}{A(x)}, \tag{1.207}$$

since $A(x) \to \infty$ by (1.187) and (1.190). The right-hand side of (1.207) is $o(x)$ and so $\mathcal{A}$ is Behrend. This completes the proof.

1.4 Witnesses

We introduce the topics to be discussed in this section with a numerical problem. Suppose that we are given non-trivial sequences $\mathcal{A}$ and $\mathcal{B}$ such that

$$\boldsymbol{\delta}\mathcal{M}(2\mathcal{A} \cup 3\mathcal{B}) = \boldsymbol{\delta}\mathcal{M}\left(\{2,3\}\right) = \frac{2}{3}. \tag{1.208}$$

Can we infer that $\mathscr{A}$ and $\mathscr{B}$ are Behrend? We answer this question affirmatively and then consider its generalizations.

$\mathscr{M}(\{2,3\})$ comprises four arithmetic progressions $\mathscr{N} = 6l + j$, $j = 0, 2, 3$, or 4. Each of the four sequences $\mathscr{M}(2\mathscr{A} \cup 3\mathscr{B}) \cap \mathscr{N}$ has upper logarithmic density $\leq \frac{1}{6}$ and the union of any three has upper density $\leq \frac{1}{2}$. By (1.208) the lower logarithmic density of each sequence is therefore $\geq \frac{1}{6}$, whence for each j, we have

$$\delta\{\mathscr{M}(2\mathscr{A} \cup 3\mathscr{B}) \cap \mathscr{N}\} = \frac{1}{6}. \tag{1.209}$$

Let $j = 2$. The multiples of $3\mathscr{B}$ contribute nothing to the intersection in (1.209) and so in the case $\mathscr{N} = 6l + 2$ we have

$$\delta\{\mathscr{M}(2\mathscr{A}) \cap \mathscr{N}\} = \frac{1}{6}. \tag{1.210}$$

We divide throughout by 2, and we deduce that

$$\delta\{\mathscr{M}(\mathscr{A}) \cap \mathscr{L}\} = \frac{1}{3}, \tag{1.211}$$

in which $\mathscr{L}$ denotes the arithmetic progression $3l + 1$. We require the following result.

Lemma 1.17 *Let $\mathscr{A}$ be any integer sequence, and q, r be coprime integers. Let $q\mathbf{Z} + r$ denote the arithmetic progression $\{n : n \equiv r (\text{mod } q)\}$. Then the sequence $\mathscr{M}(\mathscr{A}) \cap (q\mathbf{Z} + r)$ possesses logarithmic density, moreover we have the formula*

$$\delta\{\mathscr{M}(\mathscr{A}) \cap (q\mathbf{Z} + r)\} = q^{-1}\delta\mathscr{M}\left(\mathscr{A}^*(q)\right) \tag{1.212}$$

where $\mathscr{A}^*(q)$ comprises the elements of $\mathscr{A}$ prime to q. In particular, if $\mathscr{A}$ is non-trivial and the density on the left is equal to $1/q$, then $\mathscr{A}$ is Behrend.

If we assume this for the moment then we may complete the solution of our initial problem. By (1.211) and the last part of the lemma, $\mathscr{A}$ is Behrend. A similar argument shows that $\mathscr{B}$ is Behrend.

The last part of the lemma follows from (1.212). If the left-hand side equals $1/q$ then $\mathscr{A}^*(q)$ is Behrend, and so is $\mathscr{A}$. It remains to prove that $\delta\{\mathscr{M}(\mathscr{A}) \cap (q\mathbf{Z} + r)\}$ exists and that (1.212) holds. We may replace $\mathscr{A}$ by $\mathscr{A}^*(q)$ because any remaining elements of $\mathscr{A}$ contribute nothing to the intersection. Hence we may assume that $(a, q) = 1$ for every $a \in \mathscr{A}$.

Let us begin with the case when $\mathscr{A}$ is finite, say $\mathscr{A} = \{a_1, a_2, \ldots, a_N\}$. Let $1 \leq i_1 \leq i_2 \leq \cdots \leq i_h \leq N$. The integers n such that

$$n \equiv 0 (\text{mod } a_{i_g}),\ 1 \leq g \leq h; \quad n \equiv r (\text{mod } q) \tag{1.213}$$

have asymptotic density equal to $1/[a_{i_1},\dots a_{i_h}]q$ by the Chinese remainder theorem, and we employ the inclusion-exclusion principle in the usual fashion to deduce that

$$\mathbf{d}\{\mathcal{M}(\mathcal{A})\cap(q\mathbf{Z}+r)\} = q^{-1}\mathbf{d}\mathcal{M}(\mathcal{A}). \tag{1.214}$$

We may replace $\mathbf{d}$ by $\boldsymbol{\delta}$. Next, let $\mathcal{A}$ be infinite and put $\mathcal{A}^{(N)} = \{a_1, a_2, \dots, a_N\}$. Then

$$\boldsymbol{\delta}\{\mathcal{M}(\mathcal{A}^{(N)})\cap(q\mathbf{Z}+r)\} = q^{-1}\boldsymbol{\delta}\mathcal{M}(\mathcal{A}^{(N)}) \tag{1.215}$$

whence by (0.67),

$$\underline{\boldsymbol{\delta}}\{\mathcal{M}(\mathcal{A})\cap(q\mathbf{Z}+r)\} \geq q^{-1}\boldsymbol{\delta}\mathcal{M}(\mathcal{A}). \tag{1.216}$$

On the other hand we have

$$\begin{aligned}\overline{\boldsymbol{\delta}}\{\mathcal{M}(\mathcal{A})\cap(q\mathbf{Z}+r)\} &\leq \boldsymbol{\delta}\{\mathcal{M}(\mathcal{A}^{(N)})\cap(q\mathbf{Z}+r)\} + \boldsymbol{\delta}\{\mathcal{M}(\mathcal{A})\setminus\mathcal{M}(\mathcal{A}^{(N)})\\ &\leq q^{-1}\mathcal{M}(\mathcal{A}) + \varepsilon\end{aligned} \tag{1.217}$$

by (0.67), if $N \geq N_0(\varepsilon)$. This is all we need.

Theorem 1.18 *Let $\mathcal{S} = \{s_1, s_2, \dots, s_k\}$ be a (finite) primitive sequence and*

$$\boldsymbol{\delta}\mathcal{M}(s_1\mathcal{A}_1 \cup s_2\mathcal{A}_2 \cup \cdots \cup s_k\mathcal{A}_k) = \boldsymbol{\delta}\mathcal{M}(\{s_1, s_2, \dots, s_k\}). \tag{1.218}$$

Then each of the sequences $\mathcal{A}_i$, $1 \leq i \leq k$, is either trivial or Behrend.

Proof It will be sufficient to consider $\mathcal{A}_1$, which we may assume to be non-trivial. We put $s = [s_1, s_2, \dots s_k]$, $v = s\boldsymbol{\delta}\mathcal{M}(\{s_1, s_2, \dots, s_k\})$, so that the set of multiples on the left-hand side of (1.218) is contained in the union of v arithmetic progression $\mathcal{N} = s\mathbf{Z} + j_\mu$, $1 \leq \mu \leq v$. By the argument employed in our initial example we have for every μ,

$$\boldsymbol{\delta}\{\mathcal{M}(s_1\mathcal{A}_1 \cup s_2\mathcal{A}_2 \cup \cdots \cup s_k\mathcal{A}_k)\cap\mathcal{N}\} = \frac{1}{s}. \tag{1.219}$$

We put $\mathcal{N} = s\mathbf{Z} + s_1$: since $\{s_1, s_2, \dots, s_k\}$ is primitive $\mathcal{N}$ contains no multiples from $s_i\mathcal{A}_i$, $i > 1$, and we have

$$\boldsymbol{\delta}\{\mathcal{M}(s_1\mathcal{A}_1)\cap\mathcal{N}\} = \frac{1}{s} \tag{1.220}$$

and

$$\boldsymbol{\delta}\{\mathcal{M}(\mathcal{A}_1)\cap\mathcal{L}\} = \frac{s_1}{s} \tag{1.221}$$

where $\mathcal{L} = (s/s_1)\mathbf{Z} + 1$. We apply Lemma 1.17 to deduce that $\mathcal{A}_1$ is Behrend. This complete the proof.

We note that the hypothesis $(q,r) = 1$ is essential for the final conclusion of the lemma. For example, $\delta\{\mathscr{M}(\{3\}) \cap 3\mathbf{Z}\} = \frac{1}{3}$, but $\{3\}$ is not a Behrend sequence. If every element of $\mathscr{A}$ is prime to q, that is $\mathscr{A} = \mathscr{A}^*(q)$, then (1.212) holds for every r. (This can be seen by an examination of the proof of Lemma 1.17, where the extra hypothesis $(q,r) = 1$ was just used at the outset, to enable us to strike out $\mathscr{A} \setminus \mathscr{A}^*(q)$.)

Theorem 1.19 *Let $\mathscr{A}$ be a Besicovitch sequence, all of whose elements are prime to the integer q. Then for every r, $\mathbf{d}\{\mathscr{M}(\mathscr{A}) \cap (q\mathbf{Z}+r)\}$ exists and is equal to $q^{-1}\mathbf{d}\mathscr{M}(\mathscr{A})$.*

Proof When $\mathscr{A}$ is finite the proof of Lemma 1.17 already gives this. Let $\mathscr{A}$ be infinite, and $\mathscr{A}^{(N)} = \{a_1, a_2, \ldots, a_N\}$. Then we have

$$\mathbf{d}\{\mathscr{M}(\mathscr{A}^{(N)}) \cap (q\mathbf{Z}+r)\} = q^{-1}\mathbf{d}\mathscr{M}(\mathscr{A}^{(N)}) \tag{1.222}$$

and we apply (0.67) to obtain

$$\underline{\mathbf{d}}\{\mathscr{M}(\mathscr{A}) \cap (q\mathbf{Z}+r)\} \geq q^{-1}\mathbf{d}\mathscr{M}(\mathscr{A}). \tag{1.223}$$

This holds for every $r\,(\mathrm{mod}\ q)$ and so the opposite inequality applies to the upper densities by subtraction. This completes the proof.

The last part of Lemma 1.17 demonstrates a special property of the arithmetic progressions $q\mathbf{Z} + r$ with $(q,r) = 1$, and suggested to the author the following definition. The name comes from the world of law, in which a witness is a person who must testify, but only to first-hand knowledge.

Definition 1.20 *Let $\mathscr{W}$ be an integer sequence possessing positive logarithmic density. If the relation*

$$\delta\{\mathscr{M}(\mathscr{A}) \cap \mathscr{W}\} = \delta\mathscr{W} \tag{1.224}$$

implies that $\mathscr{A}$ is either trivial or Behrend, then $\mathscr{W}$ is called a witness.

Thus we may draw a conclusion about the behaviour of $\mathscr{A}$, although $\mathscr{W}$ 'sees' only a part of $\mathscr{M}(\mathscr{A})$. By Lemma 1.17, any arithmetic progression $(q\mathbf{Z}+r) \cap \mathbf{Z}^+$ is a witness. The union of two such progressions may or may not be, and a suggested exercise for the reader is to write down the necessary and sufficient condition for a finite union of arithmetic progressions to be a witness. From (1.224), the set of multiples $\mathscr{M}(\mathscr{B})$ is not a witness unless $\mathscr{B}$ is Behrend or trivial – in some sense the class of witnesses is distant from the class of sets of multiples.

In the Oberwohlfach (1994) list of problems Tenenbaum made a conjecture which, if true, would be a very interesting result, analogous to Theorem 1.18 but with $\mathscr{S}$ infinite. We conclude this chapter with a proof. We employ the notation $\mathscr{S}\mathscr{A} = \{sa : s \in \mathscr{S},\, a \in \mathscr{A}\}$.

Theorem 1.21 *Let $\mathscr{A}, \mathscr{S}, \mathscr{T}$ be sequences such that*

$$\delta\mathscr{M}(\mathscr{S} \cup \mathscr{T}) > \delta\mathscr{M}(\mathscr{T}), \tag{1.225}$$

$$\delta\mathscr{M}(\mathscr{S}\mathscr{A} \cup \mathscr{T}) = \delta\mathscr{M}(\mathscr{S} \cup \mathscr{T}). \tag{1.226}$$

Then $\mathscr{A}$ is Behrend or trivial.

We employ the following result.

Lemma 1.22 (Tenenbaum) *Let $s \in \mathbf{Z}^+$ and $\mathscr{A}, \mathscr{T}$ be arbitrary. Then*

$$\delta\mathscr{M}(\{s\} \cup \mathscr{T}) - \delta\mathscr{M}(s\mathscr{A} \cup \mathscr{T}) \ge (1 - \delta\mathscr{M}(\mathscr{A}))\left(\delta\mathscr{M}(\{s\} \cup \mathscr{T}) - \delta\mathscr{M}(\mathscr{T})\right). \tag{1.227}$$

This inequality gives an alternative proof of Theorem 1.18, not involving arithmetic progressions and therefore avoiding Lemma 1.17.

Proof of lemma Let $n \in \mathscr{M}(\{s\} \cup \mathscr{T}) \setminus \mathscr{M}(s\mathscr{A} \cup \mathscr{T})$. Then $n = sm$ where $m \notin \mathscr{M}(\mathscr{A})$. Also, $sm \notin \mathscr{M}(\mathscr{T})$ so that in the notation of §0.4, $m \notin \mathscr{M}\left(\mathscr{T}'(s)\right)$. The density of such m is

$$1 - \delta\mathscr{M}\left(\mathscr{A} \cup \mathscr{T}'(s)\right) \ge (1 - \delta\mathscr{M}(\mathscr{A}))\left(1 - \delta\mathscr{M}\left(\mathscr{T}'(s)\right)\right) \tag{1.228}$$

by Behrend's inequality (Theorem 0.12). The second factor on the right is $\mathbf{t}\left(\mathscr{T}'(s)\right)$ and we apply formula (0.78). This gives the result stated.

Proof of Theorem 1.21 We may assume that $\mathscr{S}$ is infinite, by Theorem 1.18, and primitive. We put $\mathscr{S}_\nu = \{s_1, s_2, \ldots, s_\nu\}$ for $\nu \ge 1$, and $\mathscr{S}_0 = \emptyset$. Also, for every ν, $\mathscr{S}_\nu^\sharp = \{s_{\nu+1}, s_{\nu+2}, \ldots\}$. For arbitrary $\mathscr{A}, \mathscr{T}$ and ν we may write (actually an identity between sets),

$$\begin{aligned}
&\delta\mathscr{M}(\mathscr{S} \cup \mathscr{T}) - \delta\mathscr{M}(\mathscr{S}\mathscr{A} \cup \mathscr{T}) = \\
&\quad \delta\mathscr{M}(\{s_1\} \cup \left(\mathscr{S}_1^\sharp \cup \mathscr{T}\right)) - \delta\mathscr{M}(s_1\mathscr{A} \cup \left(\mathscr{S}_1^\sharp \cup \mathscr{T}\right)) \\
&\quad + \delta\mathscr{M}(\{s_2\} \cup \left(s_1\mathscr{A} \cup \mathscr{S}_2^\sharp \cup \mathscr{T}\right)) - \delta\mathscr{M}(s_2\mathscr{A} \cup \left(s_1\mathscr{A} \cup \mathscr{S}_2^\sharp \cup \mathscr{T}\right)) \\
&\quad + \delta\mathscr{M}(\{s_3\} \cup \left(\mathscr{S}_2\mathscr{A} \cup \mathscr{S}_3^\sharp \cup \mathscr{T}\right)) - \delta\mathscr{M}(s_3\mathscr{A} \cup \left(\mathscr{S}_2\mathscr{A} \cup \mathscr{S}_3^\sharp \cup \mathscr{T}\right)) \\
&\quad \vdots \\
&\quad + \delta\mathscr{M}(\{s_\nu\} \cup (\mathscr{S}_{\nu-1}\mathscr{A} \cup \mathscr{S}_\nu^\sharp \cup \mathscr{T})) - \delta\mathscr{M}(\mathscr{S}\mathscr{A} \cup \mathscr{T})
\end{aligned} \tag{1.229}$$

The expression in the last line of this identity is equal to

$$\delta\mathcal{M}(\mathcal{S}_\nu\mathcal{A}\cup\mathcal{S}^\#_\nu\cup\mathcal{T})-\delta\mathcal{M}(\mathcal{S}\mathcal{A}\cup\mathcal{T}) \tag{1.230}$$

and this is non-negative. Every other line has the same form as the left-hand side of (1.227), and we apply the lemma, which yields

$$\begin{aligned}\delta\mathcal{M}(\mathcal{S}\cup\mathcal{T})-\delta\mathcal{M}(\mathcal{S}\mathcal{A}\cup\mathcal{T})\geq&\\ &(1-\delta\mathcal{M}(\mathcal{A}))\sum_{1\leq\mu\leq\nu}\{\delta\mathcal{M}(\mathcal{S}_{\mu-1}\mathcal{A}\cup\mathcal{S}^\#_{\mu-1}\cup\mathcal{T})\\ &-\delta\mathcal{M}(\mathcal{S}_{\mu-1}\mathcal{A}\cup\mathcal{S}^\#_{\mu}\cup\mathcal{T})\}.\end{aligned} \tag{1.231}$$

Suppose now that $\mathcal{A},\mathcal{S},\mathcal{T}$ satisfy (1.225)–(1.226), and that $\delta\mathcal{M}(\mathcal{A})<1$. The sum on the right of (1.231) is zero, and since every term in this sum is non-negative, these terms are all zero, in particular the last. That is, for every ν

$$\delta\mathcal{M}(\mathcal{S}_{\nu-1}\mathcal{A}\cup\mathcal{S}^\#_{\nu-1}\cup\mathcal{T})-\delta\mathcal{M}(\mathcal{S}_{\nu-1}\mathcal{A}\cup\mathcal{S}^\#_\nu\cup\mathcal{T})=0. \tag{1.232}$$

We apply Theorem 0.8, and deduce that for every ν the sequence

$$(\mathcal{S}_{\nu-1}\mathcal{A}\cup\mathcal{S}^\#_\nu\cup\mathcal{T})'(s_\nu) \tag{1.233}$$

is either Behrend or trivial, in which we have employed the notation introduced in (0.74). We deduce that for every ν, the sequence

$$(\mathcal{S}_{\nu-1}\cup\mathcal{S}^\#_\nu\cup\mathcal{T})'(s_\nu) \tag{1.234}$$

is either Behrend or trivial. We may assume that $\mathcal{S}$ is primitive, and since $\mathcal{S}_{\nu-1}$ is finite this implies that for every ν,

$$(\mathcal{S}^\#_\nu\cup\mathcal{T})'(s_\nu) \tag{1.235}$$

is either Behrend or trivial; applying Theorem 0.8 again we deduce that for every ν, we have

$$\delta\mathcal{M}(\mathcal{S}^\#_{\nu-1}\cup\mathcal{T})=\delta\mathcal{M}(\mathcal{S}^\#_\nu\cup\mathcal{T}). \tag{1.236}$$

Therefore

$$\delta\mathcal{M}(\mathcal{S}\cup\mathcal{T})=\delta\mathcal{M}(\mathcal{T}), \tag{1.237}$$

contradicting (1.225). We conclude that $\delta\mathcal{M}(\mathcal{A})=1$ as required.

2

Derived sequences and densities

2.1 Introduction

Let $\mathscr{A} = \{a_1, a_2, \ldots\}$ be an integer sequence and $\mathscr{M}(\mathscr{A})$ be the set of multiples of $\mathscr{A}$. Then n belongs to $\mathscr{M}(\mathscr{A})$ if and only if

$$\tau(n, \mathscr{A}) > 0 \tag{2.1}$$

where $\tau(n, \mathscr{A})$ denotes the number of divisors of n which belong to $\mathscr{A}$. For each positive integer k, there is a (possibly empty) subsequence of $\mathscr{M}(\mathscr{A})$ on which

$$\tau(n, \mathscr{A}) > k \tag{2.2}$$

and we notice that this subsequence is itself a set of multiples and as such possesses logarithmic density, by Theorem 0.2, the Davenport–Erdös theorem. For (2.2) holds if and only if there exist i_j, $0 \le j \le k$ such that $i_0 < i_1 < \cdots < i_k$ and

$$a_{i_j} | n, \quad 0 \le j \le k, \tag{2.3}$$

equivalently

$$[a_{i_0}, a_{i_1}, a_{i_2}, \ldots, a_{i_k}] | n. \tag{2.4}$$

Let us write

$$\mathscr{A}^{(k)} = \{[a_{i_0}, a_{i_1}, \ldots, a_{i_k}] : 1 \le i_0 < i_1 < \cdots < i_k\}. \tag{2.5}$$

and, to fix our ideas, order $\mathscr{A}^{(k)}$ as a non-decreasing sequence. It may contain repetitions, but we do not exclude this case. Then we see that (2.2) is equivalent to $n \in \mathscr{M}(\mathscr{A}^{(k)})$.

Definition 2.1 *The kth derived sequence of $\mathscr{A}$, denoted by $\mathscr{A}^{(k)}$, is given by (2.5) above. We write*

$$\mathbf{t}_k(\mathscr{A}) = 1 - \boldsymbol{\delta}\mathscr{M}(\mathscr{A}^{(k)}) = 1 - \underline{\mathbf{d}}\mathscr{M}(\mathscr{A}^{(k)}), \tag{2.6}$$

equivalently $t_k(\mathscr{A})$ is the logarithmic density of the sequence of integers n for which

$$\tau(n, \mathscr{A}) \leq k. \tag{2.7}$$

Sometimes we refer to these sequences as derivatives (but the reader should note that derivatives of derivatives are not generally derivatives: fortunately we are not concerned with these). We have the following immediate result, which is simply an equivalent alternative form of Corollary 0.15.

Proposition 2.2 *Every derivative of a Behrend sequence is Behrend.*

We remark, en passant, that the sequence of integers n such that $\tau(n, \mathscr{A}) = k$, which is of course $\mathscr{M}(\mathscr{A}^{(k-1)}) \setminus \mathscr{M}(\mathscr{A}^{(k)})$, also possesses logarithmic density, and we denote this by $\bar{\mathbf{t}}_k(\mathscr{A})$. We have $\bar{\mathbf{t}}_0(\mathscr{A}) = \mathbf{t}_0(\mathscr{A}) = \mathbf{t}(\mathscr{A})$: we use all three of these expressions according to the context.

This chapter contains four main parts. In the first of these, §2.2, we are concerned with inequalities satisfied by the $\mathbf{t}_k(\mathscr{A})$, particularly upper bounds for $\mathbf{t}_k(\mathscr{A})$ in terms of $\mathbf{t}(\mathscr{A})$, which are quantitative forms of Proposition 2.2. Some of this work is quite intricate and we mention at the outset that there are harmless looking problems in this area which we are unable to solve. We draw these to the reader's attention as they arise. In the whole of this section we are concerned with a single sequence $\mathscr{A}$.

In §2.3 we take two sequences $\mathscr{A}$ and $\mathscr{B}$ and we generalize Behrend's inequality in various ways. The results are lower bounds for the densities $\mathbf{t}_k(\mathscr{A} \cup \mathscr{B})$.

One of the basic difficulties in the study of sets of multiples or sieves is that, often driven back to the inclusion–exclusion principle, we are forced to write down sums involving a large number of terms of opposite sign to represent small positive quantities such as $\mathbf{t}(\mathscr{A})$. Thus for a finite sequence $\mathscr{A}$ we have the familiar formula

$$\mathbf{t}(\mathscr{A}) = 1 - \sum_i \frac{1}{a_i} + \sum_{i<j} \frac{1}{[a_i, a_j]} - \cdots, \tag{2.8}$$

which could hardly be more ill-suited to most applications, indeed it is not obvious that the right-hand side is positive. There is a parallel here with the exponential series

$$e^{-x} = 1 - x + \frac{x^2}{2!} - \cdots, \ (x > 0);$$

again the right-hand side is not obviously positive and for some purposes the 'expansion' (we are not concerned here with the definition of the exponential function) is quite useless. In §2.4 and §2.5 we describe a different method of dealing with quantities like $\mathbf{t}_k(\mathscr{A})$: this is still combinatorial in nature, but leads to formulae involving sums of positive and therefore necessarily small terms. This technique is very successful with the problem encountered in Chapter 3, for which the last sections of this chapter are an essential preparation. We do not know to what extent it applies to the open problems mentioned above, but this would appear to be a possible avenue of further study.

2.2 Upper bounds for $\mathbf{t}_k(\mathscr{A})$

In this section we seek an upper bound for $\mathbf{t}_k(\mathscr{A})$, defined by (2.6), in terms of $\mathbf{t}(\mathscr{A})$. Ideally we should like this to be simply a function of $\mathbf{t}$, that is we suppose there is no *a priori* information about $\mathscr{A}$. Our second theorem is a result of this sort but is (presumably) far from best possible. A complete solution of the problem is presented on the strong hypothesis that the elements of $\mathscr{A}$ should be pairwise coprime. This is the content of Theorem 2.9. We begin with a preliminary result.

Theorem 2.3 *We define*

$$\pi_k = \inf\{\mathbf{t}(\mathscr{A}) : |\mathscr{A}| \leq k\}. \tag{2.9}$$

Then we have

$$\pi_k = \prod_{i=1}^{k} \left(1 - \frac{1}{p_i}\right) \tag{2.10}$$

where p_i denotes the ith *prime.*

Proof We have to show that $\mathbf{t}(\mathscr{A})$ is not less than the right-hand side of (2.10). Let $P^+(a)$ denote, as usual, the greatest prime factor of a, and put

$$P = \{p : p = P^+(a) \text{ for some } a \in \mathscr{A}\}, \tag{2.11}$$

so that P is a set of at most k primes. For each p belonging to P let

$$\mathscr{A}_p = \{a : a \in \mathscr{A}, P^+(a) = p\}. \tag{2.12}$$

Then $\mathscr{M}(\mathscr{A}_p) \subseteq p\mathbf{Z}^+$ and $\mathbf{t}(\mathscr{A}_p) \geq 1 - (1/p)$. We apply Behrend's inequality, (Theorem 0.12) which yields

$$\mathbf{t}(\mathscr{A}) \geq \prod_{p \in P} \mathbf{t}(\mathscr{A}_p) \geq \prod_{p \in P} \left(1 - \frac{1}{p}\right) \tag{2.13}$$

and since $|P| \leq k$, the result follows.

It is important to notice that the functions $\mathbf{t}_k(\mathscr{A})$ are not monotonic in the strong sense that $\mathbf{t}(\mathscr{A}) < \mathbf{t}(\mathscr{B})$ would imply $\mathbf{t}_k(\mathscr{A}) < \mathbf{t}_k(\mathscr{B})$. A counterexample is $\mathscr{A} = \{3, 5, 7\}$, $\mathscr{B} = \{2, 4\}$ when we have

$$\mathbf{t}(\mathscr{A}) = \frac{16}{35}, \; \mathbf{t}(\mathscr{B}) = \frac{1}{2}, \; \mathbf{t}_1(\mathscr{A}) = \frac{92}{105}, \; \mathbf{t}_1(\mathscr{B}) = \frac{3}{4}. \tag{2.14}$$

Definition 2.4 *For each positive integer k and $\sigma \in (0, 1]$ let*

$$\varphi_k(\sigma) = \sup\{\mathbf{t}_k(\mathscr{A}) : \mathbf{t}(\mathscr{A}) \leq \sigma\}. \tag{2.15}$$

This is certainly an increasing function of σ, in the weak sense. Notice that if $\sigma \geq \pi_k$ then $\mathscr{A}$ may have only k elements, when $\tau(n, \mathscr{A}) > k$ is impossible and $\mathbf{t}_k(\mathscr{A}) = 1$. Hence $\varphi_k(\pi_k) = 1$, or

$$\pi_k \leq \sigma \leq 1 \quad \text{implies} \quad \varphi_k(\sigma) = 1. \tag{2.16}$$

A moment's thought will convince the reader that the converse implication is not obvious; indeed we are unable to say whether or not it is true.

Definition 2.5 *For each positive integer k let*

$$\rho_k = \inf\{\sigma : \varphi_k(\sigma) = 1\}. \tag{2.17}$$

Conjecture 2.6

$$\rho_k = \pi_k \text{ for all } k. \tag{2.18}$$

Evidently $\rho_k \leq \pi_k$, and we prove later that if we restrict $\mathscr{A}$ in (2.15) by requiring that the elements of $\mathscr{A}$ should be pairwise coprime then the quantity corresponding to ρ_k equals π_k. For the moment we proceed as far as we can without any side condition on $\mathscr{A}$.

Theorem 2.7 *For each positive integer k we have*

$$\sigma \leq \varphi_k(\sigma) < (k + 2)\sigma^{1/k+1}. \tag{2.19}$$

A corollary is that $\rho_k \geq 1/(k+2)^{k+1}$. This is (presumably) very far from the truth since Mertens' theorem (and the fact that $\log p_k \sim \log k$) imply

$$\pi_k \sim \frac{e^{-\gamma}}{\log k}. \tag{2.20}$$

Proof of Theorem 2.7 The lower bound is simple: we have $\mathbf{t}_k(\mathscr{A}) \geq \mathbf{t}(\mathscr{A})$ and we can find a sequence $\mathscr{A}$ (for example containing primes only) such that $\mathbf{t}(\mathscr{A}) = \sigma$. For a lower bound of the correct order of magnitude see (2.78). We turn to the upper bound. Let $\mathbf{t}(\mathscr{A}) \leq \sigma$. We are going to split $\mathscr{A}$ into $k+1$ disjoint subsequences $\mathscr{A}_j$, $0 \leq j \leq k$. If $n \in \mathscr{M}(\mathscr{A}_j)$ for all j then $\tau(n, \mathscr{A}) > k$, whence

$$\mathbf{t}_k(\mathscr{A}) \leq \sum_{j=0}^{k} \mathbf{t}(\mathscr{A}_j). \tag{2.21}$$

We may assume that $\sigma^{1/k+1} < \frac{1}{2}$ else (2.19) is trivial, and we set

$$u_j = \sigma^{(j+1)/(k+1)},\ 0 \leq j \leq k, \tag{2.22}$$

so that

$$\frac{1}{2} > u_0 > u_1 > \cdots > u_{k-1} > \sigma \geq \mathbf{t}(\mathscr{A}). \tag{2.23}$$

For each j, $(0 \leq j \leq k)$ let m_j denote the largest integer such that

$$\mathbf{t}(\{a_1, a_2, \ldots, a_{m_j}\}) > u_j. \tag{2.24}$$

Since $\mathbf{t}(\{a_1\}) = 1 - 1/a_1 \geq \frac{1}{2} > u_j > \mathbf{t}(\mathscr{A})$ by (2.23) we have $1 \leq m_j < \infty$ and put

$$\mathscr{B}_j = \{a_1, a_2, \ldots, a_{m_j}\},\ 0 \leq j < k. \tag{2.25}$$

By the maximal property of m_j we have

$$\mathbf{t}(\mathscr{B}_j \cup \{a_{m_j+1}\}) \leq u_j \tag{2.26}$$

and the Heilbronn–Rohrbach inequality (Theorem 0.9) implies

$$u_j \geq \mathbf{t}(\mathscr{B}_j)\left(1 - \frac{1}{a_{m_j+1}}\right) \geq \prod_{1 \leq i \leq m_j+1}\left(1 - \frac{1}{a_i}\right). \tag{2.27}$$

We may assume that $\mathscr{A}$ has no repeated elements for repetitions would not affect the value of $\mathbf{t}(\mathscr{A})$ and would not increase $\mathbf{t}_k(\mathscr{A})$. Hence the product on the right of (2.27) is at least $1/(m_j+2) \geq 1/(a_{m_j+1})$, whence (2.27) implies firstly that $a_{m_j+1} \geq 1/u_j$ and then that

$$\mathbf{t}(\mathscr{B}_j) \leq u_j \left(1 - \frac{1}{a_{m_j+1}}\right)^{-1} \leq \frac{u_j}{1-u_j}, \ (0 \leq j < k). \tag{2.28}$$

Now we may specify the $\mathscr{A}_j$ appearing in (2.21). We put $\mathscr{A}_0 = \mathscr{B}_0$, $\mathscr{A}_1 = \mathscr{B}_1 \setminus \mathscr{B}_0$, $\mathscr{A}_2 = \mathscr{B}_2 \setminus \mathscr{B}_1, \ldots, \mathscr{A}_k = \mathscr{A} \setminus \mathscr{B}_{k-1}$. Behrend's inequality (Theorem 0.12) gives

$$\begin{aligned} \mathbf{t}(\mathscr{B}_1) &\geq \mathbf{t}(\mathscr{B}_0)\mathbf{t}(\mathscr{A}_1),\ \mathbf{t}(\mathscr{B}_2) \\ &\geq \mathbf{t}(\mathscr{B}_1)\mathbf{t}(\mathscr{A}_2), \ldots, \mathbf{t}(\mathscr{A}) \geq \mathbf{t}(\mathscr{B}_{k-1})\mathbf{t}(\mathscr{A}_k) \end{aligned} \tag{2.29}$$

and since we have both lower bounds (2.24) and upper bounds (2.28) for the $\mathbf{t}(\mathscr{B}_j)$ we may deduce from (2.29) that

$$\begin{aligned} \mathbf{t}(\mathscr{A}_0) &\leq u_0/(1-u_0), \\ \mathbf{t}(\mathscr{A}_j) &\leq u_j u_{j-1}^{-1}/(1-u_j), \quad 1 \leq j < k, \\ \mathbf{t}(\mathscr{A}_k) &\leq \sigma u_{k-1}^{-1} \end{aligned} \tag{2.30}$$

whence by (2.22)

$$\mathbf{t}(\mathscr{A}_j) \leq \frac{\sigma^{1/(k+1)}}{1-\sigma^{1/(k+1)}}, \quad 0 \leq j < k \tag{2.31}$$

with strict inequality when $j > 0$. Thus (2.21) implies

$$\mathbf{t}_k(\mathscr{A}) < (k+1)\frac{\sigma^{1/(k+1)}}{1-\sigma^{1/(k+1)}}.$$

This is a uniform bound, and when $\sigma^{1/(k+1)} \leq 1/(k+2)$ it implies (2.19). For larger σ, (2.19) is trivial. This completes the proof.

The reader will have observed that our choice of the u_j in (2.22) is not optimal. However, it is clear that once we have adopted the strategy of this proof, simply adding the densities in (2.21), the upper bound envisaged cannot be better than $(k+1)\sigma^{1/(k+1)}$.

Theorem 2.8 *For every k, the function $\varphi_k(\sigma)$ is continuous.*

Proof By Theorem 2.7, $\varphi_k(\sigma)$ is continuous at $\sigma = 0$. Since φ_k is non-decreasing it will therefore be sufficient to show that for $\sigma > 0$ we have

$$\varphi_k(\sigma) - \varphi_k(x\sigma) \leq 1 - x, \quad 0 < x < 1. \tag{2.32}$$

We may assume $\varphi_k(\sigma) > 0$ else there is nothing to prove. Let $0 < \varepsilon < \varphi_k(\sigma)$. By definition, we can find a sequence $\mathscr{A}(\varepsilon)$ such that $\mathbf{t}(\mathscr{A}(\varepsilon)) \leq \sigma$

and $\mathbf{t}_k(\mathscr{A}(\varepsilon)) > \varphi_k(\sigma) - \varepsilon > 0$. Hence $\mathbf{t}(\mathscr{A}(\varepsilon)) > 0$ by Theorem 2.7, that is $\mathscr{A}(\varepsilon)$ is not a Behrend sequence. Therefore

$$\sum \left\{ \frac{1}{p} : p \in \mathscr{A}(\varepsilon) \right\} < \infty,$$

and we deduce that there exists a set $P(\varepsilon)$ of primes, not intersecting $\mathscr{A}(\varepsilon)$ and such that

$$\prod_{p \in P(\varepsilon)} \left(1 - \frac{1}{p}\right) = x.$$

Put $\mathscr{B}(\varepsilon) = P(\varepsilon) \cup \mathscr{A}(\varepsilon)$. By Theorem 0.12, we have $\mathbf{t}(\mathscr{B}(\varepsilon)) = \mathbf{t}(P(\varepsilon))\mathbf{t}(\mathscr{A}(\varepsilon)) = x\mathbf{t}(\mathscr{A}(\varepsilon)) \leq x\sigma$. If $\tau(n, \mathscr{A}(\varepsilon)) \leq k < \tau(n, \mathscr{B}(\varepsilon))$ then $n \in \mathscr{M}(P(\varepsilon))$, whence the density of such integers n does not exceed $1 - x$, that is $\mathbf{t}_k(\mathscr{A}(\varepsilon)) - \mathbf{t}_k(\mathscr{B}(\varepsilon)) \leq 1 - x$. This implies (2.32) because ε is arbitrarily small.

We obtain a much better result than Theorem 2.7 if we assume that the elements of $\mathscr{A}$ are pairwise coprime. In the context of sets of multiples this is a rather unwelcome hypothesis, but it leads to an interesting problem in probability theory which we are able to solve, and the final result which we state next is essentially best possible.

Theorem 2.9 *Let $\varphi_k^*(\sigma)$ be as in (2.15) with the extra hypothesis that the elements of $\mathscr{A}$ be pairwise coprime. Then for each k, $\varphi_k^*(\sigma)$ is a continuous, non-decreasing function of σ, such that for each $n \geq 1$ and $\pi_n \geq \sigma \geq \pi_{n+1}$ we have*

$$\varphi_k^*(\sigma) \leq \frac{(\sigma - \pi_{n+1})\mathbf{t}_{k,n} + (\pi_n - \sigma)\mathbf{t}_{k,n+1}}{\pi_n - \pi_{n+1}} \tag{2.33}$$

with equality at the end-points, where

$$\mathbf{t}_{k,n} = \mathbf{t}_k(\{p_1, p_2, \ldots, p_n\}) \tag{2.34}$$

is the density of the integers divisible by at most k of the first n primes. In particular, for $\pi_k \geq \sigma \geq \pi_{k+1}$ we have

$$\varphi_k^*(\sigma) \leq 1 - \left(1 - \frac{\sigma}{\pi_k}\right) \prod_{j=1}^{k} \frac{1}{p_j}. \tag{2.35}$$

Corollary 2.10 *For each $k \in \mathbf{Z}^+$ we have $\varphi_k^*(\sigma) < 1$ if and only if $\sigma < \pi_k$.*

This is immediate from (2.35), which is simply (2.33) with the values $\mathbf{t}_{k,k} = 1$, $\mathbf{t}_{k,k+1} = 1 - 1/(p_1 p_2 \ldots p_{k+1})$ inserted.

We see that the function $\varphi_k^*(\sigma)$ is determined precisely at the points π_n, and within quite narrow confines between these points. However, the exact behaviour of this function within the intervals (π_{n+1}, π_n) remains somewhat mysterious. We shall see in Theorem 2.13 below that it is constant in an interval $[\pi_n, \pi_n + \eta_{k,n}]$, where $\eta_{k,n} > 0$ for every pair $k, n \geq 1$. We do not know whether (2.33) or (2.35) holds with equality at any intermediate point.

Of course we can write down an explicit formula for $\mathbf{t}_{k,n}$ (see the proof of Theorem 2.13) and thereby derive an asymptotic formula, and bounds for this quantity. We do not undertake this here and so Theorem 2.9 suffers from the defect, in comparison with Theorem 2.7, that it does not provide us with explicit, applicable analytic bounds for $\varphi_k^*(\sigma)$. These can be very useful and we pursue the matter by an alternative, direct method in Theorem 2.12.

Lemma 2.11 *Let $Y_j\,(1 \leq j \leq m+1)$ be independent Bernoulli variables, (random variables taking the values* 0 *or* 1*) such that*

$$1 > \mathbf{Prob}(Y_1 = 1) \geq \mathbf{Prob}(Y_2 = 1) \geq \cdots \geq \mathbf{Prob}(Y_{m+1} = 1) \tag{2.36}$$

and let

$$Y = Y_1 + Y_2 + \cdots + Y_{m+1}.$$

Let X_j, $1 \leq j < \infty$ be a further sequence of independent Bernoulli variables such that

$$\mathbf{Prob}(X_1 = 1) \geq \mathbf{Prob}(X_2 = 1) \geq \mathbf{Prob}(X_3 = 1) \geq \ldots \tag{2.37}$$

$$\mathbf{Prob}(X_j = 1) \leq \mathbf{Prob}(Y_j = 1), \quad 1 \leq j \leq m \tag{2.38}$$

$$X := \textstyle\sum_j^\infty X_j < \infty \text{ almost surely} \tag{2.39}$$

$$\mathbf{Prob}(X = 0) \leq \mathbf{Prob}(Y = 0). \tag{2.40}$$

Then for every positive integer k we have

$$\mathbf{Prob}(X \leq k) \leq \mathbf{Prob}(Y \leq k). \tag{2.41}$$

Proof Put

$$\alpha_j = \mathbf{Prob}(X_j = 1), \quad \beta_j = \mathbf{Prob}(Y_j = 1). \tag{2.42}$$

Condition (2.39) is equivalent to

$$\sum_{j=1}^{\infty} \alpha_j < \infty, \tag{2.43}$$

and we begin by supposing that $\alpha_j = 0$ for $j > n$ (say). Consider the case $n \leq m$ first. We have $\alpha_j \leq \beta_j < 1$ by (2.35) and (2.38), whereas from (2.40) we have

$$\prod_{j=1}^{n}(1-\alpha_j) \leq \prod_{j=1}^{m+1}(1-\beta_j). \tag{2.44}$$

Therefore $\alpha_j = \beta_j$, $1 \leq j \leq n$, $\beta_j = 0$, $n < j \leq m+1$, that is $\alpha_j = \beta_j$ for all j, and (2.41) is trivial. We may therefore suppose that $n > m$, moreover $\alpha_{m+1} > 0$. Let $k \in \mathbf{Z}^+$ be given: we may suppose $k \leq m$ because the event $Y \leq k$ is otherwise certain and (2.41) holds. Let $\boldsymbol{\alpha}$ denote the vector $(\alpha_1, \alpha_2 \ldots, \alpha_n)$ and $P_k(\boldsymbol{\alpha}) = \mathbf{Prob}(X \leq k)$: we have to show that the supremum of this function subject to the constraints (2.37), (2.38) and (2.40) is attained when $\boldsymbol{\alpha} = \boldsymbol{\beta} := (\beta_1, \beta_2, \ldots, \beta_{m+1}, 0, 0, \ldots, 0)$. $P_k(\alpha)$ is continuous and attains its maximum: let us suppose for some $\overline{\boldsymbol{\alpha}} \neq \boldsymbol{\beta}$. Put $X^{(i)} = X - X_i$. Then

$$\begin{aligned} P_k(\alpha) &= \mathbf{Prob}(X \leq k) \\ &= \mathbf{Prob}(X^{(i)} \leq k-1) + \mathbf{Prob}(X^{(i)} = k,\ X_i = 0) \\ &= \mathbf{Prob}(X^{(i)} \leq k-1) + (1-\alpha_i)\mathbf{Prob}(X^{(i)} = k) \end{aligned} \tag{2.45}$$

whence

$$\frac{\partial P_k}{\partial \alpha_i}(\boldsymbol{\alpha}) = -\mathbf{Prob}(X^{(i)} = k). \tag{2.46}$$

Now $\overline{\alpha}_{k+1} \geq \overline{\alpha}_{m+1} > 0$, by our earlier assumption, that is at least $k+1$ of the $\overline{\alpha}_j$ are non-zero and $\mathbf{Prob}(X^{(i)} = k) \neq 0$. Hence from (2.46),

$$\frac{\partial P_k}{\partial \alpha_i}(\boldsymbol{\alpha}) < 0, \ (1 \leq i \leq n). \tag{2.47}$$

Let h be the least integer such that $\overline{\alpha}_h \neq \beta_h$. We must have $h \leq m+1$ else $X \geq Y$ almost surely and (2.41) follows. Further, let l be the greatest integer for which $\overline{\alpha}_l > 0$. We see from (2.40) that $l > h$.

We have $\overline{\alpha}_l \leq \overline{\alpha}_h$ by (2.37) and it may be that $\overline{\alpha}_l < \overline{\alpha}_h$. In this case we consider the vector $\boldsymbol{\alpha}$ in which

$$\alpha_h = \overline{\alpha}_h + \varepsilon,\ \alpha_l = \overline{\alpha}_l - \varepsilon,\ \alpha_j = \overline{\alpha}_j \text{ else}, \tag{2.48}$$

where ε is small and positive. Since $\overline{\alpha}_h < \beta_h \leq \beta_{h-1} = \alpha_{h-1}$ and $\overline{\alpha}_l > 0 = \overline{\alpha}_{l+1}$, moreover

$$\begin{aligned} \mathbf{Prob}(X = 0) &= P_0(\boldsymbol{\alpha}) \\ &= P_0(\overline{\boldsymbol{\alpha}})\left(1 - \frac{\varepsilon}{1-\overline{\alpha}_h}\right)\left(1 + \frac{\varepsilon}{1-\overline{\alpha}_l}\right) \\ &< P_0(\overline{\boldsymbol{\alpha}}) \leq \mathbf{Prob}(Y = 0) \end{aligned} \tag{2.49}$$

so that (2.40) still holds, this is permissible if ε is small enough. We have

$$P_k(\boldsymbol{\alpha}) = P_k(\overline{\boldsymbol{\alpha}}) + \varepsilon\left(\frac{\partial P_k}{\partial \alpha_h}(\overline{\boldsymbol{\alpha}}) - \frac{\partial P_k}{\partial \alpha_l}(\overline{\boldsymbol{\alpha}})\right) + O(\varepsilon^2) \tag{2.50}$$

and, by (2.46), the coefficient of ε on the right is

$$\mathbf{Prob}(X^{(l)} = k) - \mathbf{Prob}(X^{(h)} = k). \tag{2.51}$$

Now

$$\mathbf{Prob}(X^{(i)} = k) = \mathbf{Prob}(X = 0) \sum_{i_1 < i_2 < \cdots < i_k}^{(i)} \prod \frac{\alpha_{i_j}}{1 - \alpha_{i_j}} \tag{2.52}$$

where $\sum^{(i)}$ denotes that $i_j \neq i$. Since $\alpha_h > \alpha_l$ this implies that (2.51) is positive, that is $P_k(\boldsymbol{\alpha}) > P_k(\overline{\boldsymbol{\alpha}})$ if ε is sufficiently small. This is a contradiction. We deduce that $\overline{\alpha}_l = \overline{\alpha}_h$.

In this case we again vary $\overline{\boldsymbol{\alpha}}$ slightly, by increasing $\overline{\alpha}_h$ and decreasing $\overline{\alpha}_l$. As before, the constraints (2.37), (2.38) hold, but the relative movement of $P_0(\boldsymbol{\alpha})$ and $P_k(\boldsymbol{\alpha})$ is more subtle. Rather than embark on a calculation involving higher variations we make α_h and α_l functions of a parameter u in such a way that $P_0(\boldsymbol{\alpha})$ is independent of u, i.e. constant. We put

$$1 - \alpha_h = (1 - \overline{\alpha}_h)e^{-u}, \quad 1 - \alpha_l = (1 - \overline{\alpha}_l)e^{u} \tag{2.53}$$

where $u > 0$. Put $X^{(h,l)} = X - X_h - X_l$. We have

$$\begin{aligned} P_k(\alpha) &= \mathbf{Prob}(X \leq k) \\ &= \mathbf{Prob}(X^{(h,l)} \leq k - 2) \\ &\quad + \mathbf{Prob}(X^{(h,l)} = k - 1).\mathbf{Prob}(X_h + X_l = 1) \\ &\quad + \mathbf{Prob}(X^{(h,l)} = k).\mathbf{Prob}(X_h = X_l = 0). \end{aligned} \tag{2.54}$$

Only the middle term on the right depends on u; we have $\mathbf{Prob}(X^{(h,l)} = k-1) > 0$, because at least $k+1$ of the α_j are non-zero, and the right-hand factor is

$$\begin{aligned} &(1 - \alpha_l)\alpha_h + (1 - \alpha_h)\alpha_l \\ &\quad = (1 - \overline{\alpha}_l)e^u + (1 - \overline{\alpha}_h)e^{-u} - 2(1 - \overline{\alpha}_h)(1 - \overline{\alpha}_l) \\ &\quad = 2(1 - \overline{\alpha}_h)\cosh u - 2(1 - \overline{\alpha}_h)^2 \end{aligned} \tag{2.55}$$

because $\overline{\alpha}_h = \overline{\alpha}_l$. Since $\overline{\alpha}_h < 1$ by (2.35) and (2.38) this is a (strictly) increasing function of u, and this again is a contradiction. Hence $\boldsymbol{\alpha} = \boldsymbol{\beta}$ as required.

It remains to consider the case when the series (2.43) does not terminate, that is $\alpha_j > 0$ for every j. As before we may assume that for some $h \leq m+$

1 we have $\alpha_h < \beta_h$ else $X \geq Y$ almost surely and (2.41) holds. Let s be the greatest integer for which $\beta_j = \beta_h$ for $h \leq j \leq s$, so that either $s = m+1$ or $s \leq m$ and $\beta_s > \beta_{s+1}$. We decrease $\beta_h, \beta_{h+1}, \dots, \beta_s$ by a small quantity δ in the definition of the random variables Y_j, leaving the remaining variables unchanged. Thus Y becomes a new random variable $Y(\delta)$ for which

$$\mathbf{Prob}(Y(\delta) = 0) \geq \mathbf{Prob}(Y = 0)(1+\delta)^{s-h+1}. \tag{2.56}$$

Of course (2.35) is still satisfied. Let $0 < \varepsilon < \delta/2$, and $n > m$ be so large that

$$\sum_{j>n} \alpha_j < \varepsilon. \tag{2.57}$$

We write $X = X' + X''$ where $X' = X_1 + X_2 + \cdots + X_n$ and we have

$$\mathbf{Prob}(X'' \neq 0) = \mathbf{Prob}(X'' \geq 1) < \varepsilon \tag{2.58}$$

by Markoff's inequality, whence

$$\mathbf{Prob}(X' = 0) = \frac{\mathbf{Prob}(X = 0)}{\mathbf{Prob}(X'' = 0)} < \frac{\mathbf{Prob}(X = 0)}{1 - \varepsilon}. \tag{2.59}$$

Since $\varepsilon < \delta/2$, and clearly $\delta < 1$, we deduce from (2.56) and (2.59) that

$$\mathbf{Prob}(X' = 0) \leq \mathbf{Prob}(Y(\delta) = 0). \tag{2.60}$$

Since $\alpha_h < \beta_h$ we can make δ so small that (2.38) holds with the new Y_j, and we deduce from the first part of our proof that for all k,

$$\mathbf{Prob}(X' \leq k) \leq \mathbf{Prob}(Y(\delta) \leq k) \tag{2.61}$$

and clearly $\mathbf{Prob}(X \leq k) \leq \mathbf{Prob}(X' \leq k)$ since $X \geq X'$. The right-hand side of (2.61) is a continuous function of δ and (2.41) follows. This completes the proof of Lemma 2.11.

Proof of Theorem 2.9 Let $\mathscr{A} = \{a_1, a_2, \dots\}$ be any sequence of pairwise coprime elements arranged in increasing order, and such that $\mathbf{t}(\mathscr{A}) \leq \sigma$. It will be sufficient to show that for each k, $\mathbf{t}_k(\mathscr{A})$ does not exceed the right-hand side of (2.33). We may assume that $\mathbf{t}(\mathscr{A}) > 0$, for otherwise the series

$$\sum_{i=1}^{\infty} \frac{1}{a_i} \tag{2.62}$$

would be divergent, and for a sufficiently large N we should have

$$\prod_{i=1}^{N} \left(1 - \frac{1}{a_i}\right) \leq \sigma. \tag{2.63}$$

But then $\tau(n, \mathscr{A}) \geq \tau(n, \mathscr{A}')$ and $\mathbf{t}_k(\mathscr{A}) \leq \mathbf{t}_k(\mathscr{A}')$ where $\mathscr{A}' = \{a_1, a_2, \dots, a_N\}$.

Accordingly let $\mathbf{t}(\mathscr{A}) = \rho$, $0 < \rho \le \sigma$, and let $\pi_m \ge \rho \ge \pi_{m+1}$, so that $m \ge n$. We consider a sequence of Bernoulli trials in which

$$X_j = \begin{cases} 1 \text{ if } a_j | n \\ 0 \text{ else} \end{cases} \tag{2.64}$$

and we have $\mathbf{Prob}(X_j = 1) = 1/a_j$; moreover the X_j are independent because the a_j are coprime. The series (2.62) is convergent, and so conditions (2.37) and (2.39) of Lemma 2.11 are satisfied. Now we consider another sequence of Bernoulli trials in which

$$\mathbf{Prob}(Y_j = 1) = \frac{1}{p_j}, \quad 1 \le j \le m \tag{2.65}$$

where p_j is the jth prime, and

$$\mathbf{Prob}(Y_{m+1} = 1) = 1 - \frac{\rho}{\pi_m} \quad \left(\le \frac{1}{p_{m+1}}\right). \tag{2.66}$$

Then (2.38) is satisfied, because it is easy to see that $a_j \ge p_j$, moreover (2.40) holds with equality. As in Lemma 2.11 we put

$$X = \sum_{j=1}^{\infty} X_j, \; Y = \sum_{j=1}^{m+1} Y_j, \tag{2.67}$$

the left-hand series terminates if $\mathscr{A}$ is finite, and we apply the lemma to obtain

$$\mathbf{Prob}(X \le k) \le \mathbf{Prob}(Y \le k). \tag{2.68}$$

The right-hand side is a linear function of ρ, equal to $\mathbf{t}_{k,m}$ when $\rho = \pi_m$, and $\mathbf{t}_{k,m+1}$ when $\rho = \pi_{m+1}$. Hence it equals the right-hand side of (2.33) with m and ρ substituted for n and σ. We leave it to the reader to check the intuituve proposition that $\mathbf{Prob}(X \le k) = \mathbf{t}_k(\mathscr{A})$. Since the expression on the right of (2.33) is non-decreasing the conclusion follows. This completes the proof of Theorem 2.9.

Theorem 2.12 *For positive integers k and $0 < \sigma \le 1$ we have*

$$\sigma \sum_{j=0}^{k} \frac{1}{j!}\left(\log \frac{1}{\sigma}\right)^j \le \varphi_k^*(\sigma) \le e^C \sigma \sum_{j=0}^{k} \frac{1}{j!}\left(\log \frac{1}{\sigma}\right)^j, \tag{2.69}$$

where

$$C = \sum_p \left\{\frac{1}{p-1} + \log\left(1 - \frac{1}{p}\right)\right\} = .45743\ldots \tag{2.70}$$

We note that $e^C = 1.58000\ldots$ ($> \frac{79}{50}$!). This factor may, possibly, be replaced by a function of k tending to 1 as $k \to \infty$: this question could no doubt be settled by a careful analysis of the numbers $t_{k,n}$ defined by (2.34). The convergence cannot be too fast because $\varphi_k^*(\pi_k) = 1$. As it stands, (2.69) implies that $\varphi_k^*(\sigma) < 1$ only if σ is extravagantly small compared to π_k, indeed we require $\sigma < \exp\left(-k - c_0(k)\sqrt{k}\right)$ where $0 \le c_0(k) \ll 1$.

Proof of Theorem 2.12 When $\mathscr{A}$ is a sequence of pairwise coprime elements and $\mathbf{t}(\mathscr{A}) > 0$ it may be shown by straightforward combinatorial arguments that

$$\mathbf{t}_k(\mathscr{A}) = \mathbf{t}(\mathscr{A}) \sum_{j=0}^{k} S_j(\mathscr{A}) \tag{2.71}$$

where $S_j(\mathscr{A})$ denotes the jth elementary symmetric function of the numbers $1/(a_i - 1)$, $S_0(\mathscr{A}) = 1$. For $j \ge 1$, we have $S_j(\mathscr{A}) < S_1(\mathscr{A})^j/j!$, moreover

$$S_1(\mathscr{A}) = \sum_{i=1}^{\infty} \frac{1}{a_i - 1} \le -\sum_{i=1}^{\infty} \log\left(1 - \frac{1}{a_i}\right) + C \tag{2.72}$$

because $a_i \ge p_i$, the ith prime. If $\mathbf{t}(\mathscr{A}) = \rho$ this gives

$$\mathbf{t}_k(\mathscr{A}) \le \rho \sum_{j=0}^{k} \frac{1}{j!} \left(\left(\log \frac{1}{\rho}\right) + C\right)^j \tag{2.73}$$

and since $\rho \le \sigma \le \sigma e^C$ and the right-hand side of (2.73) increases with ρ, this gives the upper bound in (2.69). Next, by Lemma 13 (p.147) of Halberstam and Roth (1966) we have

$$S_j(\mathscr{A}) \ge \frac{1}{j!} S_1(\mathscr{A}) \left\{1 - \binom{j}{2} S_1(\mathscr{A})^{-2} \sum_{i=1}^{\infty} \frac{1}{(a_i - 1)^2}\right\}. \tag{2.74}$$

Let $w, z \to \infty$ together in such a way that

$$\prod_{w<p\le z} \left(1 - \frac{1}{p}\right) \to \sigma - 0, \tag{2.75}$$

and put $\mathscr{A} = \{p : w < p \le z\}$. We have

$$S_1(\mathscr{A}) = \sum_{w<p\le z} \frac{1}{p-1} \to \log \frac{1}{\sigma} \tag{2.76}$$

by (2.75), applying Cauchy's convergence criterion to the series (2.70), whence by (2.74)

$$S_j(\mathscr{A}) \geq (1 + o(1)) \frac{1}{j!} \left(\log \frac{1}{\sigma} \right)^j \tag{2.77}$$

and the lower bound in (2.69) follows from (2.71). This completes the proof of Theorem 2.12.

Evidently we also have

$$\varphi_k(\sigma) \geq \sigma \sum_{j=0}^{k} \frac{1}{j!} \left(\log \frac{1}{\sigma} \right)^j, \quad \sigma < 1 \tag{2.78}$$

but there is a substantial gap between this and the upper bound contained in Theorem 2.7. The central problem in this section is therefore to obtain an upper bound for $\varphi_k(\sigma)$ comparable with Theorems 2.9 and 2.12: of course the strongest conjecture would be that for all k, we have

$$\varphi_k(\sigma) = \varphi_k^*(\sigma), \quad 0 \leq \sigma \leq 1, \tag{2.79}$$

which in view of the corollary to Theorem 2.9, would certainly imply (2.18), $\rho_k = \pi_k$. There is a heuristic argument in favour of (2.79) as follows: if the elements of $\mathscr{A}$ are not coprime then there is a tendency for $\tau(n, \mathscr{A})$ to be large, given that it is positive. That is, for a given value of $\mathbf{t}(\mathscr{A})$ we might expect $\mathbf{t}_k(\mathscr{A})$ to be *smaller*, in the unrestricted situation.

All our upper and lower bounds for $\varphi_k(\sigma)$ and $\varphi_k^*(\sigma)$, provided by Theorems 2.7, 2.9 and 2.12 are concave functions of σ, and this might suggest that these functions are themselves concave. Our next theorem shows that this is not true of $\varphi_k^*(\sigma)$, and additionally that this function is not differentiable at the points π_n, $n \geq k$. We cannot prove any similar assertions about $\varphi_k(\sigma)$, but even if (2.79) were false we should not expect the two functions to behave completely differently.

Theorem 2.13 *For each k and $l \geq 1$ there exists $\eta_{k,l} > 0$ such that*

$$\varphi_k^*(\sigma) = \mathbf{t}_{k,l}, \; (\pi_l \leq \sigma \leq \pi_l + \eta_{k,l}), \tag{2.80}$$

where $\mathbf{t}_{k,l}$ is defined by (2.34).

Thus φ_k^* has infinitely many intervals of constancy, and cannot be concave. In particular it is not concave in any neighbourhood of $\sigma = 0$. If φ_k^* were differentiable at $\sigma = \pi_l$ then we should deduce from (2.80) that $d\varphi_k^*/d\sigma = 0$, in contradiction to (2.33) whenever $l \geq k$.

Proof of the theorem We may assume that $l > k$ because $\varphi_k^*(\sigma) = 1$ for $\sigma \geq \pi_k$. We want to find $\sigma > \pi_l$ such that $\mathbf{t}(\mathscr{A}) \leq \sigma$ implies $\mathbf{t}_k(\mathscr{A}) \leq \mathbf{t}_{k,l}$, and we begin with the restriction that $\sigma < \pi_{l-1}\left(1 - (1/p_{l+1})\right)$, where p_j denotes the jth prime. Since the upper bound for $\mathbf{t}_k(\mathscr{A})$ follows from Theorem 2.9 if $\mathbf{t}(\mathscr{A}) \leq \pi_l$, we may restrict our attention to sequences $\mathscr{A}$ (of coprime elements) such that

$$\pi_l < \mathbf{t}(\mathscr{A}) \leq \sigma < \pi_{l-1}\left(1 - \frac{1}{p_{l+1}}\right). \tag{2.81}$$

Since $\mathbf{t}(\mathscr{A}) < \pi_{l-1}$ this implies $|\mathscr{A}| \geq l$ by Theorem 2.3. As usual, we write $\mathscr{A} = \{a_1, a_2, \ldots\}$ and we have $a_l > p_l$, since $\mathbf{t}(\mathscr{A}) > \pi_l$. Hence $a_l \geq p_{l+1}$ because the elements of $\mathscr{A}$ are coprime. The right-hand inequality in (2.81) is impossible with just l elements, since the Heilbronn–Rohrbach inequality (0.80) and Theorem 2.3 would then imply in turn

$$\begin{aligned} \mathbf{t}(\mathscr{A}) &\geq \left(1 - \frac{1}{a_l}\right) \mathbf{t}(\{a_1, a_2, \ldots, a_{l-1}\}) \\ &\geq \left(1 - \frac{1}{p_{l+1}}\right) \pi_{l-1}. \end{aligned}$$

Therefore $|\mathscr{A}| \geq l + 1$. We are going to apply Lemma 2.11. Let

$$Y = \sum_{j=1}^{l+1} Y_j$$

where the Y_j are independent random variables equal to 0 or 1 (Bernoulli trials), such that

$$\begin{aligned} \mathbf{Prob}(Y_j = 1) &= p_j^{-1}, \quad j < l, \\ &= p_{l+1}^{-1}, \quad j = l, \\ \mathbf{Prob}(Y_{l+1} = 1) &= \theta p_{l+1}^{-1}, \quad 0 \leq \theta \leq 1. \end{aligned} \tag{2.82}$$

We determine $\theta = \theta(\sigma)$ in such a way that

$$\mathbf{Prob}(Y = 0) = \sigma,$$

thus $0 < \theta < \theta(\pi_l)$ where

$$\left(1 - \frac{1}{p_{l+1}}\right)\left(1 - \frac{\theta(\pi_l)}{p_{l+1}}\right) = 1 - \frac{1}{p_l}. \tag{2.83}$$

We note that $\theta(\pi_l) < 1$ by Bertrand's postulate. By Lemma 2.11 we deduce that for the sequences $\mathscr{A}$ satisfying (2.81) we have

$$\mathbf{t}_k(\mathscr{A}) \leq \mathbf{Prob}(Y \leq k). \tag{2.84}$$

We evaluate the right-hand side. Let us write $\mathbf{Prob}(Y_j = 1)/\mathbf{Prob}(Y_j = 0) = y_j$, and denote by s_h and $\bar{s}_h$ respectively, the hth elementary symmetric functions of the y_j, for $j \leq l+1$ and for $j \leq l-1$. Then we have

$$\mathbf{Prob}(Y \leq k) = \sigma\{1 + s_1 + s_2 + \cdots + s_k\}$$

and since $s_h = \bar{s}_h + (y_l + y_{l+1})\bar{s}_{h-1} + y_l y_{l+1}\bar{s}_{h-2}$ for $h \geq 1$ (with the convention that $\bar{s}_0 = 1$ and $\bar{s}_{-1} = 0$), this becomes

$$\mathbf{Prob}(Y \leq k) = \sigma\{\overline{S}_k + (y_l + y_{l+1})\overline{S}_{k-1} + y_l y_{l+1}\overline{S}_{k-2}\} \tag{2.85}$$

where $\overline{S}_h = \bar{s}_0 + \bar{s}_1 + \cdots + \bar{s}_h$. We denote the right-hand side of (2.85) by $F(\sigma)$ – notice that within the curly brackets only y_{l+1} depends on σ – and in view of (2.84) we have now shown that

$$\varphi_k^*(\sigma) \leq \max\left(t_{k,l}, F(\sigma)\right), \quad \sigma < \pi_{l-1}(1 - p_{l+1}^{-1}). \tag{2.86}$$

Since $F(\sigma)$ is continuous, our result will follow if we show that $F(\pi_l) < t_{k,l}$. Recall that $\mathbf{t}_{k,l}$ is the density of the integers divisible by at most k of the first l primes. We may write

$$\mathbf{t}_{k,l} = \mathbf{Prob}(Y' \leq k)$$

where

$$Y' = \sum_{j=1}^{l} Y_j', \ \mathbf{Prob}(Y_j' = 1) = \frac{1}{p_j}$$

and the Y_j' are independent Bernoulli trials. By (2.82) we have $\mathbf{Prob}(Y_j' = 1) = \mathbf{Prob}(Y_j = 1)$, $\quad j < l$, whence by a similar calculation to that which led to (2.85) we have

$$\mathbf{t}_{k,l} = \pi_l\{\overline{S}_k + z\overline{S}_{k-1}\}, \quad z = \frac{1}{p_l - 1}. \tag{2.87}$$

We put $\sigma = \pi_l$ in (2.85). By definition, $y_l = 1/(p_{l+1} - 1)$ and $y_{l+1} = \theta(\pi_l)/(p_{l+1} - \theta(\pi_l))$ where $\theta(\pi_l)$ is determined by (2.83). We notice that we may write (2.83) in the form

$$(1 + y_l)^{-1}(1 + y_{l+1})^{-1} = (1 + z)^{-1}$$

so that in fact

$$y_l + y_{l+1} + y_l y_{l+1} = z. \tag{2.88}$$

Since $\bar{s}_{k-1} > 0$ we have $\overline{S}_{k-1} > \overline{S}_{k-2}$, whence (2.85), (2.87) and (2.88) imply that $F(\pi_l) < \mathbf{t}_{k,l}$ as required. This completes the proof.

We may pursue this matter a little further. Let $\eta_{k,l}$ be taken to be as large as possible in (2.80), that is $\varphi_k^*(\sigma) > \mathbf{t}_{k,l}$ for $\sigma > \pi_l + \eta_{k,l}$. Then we obtain a lower bound for $\eta_{k,l}$ by solving the equation $F(\sigma) = t_{k,l}$. We have, for $l > k \geq 1$,

$$\eta_{k,l} \geq \frac{\bar{s}_{k-1}}{(p_{l+1}-1)\bar{s}_k + \bar{s}_{k-1}} \left\{ \pi_{l-1}\left(1 - \frac{1}{p_{l+1}}\right) - \pi_l \right\}. \tag{2.89}$$

Theorem 2.13 shows that the function $\varphi_k^*(\sigma)$ is not concave. However, we do have the following result concerning the $\mathbf{t}_{k,l}$.

Theorem 2.14 *For every k and l we have*

$$\mathbf{t}_{k,l} \geq \frac{\pi_l - \pi_{l+1}}{\pi_{l-1} - \pi_{l+1}} \mathbf{t}_{k,l-1} + \frac{\pi_{l-1} - \pi_l}{\pi_{l-1} - \pi_{l+1}} \mathbf{t}_{k,l+1}, \tag{2.90}$$

moreover the inequality is strict if and only if $l \geq k$. Thus the function $\varphi_k^*(\sigma)$, if it be restricted to the points $\sigma = \pi_l$, is concave.

Proof Put $y_j = 1/(p_j - 1)$ for every j, and let $\bar{s}_h$ denote the hth elementary symmetric function of $y_1, y_2, \ldots, y_{l-1}$, and $\overline{S}_h = \bar{s}_0 + \bar{s}_1 + \cdots + \bar{s}_h$. As in the proof of the previous theorem, it is understood that $\bar{s}_0 = 1$, $\bar{s}_{-1} = 0$. We have

$$\begin{aligned} \mathbf{t}_{k,l-1} &= \pi_{l-1}\overline{S}_k, \\ \mathbf{t}_{k,l} &= \pi_l(\overline{S}_k + y_l\overline{S}_{k-1}), \\ \mathbf{t}_{k,l+1} &= \pi_{l+1}\{\overline{S}_k + (y_l + y_{l+1})\overline{S}_{k-1} + y_l y_{l+1}\overline{S}_{k-2}\}. \end{aligned}$$

Let $u_{k,l}$ denote the difference between the left- and right-hand sides of (2.90). We find after some algebra, that

$$u_{k,l} = \frac{\bar{s}_{k-1}}{p_l(p_l + p_{l+1} - 1)} \geq 0,$$

moreover we have $\bar{s}_{k-1} > 0$ when $l \geq k$, as required. This completes the proof.

These theorems provide us with a fairly detailed description of the function φ_k^*. Our information about φ_k is, by comparison, vague, and fresh progress in this direction would be of great interest.

2.3 Generalized Behrend inequalities

We consider two sequences $\mathscr{A}$ and $\mathscr{B}$, and the densities $\mathbf{t}_k(\mathscr{A} \cup \mathscr{B})$, and we are concerned with inequalities involving these densities and the densities $\mathbf{t}_g(\mathscr{A})$, $\mathbf{t}_h(\mathscr{B})$, for various combinations of g, h and k. As in the classical

Behrend inequality (Theorem 0.12) $\mathbf{t}(\mathscr{A} \cup \mathscr{B}) \geq \mathbf{t}(\mathscr{A})\mathbf{t}(\mathscr{B})$, our results will be for the most part lower bounds for $\mathbf{t}_k(\mathscr{A} \cup \mathscr{B})$.

We shall be able to concentrate on the finite case, that is $|\mathscr{A}|$, $|\mathscr{B}| < \infty$; because firstly the t_k were defined as logarithmic densities, so that we can apply the results about sequential density from Chapter 0 to the derivatives $\mathscr{A}^{(k)}$, $\mathscr{B}^{(k)}$ and $(\mathscr{A} \cup \mathscr{B})^{(k)}$ in the infinite case, and secondly because essentially all the difficulties with which we have to contend already occur when $\mathscr{A}$ and $\mathscr{B}$ are finite. An exception to this rule already arose in Theorem 0.12 when we considered the cases of equality – for example see (0.103) – but it is enough at this stage to be aware that such things can happen. The tail might wag a little at infinity.

We begin with a simple and intuitive result concerning the densities

$$\bar{\mathbf{t}}_k(\mathscr{A}) = \delta\{n : \tau(n, \mathscr{A}) = k\}.$$

Recall that $\mathscr{A}$ and $\mathscr{B}$ are said to be coprime if $(a, b) = 1$ for all $a \in \mathscr{A}, b \in \mathscr{B}$.

Proposition 2.15 *Let $\mathscr{A}$ and $\mathscr{B}$ be non-trivial and coprime. Then for every k,*

$$\bar{\mathbf{t}}_k(\mathscr{A} \cup \mathscr{B}) = \sum_{g+h=k} \bar{\mathbf{t}}_g(\mathscr{A})\bar{t}_h(\mathscr{B}). \tag{2.91}$$

We give two proofs. Each uses ideas which we need later. In view of our previous remarks we may assume that $\mathscr{A}$ and $\mathscr{B}$ are finite.

Proof (i) The hypotheses imply that $\mathscr{A}$ and $\mathscr{B}$ are disjoint and so we have

$$\tau(n, \mathscr{A} \cup \mathscr{B}) = \tau(n, \mathscr{A}) + \tau(n, \mathscr{B}).$$

We may therefore suppose that $\tau(n, \mathscr{A}) = g$ and $\tau(n, \mathscr{B}) = h$, where $g + h = k$, and sum over the possible values of g and h. Thus (2.91) expresses the fact that, in probabilistic terminology, the events $\tau(n, \mathscr{A}) = g$ and $\tau(n, \mathscr{B}) = h$ are independent.

Let $A = \text{l.c.m.}[a : a \in \mathscr{A}]$ and B be defined similarly. Then $\tau(n, \mathscr{A}) = \tau((n, A), \mathscr{A})$, that is the value of $\tau(n, \mathscr{A})$ depends only on the congruence class (mod A) to which n belongs. There are $A\bar{\mathbf{t}}_g(\mathscr{A})$ classes in which $\tau(n, \mathscr{A}) = g$ and similarly $B\bar{\mathbf{t}}_h(\mathscr{B})$ classes (mod B) in which $\tau(n, \mathscr{B}) = h$. By the Chinese Remainder Theorem, there are $AB\bar{\mathbf{t}}_g(\mathscr{A})\bar{\mathbf{t}}_h(\mathscr{B})$ congruence classes (mod AB) in which both events occur, because A and B are relatively prime. This is all we need.

Proof (ii) We employ the inclusion–exclusion principle to write down a formula for $\bar{\mathfrak{t}}_k(\mathscr{A})$. This is

$$\bar{\mathfrak{t}}_k(\mathscr{A}) = \sum_{l \geq k} (-1)^{l-k} \binom{l}{k} f_l(\mathscr{A}) \tag{2.92}$$

where

$$f_l(\mathscr{A}) = \sum_{i_1 < i_2 < \cdots < i_l} [a_{i_1}, a_{i_2}, \ldots, a_{i_l}]^{-1}.$$

We may drop the summation condition $l \geq k$ if we define the binomial coefficient to be 0 when $l < k$: also $f_l(\mathscr{A}) = 0$ when $l > |\mathscr{A}|$. Since $\mathscr{A}$ and $\mathscr{B}$ are disjoint and coprime we have

$$f_l(\mathscr{A} \cup \mathscr{B}) = \sum_{r+s=l} f_r(\mathscr{A}) f_s(\mathscr{B}). \tag{2.93}$$

By (2.92), the right-hand side of (2.91) is

$$\sum_{g+h=k} \sum_r \sum_s (-1)^{r+s-g-h} \binom{r}{g} \binom{s}{h} f_r(\mathscr{A}) f_s(\mathscr{B})$$

and since

$$\sum_{g+h=k} \binom{r}{g} \binom{s}{h} = \binom{r+s}{k}$$

we may perform the outer summation to obtain

$$\sum_r \sum_s (-1)^{r+s-k} \binom{r+s}{k} f_r(\mathscr{A}) f_s(\mathscr{B}).$$

We employ (2.92) and (2.93) to see that this is $\bar{\mathfrak{t}}_k(\mathscr{A} \cup \mathscr{B})$. This completes the proof.

Notice that whenever $\mathscr{A}$ and $\mathscr{B}$ are disjoint the left-hand side of (2.93) is at least as great as the right-hand side, but this inequality seems difficult to use. Similarly we do not obtain an inequality from the Chinese Remainder Theorem in the first proof: recall that if $(A, B) = d > 1$ then the congruences $n \equiv u \pmod{A}$, $n \equiv v \pmod{B}$ are consistent, and define a unique congruence class (mod $[A, B]$), if and only if $d | (u - v)$. In the situation which arises in Proof (i) we should not in general know whether this condition held.

The next result is our first inequality of Behrend type. It generalizes the classical inequality but falls short of our expectations in one respect.

Proposition 2.16 *For all $\mathscr{A}, \mathscr{B}, g$ and h we have*

$$\mathbf{t}_{g+h}(\mathscr{A} \cup \mathscr{B}) \geq \mathbf{t}_g(\mathscr{A})\mathbf{t}_h(\mathscr{B}).$$

Proof We may assume, by the limiting argument described previously, that $\mathscr{A}$ and $\mathscr{B}$ are finite. We shall employ Lemma 0.18, and we define, for any $\mathscr{A}$, finite or not, and $g \geq 0$,

$$\theta^{(g)}_{\mathscr{A}}(n) = \begin{cases} 1 & \text{if } \tau(n, \mathscr{A}) \leq g, \\ 0 & \text{else.} \end{cases} \tag{2.94}$$

This function is multiplicatively non-increasing, that is $n|n' \Rightarrow \theta^{(g)}_{\mathscr{A}}(n) \geq \theta^{(g)}_{\mathscr{A}}(n')$. Let $A = \text{l.c.m.}[a : a \in \mathscr{A}]$ and B be defined similarly. Then as in (0.107) we have

$$\mathbf{t}_g(\mathscr{A}) = E(\theta^{(g)}_{\mathscr{A}}; M) = \frac{1}{M}\sum_{d|M} \varphi\left(\frac{M}{d}\right)\theta^{(g)}_{\mathscr{A}}(d) \tag{2.95}$$

for any multiple M of A. This is because $\tau(n, \mathscr{A}) = \tau((n, M), \mathscr{A})$ and there are $\varphi(M/d)$ congruence classes (mod M) in which $(n, M) = d$. We put $M = [A, B]$, and apply Lemma 0.18 to obtain

$$E\left(\theta^{(g)}_{\mathscr{A}}\theta^{(h)}_{\mathscr{B}}; M\right) \geq E\left(\theta^{(g)}_{\mathscr{A}}; M\right) E\left(\theta^{(h)}_{\mathscr{B}}; M\right). \tag{2.96}$$

Since $\tau(n, \mathscr{A} \cup \mathscr{B}) \leq \tau(n, \mathscr{A}) + \tau(n, \mathscr{B})$ we have

$$\theta^{(g+h)}_{\mathscr{A}\cup\mathscr{B}}(n) \geq \theta^{(g)}_{\mathscr{A}}(n)\theta^{(h)}_{\mathscr{B}}(n) \tag{2.97}$$

and our result is an assembly of (2.95) applied to $\mathscr{A}, \mathscr{B}$ and $\mathscr{A} \cup \mathscr{B}$, (2.96) and (2.97). This completes the proof.

Proposition 2.16 includes Behrend's inequality (Theorem 0.12) as a special case $g = h = 0$. However the reader will have noticed that we have not mentioned the cases of equality. Of course there are such cases, a trivial example being given by $g \geq |\mathscr{A}|$, $h \geq |\mathscr{B}|$, but they are necessarily a little artificial, because (2.97) is not generally an equation except when $g = h = 0$.

Let us return to the case when $\mathscr{A}$ and $\mathscr{B}$ are non-trivial and coprime. By Proposition 2.15 we have

$$\begin{aligned} \mathbf{t}_{g+h}(\mathscr{A} \cup \mathscr{B}) &= \sum_{i+j \leq g+h} \bar{\mathbf{t}}_i(\mathscr{A})\bar{\mathbf{t}}_j(\mathscr{B}) \\ &= \mathbf{t}_g(\mathscr{A})\mathbf{t}_h(\mathscr{B}) + \sum_{i=g+1}^{g+h} \bar{\mathbf{t}}_i(\mathscr{A})\mathbf{t}_{g+h-i}(\mathscr{B}) \\ &\quad + \sum_{j=h+1}^{g+h} \mathbf{t}_{g+h-j}(\mathscr{A})\bar{\mathbf{t}}_j(\mathscr{B}). \end{aligned}$$

The last two sums are an excess, moreover the t-factors are non-zero, except in the trivial case when $\mathscr{A}$ or $\mathscr{B}$ is Behrend. However, it is possible to arrange that $\bar{\mathbf{t}}_i(\mathscr{A}) = 0$ in a non-trivial, albeit somewhat artificial, fashion. We explore this phenomenon briefly here, to warn the reader that such odd cases can arise. Let $p_1, p_2, \ldots, p_m$ be distinct primes and, for fixed $l < m$, let

$$\mathscr{A} = \{p_{i_1} p_{i_2} \ldots p_{i_l} : 1 \le i_1 < i_2 < \cdots < i_l \le m\}.$$

If n is divisible by (exactly) k of these primes then

$$\tau(n, \mathscr{A}) = \binom{k}{l},$$

with the usual convention if $k < l$. Thus $\tau(n, \mathscr{A})$ has omitted values. A simple example is given by $\mathscr{A} = \{6, 10, 15\}$ when $\tau(n, \mathscr{A}) \neq 2$. If p, q, r, p', q', r' are distinct primes and $\mathscr{A} = \{pq, qr, rp\}$, $\mathscr{B} = \{p'q', q'r', r'p'\}$ then indeed

$$\mathbf{t}_2(\mathscr{A} \cup \mathscr{B}) = \mathbf{t}_1(\mathscr{A})\mathbf{t}_1(\mathscr{B})$$

with none of the three densities equal to 0 or 1.

We should like to have a generalization of Behrend's inequality which maintains the property that there is equality when $\mathscr{A}$ and $\mathscr{B}$ are coprime. A possible strategy is to consider identities like (2.91), and see whether they generalize to inequalities. If we intend to employ Lemma 0.18, which involves multiplicatively non-increasing functions, it would seem that we must begin with an identity clear of the densities $\bar{\mathbf{t}}_k$, $(k > 0)$. We obtain an identity of this type from Proposition 2.15 after two summations which yield: for $\mathscr{A}$ and $\mathscr{B}$ non-trivial and coprime,

$$\sum_{h=0}^{k} \mathbf{t}_h(\mathscr{A} \cup \mathscr{B}) = \sum_{h=0}^{k} \mathbf{t}_h(\mathscr{A})\mathbf{t}_{k-h}(\mathscr{B}). \tag{2.98}$$

Examination of Theorem 0.12 suggests that even in the finite case coprimality is unnecessary for this to hold: for $k = 0$ the relevant extra condition is that $\mathscr{A}$ and $\mathscr{B}$ should be primitive. An appropriate generalization is as follows.

Definition 2.17 *The sequence $\mathscr{A}$ is said to be k-primitive if $\mathscr{A}$ has at least $k+1$ elements, and for any indices $i_0 < i_1 < i_2 < \cdots < i_k$ the least common multiple $[a_{i_0}, a_{i_1}, \ldots, a_{i_k}]$ is not divisible by any other element of $\mathscr{A}$ (if such exists).*

Thus a 0-primitive sequence is primitive. We note that any sequence of exactly $k+1$ terms is k-primitive. If $|\mathscr{A}| \geq k+2$, the condition $a_i \nmid [a_{i_0}, a_{i_1}, \ldots, a_{i_k}]$, $i \neq i_h$ for any $h \leq k$, may be written in the alternative form

$$\frac{[a_i, a_{i_0}, a_{i_1}, \ldots, a_{i_k}]}{[a_{i_0}, a_{i_1}, \ldots, a_{i_k}]} > 1 \tag{2.99}$$

which leads to a useful description of k-primality in terms of total decomposition sets in the case when $\mathscr{A}$ is finite. If $\{d(S)\}$ denotes the total decomposition set of $\mathscr{A}$ then $\mathscr{A}$ is k-primitive if and only if for any $k+2$ distinct indices $i, i_0, i_1 \ldots i_k$ we can find $S \subseteq \{1, 2, \ldots, |\mathscr{A}|\}$ such that $i \in S$, $i_h \notin S$, $0 \leq h \leq k$, and $d(S) > 1$. We leave the proof to the reader.

Theorem 2.18 *For all $\mathscr{A}$ and $\mathscr{B}$, we have*

$$\sum_{h=0}^{k} \mathbf{t}_h(\mathscr{A} \cup \mathscr{B}) \geq \sum_{h=0}^{k} \mathbf{t}_h(\mathscr{A}) \mathbf{t}_{k-h}(\mathscr{B}) \tag{2.100}$$

with equality if $\mathscr{A}$ and $\mathscr{B}$ are non-trivial and coprime. If $\mathscr{A}$ and $\mathscr{B}$ are finite sequences, and if there exists an h, $0 \leq h \leq k$, such that $\mathscr{A}$ is h-primitive and $\mathscr{B}$ is $(k-h)$-primitive, then it is necessary for equality in (2.100) *that $\mathscr{A}$ and $\mathscr{B}$ should be coprime.*

Proof We begin with the inequality: as usual we may assume $\mathscr{A}$ and $\mathscr{B}$ finite, moreover we let $A = \text{l.c.m.}[a : a \in \mathscr{A}]$, B be defined similarly, and $M = [A, B]$. We let $\theta_{\mathscr{A}}^{(g)}, \theta_{\mathscr{B}}^{(g)}$ be as in (2.94) so that in the notation of Lemma 0.18, the right-hand side of (2.100) equals

$$\sum_{h=0}^{k} E(\theta_{\mathscr{A}}^{(h)}; M) E(\theta_{\mathscr{B}}^{(k-h)}; M). \tag{2.101}$$

Put

$$\Theta_{\mathscr{A} \cup \mathscr{B}}^{(k)}(n) = (k + 1 - \tau(n, \mathscr{A} \cup \mathscr{B}))^+ .$$

For all n, we have $\tau(n, \mathscr{A} \cup \mathscr{B}) \leq \tau(n, \mathscr{A}) + \tau(n, \mathscr{B})$ whence

$$\Theta_{\mathscr{A} \cup \mathscr{B}}^{(k)}(n) \geq \sum_{h=0}^{k} \theta_{\mathscr{A}}^{(h)}(n) \theta_{\mathscr{B}}^{(k-h)}(n). \tag{2.102}$$

We also have

$$\sum_{h=0}^{k} \mathbf{t}_h(\mathscr{A} \cup \mathscr{B}) = E\left(\Theta_{\mathscr{A} \cup \mathscr{B}}^{(k)}; M\right)$$

and in view of (2.102) this is

$$\geq \sum_{h=0}^{k} E\left(\theta_{\mathscr{A}}^{(h)}\theta_{\mathscr{B}}^{(k-h)}; M\right).$$

We apply Lemma 0.18 to deduce that each term in this sum is at least as great as the corresponding term in the sum (2.101), and (2.100) follows. This proves the first part of the theorem.

Let the hypotheses of the second part hold and in addition suppose that $\mathscr{A}$ and $\mathscr{B}$ are not coprime. We have to show that there is strict inequality in (2.100). It will be sufficient to show that for some h, we have

$$E\left(\theta_{\mathscr{A}}^{(h)}\theta_{\mathscr{B}}^{(k-h)}; M\right) > E\left(\theta_{\mathscr{A}}^{(h)}; M\right) E\left(\theta_{\mathscr{B}}^{(k-h)}; M\right) \tag{2.103}$$

and we consider the h specified in the statement of the theorem. We claim, in the terminology of Lemma 0.18, that $\theta_{\mathscr{A}}^{(h)}$ and $\theta_{\mathscr{B}}^{(k-h)}$ do not split M, whence (2.103). By hypothesis A and B have a common prime factor p. Let $p^{\alpha} \| A$, $p^{\beta} \| B$. Let $a^{(0)}, a^{(1)}, \ldots, a^{(h)}$ be elements of $\mathscr{A}$ such that $p^{\alpha} | a^{(0)}$, and put $d = [a^{(0)}, a^{(1)}, \ldots, a^{(h)}]$. Then $\tau(d, \mathscr{A}) \geq h + 1$. On the other hand $\tau\left((d, Ap^{-\alpha}), \mathscr{A}\right) \leq h$, for $a^{(0)} \nmid (d, Ap^{-\alpha})$ and no other element of $\mathscr{A}$ divides d. Thus $\theta_{\mathscr{A}}^{(h)}(d) < \theta_{\mathscr{A}}^{(h)}\left((d, Ap^{-\alpha})\right)$. By a similar argument we can find e such that $\theta_{\mathscr{B}}^{(k-h)}(e) < \theta_{\mathscr{B}}^{(k-h)}\left((e, Bp^{-\beta})\right)$, moreover d and e are divisors of M. If $M = FG$ where $(F, G) = 1$, either $p \nmid F$ or $p \nmid G$, and either $\theta_{\mathscr{A}}^{(h)}(d) < \theta_{\mathscr{A}}^{(h)}((d, F))$ or $\theta_{\mathscr{B}}^{(k-h)}(e) < \theta_{\mathscr{B}}^{(k-h)}((e, G))$. Therefore the functions $\theta_{\mathscr{A}}^{(h)}$ and $\theta_{\mathscr{B}}^{(k-h)}$ do not split M, whereby our claim is substantiated. This completes the proof.

2.4 Multilinear functions

When $\mathscr{A}$ is finite we can write down explicit formulae for the densities $\mathbf{t}_k(\mathscr{A})$ by means of the inclusion–exclusion principle; for example (2.8) is the familiar expression for $\mathbf{t}(\mathscr{A})$. These formulae are almost never applied because of two drawbacks: the alternating signs and the occurrence of the l.c.m.'s in the denominators. Partial sums of (2.8) give, alternately, upper and lower bounds for $\mathbf{t}(\mathscr{A})$, by similar considerations to those employed in the early stages of Brun's sieve, and of the two snags the l.c.m.'s are usually the more crippling.

We know from Chapter 0 that any finite sequence $\mathscr{A} = \{a_1, a_2, \ldots, a_n\}$ of positive integers possesses a *total decomposition set* $\{d(S)\}$ of $2^n - 1$ positive integers $d(S)$, indexed by the non-empty subsets $S \subseteq \{1, 2, 3, \ldots, n\}$.

The h.c.f. and l.c.m. of any sub-sequence of $\mathscr{A}$ have a representation as a product of certain $d(S)$: we do not need to write these representations down here since all we require at this point, as a glance at (0.105), (0.106) will confirm, is that each $d(S)$ which occurs in such a product does so *linearly*, that is the S are distinct. The $d(S)$ may not be distinct but this is irrelevant to the present discussion in which we regard the numbers $d(S)$ as variables.

Let $f(x_1, x_2, \ldots, x_m)$ be a polynomial and suppose

$$\frac{\partial^2 f}{\partial x_i^2} \equiv 0, \quad 1 \le i \le m.$$

We call such a polynomial *multilinear*, and we see from the foregoing observations that $t(\mathscr{A})$ and $t^{(k)}(\mathscr{A})$ are multilinear functions of the $2^n - 1$ variables $x(S) = d(S)^{-1}$. For example, if $\mathscr{A} = \{a_1, a_2\}$ and the total decomposition set is $d(\{1\})$, $d(\{2\})$, $d(\{1,2\})$ so that

$$\begin{aligned} a_1 &= d(\{1\})d(\{1,2\}), \\ a_2 &= d(\{2\})d(\{1,2\}), \\ (a_1, a_2) &= d(\{1,2\}), \\ [a_1, a_2] &= d(\{1\})d(\{2\})d(\{1,2\}) \end{aligned}$$

then

$$t(\mathscr{A}) = 1 - x(\{1\})x(\{1,2\}) - x(\{2\})x(\{1,2\}) + x(\{1\})x(\{2\})x(\{1,2\}). \quad (2.104)$$

We now have an expression which we can think about analytically, for example it may be differentiated; and maybe we can establish inequalities in this way.

Now (2.104) suffers from appalling notation, it still contains terms of opposite sign, and we now have $2^n - 1$ variables instead of n. In fact this last remark is the tip of an iceberg because we shall have to consider not just the subsets $S \subseteq \{1, 2, \ldots, n\}$ but all the possible families of such subsets. We overcome the first difficulty by writing

$$d(\{1,2\}) = d_{12}, \; x(\{1,2,\}) = x_{12} \;\text{ etc.},$$

and it is the object of this section to overcome the second difficulty of alternate signs.

If we consider (2.104) again we can make two observations. We know that the variables $x(S)$ on the right satisfy $0 < x(S) \le 1$, (it will be convenient to allow $0 \le x(S) \le 1$ and to regard $x(S)$ as a *continuous* variable), and we also know that $\mathbf{t}(\mathscr{A}) \ge 0$. (For number theoretic reasons, of course $\mathbf{t}(\mathscr{A}) > 0$ if $\mathscr{A}$ is finite and the a_j exceed 1: we allow our functions to be zero in this section, moreover a sequence $\mathscr{A}$ containing 1's and repetitions still possesses a total decomposition set.)

When a function is non-negative it is not unreasonable to ask it to be written in such a way that this property is obvious. Thus if $0 \le x, y \le 1$ we can write

$$1 - x - y + xy = (1-x)(1-y). \tag{2.105}$$

A similar example, with $0 \le x, y, z \le 1$, is

$$\begin{aligned} 2 - x - y - z + xyz &= 2(1-x)(1-y)(1-z) + (1-x)(1-y)z \\ &\quad + (1-x)(1-z)y + (1-y)(1-z)x. \end{aligned} \tag{2.106}$$

In this case it is not so obvious that the left-hand side is non-negative.

The multilinear functions (we restrict our attention to polynomials) of m variables may be viewed as a vector space of dimension 2^m, with basis

$$x_1^{\varepsilon_1} x_2^{\varepsilon_2} \dots x_m^{\varepsilon_m}, \ \varepsilon_i = 0 \text{ or } 1, \quad 1 \le i \le m.$$

We are concerned with real functions and so the field of scalars is $\mathbf{R}$. Guided by examples (2.105) and (2.106), we seek another basis with the property that any non-negative function is a linear combination of the base functions, with non-negative coefficients. This is

$$\prod_{i=1}^{m} (1-x_i)^{1-\varepsilon_i} x_i^{\varepsilon_i}, \ \varepsilon_i = 0 \text{ or } 1, \quad 1 \le i \le m. \tag{2.107}$$

We have to show that this is a basis and that it has the positivity property required. Clearly there are 2^m polynomials (2.107); moreover they are independent, since if we set $x_i = \varepsilon_i$, $1 \le i \le m$, all the polynomials are zero with one exception, equal to 1. Hence (2.107) is a basis. Furthermore, the coefficient $\lambda(\varepsilon_1, \varepsilon_2, \dots, \varepsilon_m)$ of each base polynomial in (2.107) may be determined by setting $x_i = \varepsilon_i$, $1 \le i \le m$, that is

$$\lambda(\varepsilon_1, \varepsilon_2, \dots, \varepsilon_m) = f(\varepsilon_1, \varepsilon_2, \dots, \varepsilon_m) \tag{2.108}$$

where f is the multilinear function to be expanded. We have used the fact that such a function is non-negative throughout the m-dimensional cube $[0, 1]^m$ if and only if it is non-negative at the corners of the cube.

The reader will have noticed that the right-hand side of (2.106) may be simplified, for example to

$$(1-x)(1-y) + (1-y)(1-z) + (1-x)(1-z)y$$

which has the same 'obviously positive' property. Such an expansion is not unique, indeed there are clearly 3^m products in which each variable contributes one of the factors $1, x$, or $1-x$, and we make no use of such expansions here, tempting as such reductions may occasionally be.

We refer to the polynomials (2.107) as the *elementary multilinear functions (of m variables)* and we have proved the following result.

Theorem 2.19 *Every multilinear function of the m variables* $x_1, x_2, \ldots, x_m$ *has a unique expansion as a linear combination of the* 2^m *elementary multilinear funtions (2.107), moreover (2.108) is a formula for the coefficients. The coefficients are all non-negative if and only if the function is non-negative throughout the cube* $[0,1]^m$.

2.5 Formulae for the densities $\mathbf{t}_k(\mathscr{A})$

In this section we develop the ideas introduced in §2.4. We require that $\mathscr{A}$ be finite, but it need not be monotonic and we allow repetitions and 1's amongst the elements. Let $\mathscr{A} = \{a_1, a_2, \ldots, a_n\}$, and $\{d(S)\}$ denote the total decomposition set of $\mathscr{A}$; thus S is any non-empty subset of $S_0 := \{1, 2, \ldots, n\}$ and the cardinality of $\{d(S)\}$ is $m = 2^n - 1$. We introduce m real variables $x(S)$, $0 \leq x(S) \leq 1$ and write down multilinear functions of these variables taking the values $\mathbf{t}^{(k)}(\mathscr{A})$ when

$$x(S) = d(S)^{-1} \text{ for all } S \subseteq S_0. \tag{2.109}$$

Furthermore, we expand these functions as linear combinations of the 2^m elementary functions defined by (2.107).

It may be helpful to work out a simple example in ad hoc fashion to begin with, so that we can visualize the shape of the formulae to which we are heading. Let $\mathscr{A} = \{a_1, a_2\}$ and the total decomposition set be $\{d_1, d_2, d_{12}\}$ – we streamline our notation when convenient by indexing with suffices instead of sets – where $d_{12} = (a_1, a_2)$, and $a_1 = d_1 d_{12}$, $a_2 = d_2 d_{12}$. Now consider the polynomial

$$t_0 = 1 - x_1 x_{12} - x_2 x_{12} + x_1 x_2 x_{12}.$$

When we make the substitution (2.99) we obtain $t_0 = t(\mathscr{A})$, and of course t_0 is multilinear. In this example $m = 3$, the suffix 12 corresponding to $S = \{1, 2\}$ replacing 3. There are eight elementary functions (2.107): if we denote these by $f(\underline{x}, \underline{\varepsilon})$ where $\underline{x} = (x_1, x_2, x_{12})$, $\underline{\varepsilon} = (\varepsilon_1, \varepsilon_2, \varepsilon_{12})$ then by

Theorem 2.19 we can write

$$t_0 = \sum_{\underline{\varepsilon}} \lambda(\underline{\varepsilon}) f(\underline{x}, \underline{\varepsilon})$$

where, by (2.108),

$$\lambda(\underline{\varepsilon}) = t_0(\varepsilon_1, \varepsilon_2, \varepsilon_{12}).$$

Evaluating these eight coefficients we find that

$$\begin{aligned} t_0 &= (1-x_1)(1-x_2)(1-x_{12}) + (1-x_1)(1-x_2)x_{12} \\ &\quad + (1-x_1)(1-x_{12})x_2 + (1-x_2)(1-x_{12})x_1 \\ &\quad + (1-x_{12})x_1x_2, \end{aligned}$$

whence

$$\begin{aligned} t(\mathscr{A}) &= \left(1-\frac{1}{d_1}\right)\left(1-\frac{1}{d_2}\right)\left(1-\frac{1}{d_{12}}\right) + \left(1-\frac{1}{d_1}\right)\left(1-\frac{1}{d_2}\right)\frac{1}{d_{12}} \\ &\quad + \left(1-\frac{1}{d_1}\right)\left(1-\frac{1}{d_{12}}\right)\frac{1}{d_2} + \left(1-\frac{1}{d_2}\right)\left(1-\frac{1}{d_{12}}\right)\frac{1}{d_1} \\ &\quad + \left(1-\frac{1}{d_{12}}\right)\frac{1}{d_1d_2}. \end{aligned} \tag{2.110}$$

Notice that the coefficients equal either 0 or 1 and that there are five out of eight equal to 1. It would be foolhardy to extrapolate too far from a simple example, nevertheless the reader might also observe a characteristic of these five terms not shared by the missing three: in each term *both* integers 1 and 2 appear somewhere within suffices of variables *inside* brackets.

We require some notation and terminology. We put $S_0 = \{1, 2, \ldots, n\}$ throughout, R, S, T etc. are subsets of S_0. Curly letters, $\mathscr{F}, \mathscr{G}$ etc. denote families of such subsets and we define

$$\operatorname{Span}\mathscr{F} = \bigcup\{S : S \in \mathscr{F}\}. \tag{2.111}$$

We say that $\mathscr{F}$ is *complete* if $\operatorname{Span}\mathscr{F} = S_0$; more generally we define the *deficiency* of $\mathscr{F}$ as

$$\delta(\mathscr{F}) = n - |\operatorname{Span}\mathscr{F}|. \tag{2.112}$$

Cardinalities of sets or families are written $|S|$ or $|\mathscr{F}|$, thus $|\mathscr{F}|$ is the number of sets in $\mathscr{F}$. We write

$$\mathscr{X}(\mathscr{F}) = \begin{cases} 1 & \text{if } \mathscr{F} \text{ is complete,} \\ 0 & \text{else.} \end{cases} \tag{2.113}$$

It is understood, unless stated otherwise, that subsets of S_0 are non-empty. Families may be empty, so there are $m = 2^n - 1$ subsets and 2^m families. Summation over sets or families with no further instructions are over these ranges. A typical elementary multilinear function as in (2.107) therefore has the form

$$\prod_{S\in\mathscr{F}} (1 - x(S)) \prod_{R\notin\mathscr{F}} x(R). \tag{2.114}$$

We can now state our first theorem in this section.

Theorem 2.20 *Let* $\mathscr{A}$ *be finite, with total decomposition set* $\{d(S)\}$. *Then*

$$t(\mathscr{A}) = \sum_{\mathscr{F}} \mathscr{X}(\mathscr{F}) \prod_{S\in\mathscr{F}} \left(1 - \frac{1}{d(S)}\right) \prod_{R\notin\mathscr{F}} \frac{1}{d(R)}. \tag{2.115}$$

Proof We follow the steps which led to (2.110). For $T \subseteq S_0 = \{1, 2, 3, \ldots, |\mathscr{A}|\}$ we have, by (0.106)

$$\text{l.c.m.}[a_i : i \in T] = \prod\{d(S) : S \cap T \neq \emptyset\}. \tag{2.116}$$

In view of the expression (2.8) for $\mathbf{t}(\mathscr{A})$, we consider the polynomial

$$t_0 = \sum_T (-1)^{|T|} \prod\{x(S) : S \cap T \neq \emptyset\} \tag{2.117}$$

where the sum includes $T = \emptyset$; the product is 1. Let $\lambda(\mathscr{F})$ denote the coefficient of the elementary function (2.114) in the expression afforded by Theorem 2.12: we have to show that $\lambda(\mathscr{F}) = \mathscr{X}(\mathscr{F})$. By (2.108) we obtain $\lambda(\mathscr{F})$ by evaluating t_0 at the point

$$x(S) = \begin{cases} 0 & \text{if } S \in \mathscr{F} \\ 1 & \text{else.} \end{cases} \tag{2.118}$$

whence by (2.117),

$$\lambda(\mathscr{F}) = \sum_T \{(-1)^{|T|} : S \in \mathscr{F} \Rightarrow S \cap T = \emptyset\}, \tag{2.119}$$

since the product on the right of (2.117) is zero unless $x(S) = 1$, i.e. $S \notin \mathscr{F}$, whenever S intersects T. We can rewrite (2.119) in the form

$$\lambda(\mathscr{F}) = \sum_T \{(-1)^{|T|} : T \subseteq S_0 \setminus \text{Span}\mathscr{F}\}, \tag{2.120}$$

where $\setminus$ denotes set subtraction, so that $S_0 \setminus Q$ is the complement of Q.

If $\mathscr{F}$ is complete, the complement of Span$\mathscr{F}$ is empty and $\lambda(\mathscr{F}) = 1$. If $\mathscr{F}$ has deficiency k, the sum on the left of (2.120) is

$$\sum_{h=0}^{k} \binom{k}{h} (-1)^h = 0$$

whence $\lambda(\mathscr{F}) = \mathscr{X}(\mathscr{F})$ as required. This completes the proof of Theorem 2.20. Next we extend this result to a formula for the density $\mathbf{t}_k(\mathscr{A})$.

Theorem 2.21 *Let $\mathscr{A}$ be finite, with total decomposition set $\{d(S)\}$. Then we have*

$$\mathbf{t}_k(\mathscr{A}) = \sum_{\mathscr{F}} \mathscr{X}_k(\mathscr{F}) \prod_{S \in \mathscr{F}} \left(1 - \frac{1}{d(S)}\right) \prod_{R \notin \mathscr{F}} \frac{1}{d(R)} \tag{2.121}$$

where $\mathscr{X}_k$ is the characteristic function of the families $\mathscr{F}$ with deficiency $\delta(\mathscr{F}) \le k$.

The deficiency of a family $\mathscr{F}$ of subsets $S \subseteq \{1, 2, \ldots, |\mathscr{A}|\}$ was defined by (2.111) and (2.112). By (2.112), $\delta(\mathscr{F}) \le |\mathscr{A}|$ with equality if and only if $\mathscr{F}$ is empty (recall that subsets S are not allowed to be empty but families of subsets are so allowed). This corresponds to the fact that $t_k(\mathscr{A}) = 1$ whenever $k \ge |\mathscr{A}|$. If $k = |\mathscr{A}| - 1$, only the empty family is excluded from the sum in (2.121), and we obtain

$$\mathbf{t}_k(\mathscr{A}) = 1 - \prod_{R} \frac{1}{d(R)} = 1 - \text{l.c.m.}[a : a \in \mathscr{A}]^{-1}$$

by (0.106).

Notice that to obtain $\bar{\mathbf{t}}_k(\mathscr{A})$ instead of $\mathbf{t}_k(\mathscr{A})$ in (2.121) we need only replace $\mathscr{X}_k(\mathscr{F})$ by $\overline{\mathscr{X}}_k(\mathscr{F})$, the characteristic function of the families with deficiency k. Notice that such families exist for all $k \le |\mathscr{A}|$, even though we may have $\bar{\mathbf{t}}_k(\mathscr{A}) = 0$ (see the discussion following (2.97) in which non-trivial examples are constructed). An exercise for the reader is to see how the new formula for $\bar{\mathbf{t}}_k(\mathscr{A})$ can be correct in such cases.

Proof of the theorem It is convenient to work with $\bar{\mathbf{t}}_k(\mathscr{A})$ rather than $\mathbf{t}_k(\mathscr{A})$, and we begin with the formula for this function provided by the inclusion–exclusion principle, which is

$$\bar{\mathbf{t}}_k(\mathscr{A}) = \sum_{l \ge k} (-1)^{l-k} \binom{l}{k} \sum_{i_1 < i_2 < \cdots < i_l} [a_{i_1}, a_{i_2}, \ldots, a_{i_l}]^{-1}. \tag{2.122}$$

We recall formula (0.106) which states that

$$[a_i : i \in T] = \prod\{d(S) : S \cap T \neq \emptyset\}$$

and we put $x(S) = d(S)^{-1}$ so that (2.122) becomes

$$\bar{\mathbf{t}}_k(\mathscr{A}) = \sum_T c_k\left(|T|\right) \prod\{x(S) : S \cap T \neq \emptyset\}. \tag{2.123}$$

where

$$c_k(l) = (-1)^{l-k}\binom{l}{k}$$

with the usual convention that $c_k(l) = 0$ if $l < k$. Let us think of the $x(S)$ as independent variables, $0 \leq x(S) \leq 1$, and consider the polynomial on the right-hand side of (2.123), denoting this by $\bar{\mathbf{t}}_k$. We may expand $\bar{\mathbf{t}}_k$ as a series of elementary multilinear functions, viz

$$\bar{\mathbf{t}}_k = \sum_{\mathscr{F}} \lambda(\mathscr{F}) \prod_{S \in \mathscr{F}} (1 - x(S)) \prod_{R \notin \mathscr{F}} x(R) \tag{2.124}$$

and we claim that $\lambda(\mathscr{F}) = \overline{\mathscr{X}}_k(\mathscr{F})$. We evaluate $\lambda(\mathscr{F})$ by choosing $\{x(S)\}$ as in (2.118) and we obtain, similarly to (2.120),

$$\lambda(\mathscr{F}) = \sum_T \{c_k\left(|T|\right) : T \subseteq S_0 \setminus \mathrm{Span}\mathscr{F}\}. \tag{2.125}$$

Let the deficiency $\delta(\mathscr{F}) < k$. Then $T \subseteq S_0 \setminus \mathrm{Span}\mathscr{F}$ implies $|T| < k$ and $c_k\left(|T|\right) = 0$. Now let $\delta(\mathscr{F}) = r \geq k$. The sum on the right of (2.125) is

$$\begin{aligned}
\sum_{l \leq r} c_k(l)\binom{r}{l} &= \sum_{l=k}^{r}(-1)^{l-k}\binom{l}{k}\binom{r}{l} \\
&= \binom{r}{k}\sum_{l=k}^{r}(-1)^{l-k}\binom{r-k}{l-k} \\
&= \begin{cases} 1 & \text{if } r = k \\ 0 & \text{if } r > k. \end{cases}
\end{aligned}$$

Therefore $\lambda(\mathscr{F}) = \overline{\mathscr{X}}_k(\mathscr{F})$ and by (2.124),

$$\bar{\mathbf{t}}_k = \sum_{\mathscr{F}} \overline{\mathscr{X}}_k(\mathscr{F}) \prod_{S \in \mathscr{F}} (1 - x(S)) \prod_{R \notin \mathscr{F}} x(R).$$

When $x(S) = d(S)^{-1}$ we have $\bar{\mathbf{t}}_k = \bar{\mathbf{t}}_k(\mathscr{A})$, and this is equivalent to (2.121).

3

Oscillation

3.1 Introduction

Let $\mathcal{M} = \mathcal{M}(\mathcal{A})$ be a set of multiples and

$$M(x) = \text{card}\{m : m \leq x, m \in \mathcal{M}\} \tag{3.1}$$

be the counting function of $\mathcal{M}$. We are interested in the oscillations of the function $M(x)$, in other words in irregularities in the distribution of $\mathcal{M}$, specifically in showing that usually these are not too small.

We distinguish two cases according as $\mathcal{A}$, or rather $\mathcal{P}(\mathcal{A})$ if $\mathcal{A}$ is not assumed to be primitive, is finite or not. In the finite case $\mathcal{M}(\mathcal{A})$ is a finite union of arithmetic progressions and is of course Besicovitch, moreover the discrepancy

$$E(x, \mathcal{A}) := M(x) - \mathbf{d}M(\mathcal{A})x \tag{3.2}$$

is a periodic function of x. We are interested in the extreme, or suitable mean, values of $E(x, \mathcal{A})$ over its period, and we seek *lower* bounds or what are often called Ω-theorems.

In the infinite case $\mathcal{A}$ may not be Besicovitch and $|E(x, \mathcal{A})| \gg x$ for arbitrarily large x. We might instead consider the logarithmic discrepancy, that is

$$\sum\left\{\frac{1}{m} : m \leq x, m \in \mathcal{M}(\mathcal{A})\right\} - \boldsymbol{\delta}\mathcal{M}(\mathcal{A}) \log x \tag{3.3}$$

but this idea is not taken up in the present work. Instead we consider, in §3.5 below, the problem of what we call *perfect sequences,* for which $M(x) = \boldsymbol{\delta}\mathcal{M}(\mathcal{A})x + O(1)$. This highly restrictive looking condition leads to all sorts of ramifications.

We assume until then that $\mathscr{A}$ is finite and we write $\mathscr{A} = \{a_1, a_2, \ldots, a_n\}$. We do not demand initially that $\mathscr{A}$ should be primitive; indeed we allow repetitions and 1's. We recall that $\mathscr{A}$ is non-trivial if there are no 1's. We denote the (minimal) period of $E(x, \mathscr{A})$ by X_0 so that always

$$X_0 | [a_1, a_2, \ldots, a_n]. \tag{3.4}$$

Note that $X_0 = [a_1, a_2, \ldots, a_n]$ if (but not only if) $\mathscr{A}$ is primitive.

It is convenient to make a slight formal change. Let $\overline{M}(x) = [x] - M(x)$ denote the counting function of $\mathscr{T}(\mathscr{A})$, that is the set of integers which remain after $\mathbf{Z}^+$ is sifted by $\mathscr{A}$, and put

$$\overline{E}(x, \mathscr{A}) = \overline{M}(x) - \mathbf{t}(\mathscr{A})x. \tag{3.5}$$

We work with $\overline{E}(x, \mathscr{A})$ rather than $E(x, \mathscr{A})$ because some of our various formulae are cleaner: the change is unimportant because

$$E(x, \mathscr{A}) + \overline{E}(x, \mathscr{A}) = [x] - x. \tag{3.6}$$

We note, en passant, that $\overline{E}(x, \mathscr{A})$ is an odd function in the sense that

$$\overline{E}(X_0 - x, \mathscr{A}) = -\overline{E}(x, \mathscr{A}). \tag{3.7}$$

The restriction to finite $\mathscr{A}$ is a serious one and so we insist on the most general results which we can achieve within this framework.

The obvious measure of the oscillations of $\overline{M}(x)$ is

$$\max_x \left|\overline{E}(x, \mathscr{A})\right| = \max_x \overline{E}(x, \mathscr{A}) \tag{3.8}$$

but this is not very easy to handle. We define

$$\langle \mathscr{A}, \mathscr{A} \rangle = 12 X_0^{-1} \int_0^{X_0} \overline{E}(x, \mathscr{A})^2 dx \tag{3.9}$$

and we begin with the following simple result.

Theorem 3.1 *Provided $\mathscr{A}$ is non-trivial, we have*

$$\langle \mathscr{A}, \mathscr{A} \rangle \geq 1. \tag{3.10}$$

Proof Since $\mathscr{A}$ is non-trivial the complement $\mathscr{T}(\mathscr{A})$ of $\mathscr{M}(\mathscr{A})$ is non-empty: we denote the elements of $\mathscr{T}(\mathscr{A})$ not exceeding X_0 by $b_1, b_2, \ldots, b_N$. Then $N = \mathbf{t}(\mathscr{A})X_0$ and we have, splitting the range of integration at the points b_j, that

$$\int_0^{X_0} \overline{E}(x,\mathscr{A})^2 dx$$
$$= (b_2 - b_1) + 4(b_3 - b_2) + 9(b_4 - b_3) + \cdots + N^2(X_0 - b_N)$$
$$-\mathbf{t}(\mathscr{A})\{(b_2^2 - b_1^2) + 2(b_3^2 - b_2^2) + \cdots + N(X_0^2 - b_N^2)\} + \frac{1}{3}\mathbf{t}(\mathscr{A})^2 X_0^3$$
$$= \frac{1}{12}X_0 + \mathbf{t}(\mathscr{A})^{-1}\sum_{j=1}^{N}\left(\mathbf{t}(\mathscr{A})b_j - j + \frac{1}{2}\right)^2, \tag{3.11}$$

whence

$$\langle \mathscr{A}, \mathscr{A}\rangle = 1 + 12N^{-1}\sum_{j=1}^{N}\left(\mathbf{t}(\mathscr{A})b_j - j + \frac{1}{2}\right)^2. \tag{3.12}$$

This completes the proof.

In general formula (3.12) is not very helpful because we have too little information about the b_j. Our next task is to write down a (somewhat cumbersome) expression for $\langle \mathscr{A}, \mathscr{A}\rangle$ in terms of the elements a_i of $\mathscr{A}$. By the inclusion–exclusion principle we have

$$\overline{M}(x) = [x] - \sum_i \left[\frac{x}{a_i}\right] + \sum_{i<j}\left[\frac{x}{[a_i, a_j]}\right] - \cdots \tag{3.13}$$

and we define, for real x,

$$\rho(x) = [x] - x + \frac{1}{2}. \tag{3.14}$$

We then have

$$\overline{E}(x,\mathscr{A}) = \rho(x) - \sum_i \rho\left(\frac{x}{a_i}\right) + \sum_{i<j}\rho\left(\frac{x}{[a_i,a_j]}\right) - \cdots; \tag{3.15}$$

of course the $\frac{1}{2}$'s all cancel. We require a formula of J.Franel (1924), Kluyver (1903) (see Landau, *Vorlesungen über Zahlentheorie* Satz 484).

Lemma 3.2 *Let a, b be positive integers. Then*

$$\int_0^1 \rho(ax)\rho(bx)dx = \frac{1}{12}\frac{(a,b)}{[a,b]}, \tag{3.16}$$

equivalently if X be any multiple of $[a, b]$,

$$X^{-1}\int_0^X \rho\left(\frac{x}{a}\right)\rho\left(\frac{x}{b}\right)dx = \frac{1}{12}\frac{(a,b)}{[a,b]}. \tag{3.17}$$

We obtain (3.17), with $X = ab$, from (3.16) by the substitution $x \to x/ab$, and then we note that the integrand in (3.17) has period $[a, b]$. One way to prove (3.16) is by means of the Fourier series

$$\rho(x) \sim \sum_{r \neq 0} \frac{e(rx)}{2\pi ri}, \quad e(\theta) = e^{2\pi i\theta} \tag{3.18}$$

but Landau avoids this. There has been some interest in these integrals recently, see Greaves, Hall, Huxley and Wilson (1993), Wilson (1994).

Proposition 3.3 *We have*

$$\begin{aligned}\langle \mathscr{A}, \mathscr{A} \rangle \;=\; & \sum_{\varepsilon}{}^{*} \sum_{\delta}{}^{*} (-1)^{\varepsilon_1 + \cdots + \varepsilon_n + \delta_1 + \cdots + \delta_n} \\ & \times \frac{([a_1^{\varepsilon_1}, a_2^{\varepsilon_2}, \ldots, a_n^{\varepsilon_n}], [a_1^{\delta_1}, a_2^{\delta_2}, \ldots, a_n^{\delta_n}])}{[[a_1^{\varepsilon_1}, a_2^{\varepsilon_2}, \ldots, a_n^{\varepsilon_n}], [a_1^{\delta_1}, a_2^{\delta_2}, \ldots, a_n^{\delta_n}]]},\end{aligned} \tag{3.19}$$

where every ε_i, δ_i takes the value 0 or 1 and $\sum^*$ denotes that summation is over all 2^n choices.

This is an immediate consequence of (3.15), the definition of $\langle \mathscr{A}, \mathscr{A} \rangle$ and (3.17). The next result breaks our rule about generality but is a useful precursor of developments to follow.

Corollary 3.4 *When the elements of $\mathscr{A}$ are pairwise coprime, we have*

$$\langle \mathscr{A}, \mathscr{A} \rangle = 2^n \mathbf{t}(\mathscr{A}). \tag{3.20}$$

This formula is implicit in Perelli and Zannier (1989) and Erdös (1946).

Proof When the a_i are pairwise coprime the interior l.c.m.'s on the right of (3.19) are simply products. The ratio h.c.f./l.c.m. may then be viewed as a symmetric quotient, that is the analogue for products of the symmetric difference of sets. The factor a_i appears in both numerator and denominator if $\varepsilon_i = \delta_i = 1$, and so cancels; so it survives as a factor in the denominator if and only if $\varepsilon_i \neq \delta_i$. We put $\Delta_i = |\varepsilon_i - \delta_i|$ for each i and we have $\Delta_i = 0$ or 1: in either case there are two choices of the pair $\{\varepsilon_i, \delta_i\}$ giving the same value of Δ_i. Since $\Delta_i \equiv \varepsilon_i + \delta_i \pmod 2$ we deduce from (3.19) that

$$\langle \mathscr{A}, \mathscr{A} \rangle = 2^n \sum_{\Delta}{}^{*} \frac{(-1)^{\Delta_1 + \cdots + \Delta_n}}{a_1^{\Delta_1} a_2^{\Delta_2} \cdots a_n^{\Delta_n}} = 2^n \mathbf{t}(\mathscr{A}). \tag{3.21}$$

The notation $\langle , \rangle$ arose from the idea of an inner product. If $\mathscr{A}$ and $\mathscr{B}$ are finite the function

$$\overline{E}(x, \mathscr{A})\overline{E}(x, \mathscr{B}) \tag{3.22}$$

is periodic with period say X. We define

$$\langle \mathscr{A}, \mathscr{A} \rangle = 12X^{-1} \int_0^X \overline{E}(x, \mathscr{A})\overline{E}(x, \mathscr{B})dx. \tag{3.23}$$

This has not been developed very far. We leave the proof of the following proposition, which indicates some of the possibilities, to the reader.

Proposition 3.5 *Let* $\mathscr{A} = \{a_1, a_2, \ldots, a_m\}$, $\mathscr{B} = \{b_1, b_2, \ldots, b_n\}$ *be such that* $(a_i, b_j) = 1$ *for all* i *and* j. *Then we have*

$$\langle \mathscr{A}, \mathscr{A} \rangle = \mathbf{t}(\mathscr{A})\mathbf{t}(\mathscr{B}) \tag{3.24}$$

$$\langle \mathscr{A} \cup \mathscr{B}, \mathscr{A} \rangle = \langle \mathscr{A}, \mathscr{A} \rangle \mathbf{t}(\mathscr{B}) \tag{3.25}$$

$$\langle \mathscr{A} \cup \mathscr{B}, \mathscr{A} \cup \mathscr{B} \rangle = \langle \mathscr{A}, \mathscr{A} \rangle \langle \mathscr{B}, \mathscr{B} \rangle \tag{3.26}$$

We want to make progress without such coprimality conditions. The type of sequences $\mathscr{A}$ which we have in mind are

$$\begin{aligned} \mathscr{A}_1 &= \{a+1, a+2, a+3, \ldots, 2a\}, \\ \mathscr{A}_2 &= \{p_1 p_2 \ldots p_h \,:\, p_1 < p_2 < \cdots < p_h \le y\}. \end{aligned} \tag{3.27}$$

Our results yield in these particular cases:

$$\begin{aligned} \langle \mathscr{A}_1, \mathscr{A}_1 \rangle &\ge \frac{1}{7} a^\beta, \quad \beta = 1 - \frac{\log(\pi^2/6)}{\log 2} = .28197\ldots, \\ \langle \mathscr{A}_2, \mathscr{A}_2 \rangle &\ge 2^{\pi(y)-h} \prod_{p \le y} \left(1 - \frac{1}{p}\right) \sum_{g<h} \frac{\left(\sum_{p \le y} \frac{1}{p}\right)^g}{g!} \end{aligned} \tag{3.28}$$

The second inequality above is non-trivial whenever $\pi(y) - h$ is large. notice that when $h = 1$, $\overline{M}(x) = \Phi(x, y)$ is Buchstab's function. If we write

$$\overline{E}(x, \mathscr{A}_2) = \Phi(x, y) - x \prod_{p \le y} \left(1 - \frac{1}{p}\right) =: \eta(x, y) x \prod_{p \le y} \left(1 - \frac{1}{p}\right)$$

then we have, for each fixed $u > 1$ that

$$\lim_{y \to \infty} \eta(y^u, y) = e^\gamma \omega(u) - 1$$

where γ is Euler's constant and ω is defined by a differential–difference equation. The right-hand side oscillates about, and converges to 0 as

$u \to \infty$; (cf. Maier (1985), Friedlander and Granville (1989), Hildebrand and Maier (1989). It would be of interest to know whether (3.28) is applicable to this area of research).

We employ the machinery involving total decomposition sets introduced in §2.4–5 to cope with the l.c.m.'s appearing in the formula (3.19) for $\langle \mathscr{A}, \mathscr{A} \rangle$. Let $\mathscr{F}$ be a family of non-empty subsets $S \subseteq S_0 = \{1, 2, \ldots, n\}$, where $n = |\mathscr{A}|$. For each subfamily $\mathscr{E}$ of $\mathscr{F}$ (including the empty family) we define

$$a(\mathscr{E}, \mathscr{F}) = \sum_{\mathscr{G}} \left\{ (-1)^{|\mathscr{G}|} \mathscr{X}(\mathscr{G}) : \mathscr{E} \subseteq \mathscr{G} \subseteq \mathscr{F} \right\} \tag{3.29}$$

where $\mathscr{X}$, as in (2.113), is the characteristic function of the complete families. We also define

$$b(\mathscr{F}) = \sum_{\mathscr{E}} \{ |a(\mathscr{E}, \mathscr{F})| : \mathscr{E} \subseteq \mathscr{F} \}, \tag{3.30}$$

$$c(\mathscr{F}) = \sum_{\mathscr{E}} \{ a(\mathscr{E}, \mathscr{F})^2 : \mathscr{E} \subseteq \mathscr{F} \}. \tag{3.31}$$

The main result of this section is the formula for $\langle \mathscr{A}, \mathscr{A} \rangle$ which follows.

Theorem 3.6 *Let $\mathscr{A} = \{a_1, a_2, \ldots, a_n\}$ and $\{d(S)\}$ be the total decomposition set of $\mathscr{A}$. Then*

$$\langle \mathscr{A}, \mathscr{A} \rangle = \sum_{\mathscr{F}} c(\mathscr{F}) \prod_{S \in \mathscr{F}} \left(1 - \frac{1}{d(S)}\right) \prod_{R \notin \mathscr{F}} \frac{1}{d(R)} \tag{3.32}$$

with $c(\mathscr{F})$ as in (3.31).

This formula has a similar shape to (2.115), for $\mathbf{t}(\mathscr{A})$. As an example we write out (3.32) in full in the case $\mathscr{A} = \{a_1, a_2\}$, as before using the shorthand $\{d_1, d_2, d_{12}\}$ for $\{d(S)\}$. We have

$$\begin{aligned} \langle \mathscr{A}, \mathscr{A} \rangle = {} & 4\left(1 - \frac{1}{d_1}\right)\left(1 - \frac{1}{d_2}\right)\left(1 - \frac{1}{d_{12}}\right) + 4\left(1 - \frac{1}{d_1}\right)\left(1 - \frac{1}{d_2}\right)\frac{1}{d_{12}} \\ & + 2\left(1 - \frac{1}{d_1}\right)\left(1 - \frac{1}{d_{12}}\right)\frac{1}{d_2} + 2\left(1 - \frac{1}{d_2}\right)\left(1 - \frac{1}{d_{12}}\right)\frac{1}{d_1} \\ & + 2\left(1 - \frac{1}{d_{12}}\right)\frac{1}{d_1 d_2}. \end{aligned} \tag{3.33}$$

Comparing this with (2.110), we observe that the same five elementary multilinear functions appear on the right-hand side: only the coefficients have altered. An immediate deduction from (2.110) and (3.33) is that for $n = 2$,

$$2\mathbf{t}(\mathscr{A}) \le \langle \mathscr{A}, \mathscr{A} \rangle \le 4\mathbf{t}(\mathscr{A}). \tag{3.34}$$

Moreover it is not difficult to isolate the cases of equality. Equality on the left is equivalent to $d_1 = 1$ or $d_2 = 1$, that is $a_1|a_2$ or $a_2|a_1$. This means that $\mathscr{A}$ is not primitive. Equality on the right is equivalent to $d_{12} = 1$, that is $(a_1, a_2) = 1$, when (3.33) is a special case of Proposition 3.3. It will emerge that the left-hand inequality in (3.34) generalizes to all n. However, it is not true that for all n

$$\langle \mathscr{A}, \mathscr{A} \rangle \le 2^n\mathbf{t}(\mathscr{A}). \tag{3.35}$$

This holds if and only if $n \le 3$; for $n > 3$ the pairwise coprime case is not extremal. We must take care not to extrapolate too far.

Proof of Theorem 3.6 We begin with formula (3.19). Let Q and T (which we allow to be empty) be the subsets of $\{1, 2, \ldots, n\}$ on which, respectively $\varepsilon_i = 1$, $\delta_j = 1$. We recall from (0.112) that

$$\text{l.c.m.}[a_i : i \in Q] = \prod \{d(R) : R \cap Q \neq \emptyset\} \tag{3.36}$$

whence we have

$$\begin{aligned} \langle \mathscr{A}, \mathscr{A} \rangle &= \sum_Q \sum_T (-1)^{|Q|+|T|} \frac{(\text{l.c.m.}[a_i : i \in Q], \text{l.c.m.}[a_j : j \in T])}{[\text{l.c.m.}[a_i : i \in Q], \text{l.c.m.}[a_j : j \in T]]} \\ &= \sum_Q \sum_T (-1)^{|Q|+|T|} \\ &\quad \times \frac{(\prod\{d(R) : R \cap Q \neq \emptyset\}, \prod\{d(S) : S \cap T \neq \emptyset\})}{[\prod\{d(R) : R \cap Q \neq \emptyset\}, \prod\{d(S) : S \cap T \neq \emptyset\}]}. \end{aligned} \tag{3.37}$$

Let P be a subset of $S_0 = \{1, 2, \ldots, n\}$ which intersects both Q and T. Then $d(P)$ is a factor of both the numerator and denominator on the right of (3.37) and so cancels. On the other hand let R and S be subsets such that

$$R \cap Q \neq \emptyset,\ R \cap T = \emptyset,\ S \cap T \neq \emptyset,\ S \cap Q = \emptyset. \tag{3.38}$$

Then we cannot have $R \subseteq S$ or $S \subseteq R$. For example if $R \subseteq S$ then $R \cap Q \neq \emptyset$ implies $S \cap Q \neq \emptyset$, contradicting (3.38). We deduce from Theorem 0.21 that for R, S satisfying (3.38) we have $d(R)$ and $d(S)$ relatively prime.

Now consider the quotient on the right of (3.37). Once the factors

$d(P)$ described above have been cancelled, the h.c.f. remaining in the numerator is equal to 1. We define

$$\nabla(Q,T) = \{W : W \text{ intersects one and only one of } Q \text{ and } T\} \quad (3.39)$$

and we deduce that

$$\langle \mathscr{A}, \mathscr{A} \rangle = \sum_Q \sum_T (-1)^{|Q|+|T|} \prod \{d(W)^{-1} : W \in \nabla(Q,T)\}. \quad (3.40)$$

Thus $\langle \mathscr{A}, \mathscr{A} \rangle$ is a multilinear function of the variables $d(S)^{-1}$. We consider the corresponding polynomial

$$f = \sum_Q \sum_T (-1)^{|Q|+|T|} \prod \{x(W) : W \in \nabla(Q,T)\} \quad (3.41)$$

and by Theorem 2.12, f has an expansion as a linear combination of elementary functions. Let $\lambda(\mathscr{F})$ be the coefficient of

$$\prod_{S \in \mathscr{F}} (1 - x(S)) \prod_{R \notin \mathscr{F}} x(R) \quad (3.42)$$

so that we have to show that $\lambda(\mathscr{F}) = c(\mathscr{F})$. To obtain $\lambda(\mathscr{F})$ we evaluate f at the point

$$x(S) = 0, \ (S \in \mathscr{F}), \ = 1, \ (S \notin \mathscr{F}). \quad (3.43)$$

Hence

$$\begin{aligned} \lambda(\mathscr{F}) &= \sum_Q \sum_T \{(-1)^{|Q|+|T|} : \mathscr{F} \cap \nabla(Q,T) = \emptyset\} \\ &= \sum_Q \sum_T \{(-1)^{|Q|+|T|} : \Delta(Q,T) \supseteq \mathscr{F}\} \end{aligned} \quad (3.44)$$

where

$$\Delta(Q,T) = \{W : W \text{ intersects both or neither of } Q \text{ and } T\} \quad (3.45)$$

is the complementary family to $\nabla(Q,T)$. We have $\mathscr{F} \subseteq \Delta(Q,T)$ if and only if $\mathscr{F}$ possesses a subfamily $\mathscr{E}$ such that $Q, T \in \mathscr{H}(\mathscr{E}, \mathscr{F})$, where $\mathscr{H}(\mathscr{E}, \mathscr{F})$ denotes the family of sets H which intersect *no* $W \in \mathscr{E}$ and *every* $W \in \mathscr{F} \setminus \mathscr{E}$. Therefore

$$\lambda(\mathscr{F}) = \sum_{\mathscr{E} \subseteq \mathscr{F}} \left(\sum_Q \{(-1)^{|Q|} : Q \in \mathscr{H}(\mathscr{E}, \mathscr{F})\} \right)^2, \quad (3.46)$$

and, by (3.31), our proof will be complete if we can show that the sum over Q in (3.46) equals $\pm a(\mathscr{E}, \mathscr{F})$. We employ the inclusion–exclusion principle to evaluate this sum. For any family $\mathscr{G}$ of subsets of S_0 let

$\mathscr{K}(\mathscr{G})$ comprise the sets H which do not intersect any member of $\mathscr{G}$, that is H is a subset of the complement of Span$\mathscr{G}$, defined in (2.111). As in (2.114) and (2.115) we have

$$\sum_{H}\{(-1)^{|H|} : H \in \mathscr{K}(\mathscr{G})\} = \mathscr{X}(\mathscr{G}). \tag{3.47}$$

The inner sum in (3.46) is

$$\begin{aligned}\sum_{Q}\{(-1)^{|Q|} : Q \in \mathscr{H}(\mathscr{E},\mathscr{F})\} &= \sum_{\mathscr{E}\subseteq\mathscr{G}\subseteq\mathscr{F}} (-1)^{|\mathscr{G}|-|E|}\sum_{Q}\{(-1)^{|Q|} : Q \in \mathscr{K}(\mathscr{G})\} \\ &= \sum_{\mathscr{E}\subseteq\mathscr{G}\subseteq\mathscr{F}} (-1)^{|\mathscr{G}|-|E|}\mathscr{X}(\mathscr{G}) = (-1)^{|\mathscr{E}|}a(\mathscr{E},\mathscr{F}) \end{aligned}\tag{3.48}$$

by (3.47) and (3.29); and (3.46), (3.48) give

$$\lambda(\mathscr{F}) = \sum\{a(\mathscr{E},\mathscr{F})^2 : \mathscr{E} \subseteq \mathscr{F}\} = c(\mathscr{F}) \tag{3.49}$$

as required. Hence the polynomial in (3.41) is

$$f = \sum_{\mathscr{F}} c(\mathscr{F}) \prod_{S\in\mathscr{F}} (1 - x(S)) \prod_{R\notin\mathscr{F}} x(R) \tag{3.50}$$

and we substitute $x(S) = d(S)^{-1}$, when $f = \langle\mathscr{A},\mathscr{A}\rangle$ by (3.40). This proves Theorem 3.6.

In the next section we investigate the coefficients $c(\mathscr{F})$. We conclude this section by stating an analogue of Theorem 3.6 for the derived sequences $\mathscr{A}^{(k)}$ defined in Chapter 2. We recall from (2.112) the definition of the deficiency $\delta(\mathscr{F})$ of a family $\mathscr{F}$. As in Theorem 2.21 we put

$$\mathscr{X}_k(\mathscr{F}) = \begin{cases} 1 & \text{if } \delta(\mathscr{F}) \le k, \\ 0 & \text{else.} \end{cases} \tag{3.51}$$

Now we generalize (3.29)–(3.31) by defining

$$a_k(\mathscr{E},\mathscr{F}) = \sum_{\mathscr{G}}\{(-1)^{|\mathscr{G}|}\mathscr{X}_k(\mathscr{G}) : \mathscr{E} \subseteq \mathscr{G} \subseteq \mathscr{F}\}, \tag{3.52}$$

$$b_k(\mathscr{F}) = \sum_{\mathscr{E}}\{|a_k(\mathscr{E},\mathscr{F})| : \mathscr{E} \subseteq \mathscr{F}\}, \tag{3.53}$$

$$c_k(\mathscr{F}) = \sum_{\mathscr{E}}\{a_k(\mathscr{E},\mathscr{F})^2 : \mathscr{E} \subseteq \mathscr{F}\}. \tag{3.54}$$

Theorem 3.7 *Let $\mathscr{A} = \{a_1, a_2, \dots, a_n\}$ and $\{d(S)\}$ be the total decomposition set of $\mathscr{A}$. Let $\mathscr{A}^{(k)}$ be the derived sequence defined by (2.5). Then*

$$\langle\mathscr{A}^{(k)},\mathscr{A}^{(k)}\rangle = \sum_{\mathscr{F}} c_k(\mathscr{F}) \prod_{S\in\mathscr{F}} \left(1 - \frac{1}{d(S)}\right) \prod_{R\notin\mathscr{F}} \frac{1}{d(R)}. \tag{3.55}$$

The proof is similar to that of Theorem 3.6 and is left to the reader. Notice that we could apply Theorem 3.6 to $\mathscr{A}^{(k)}$ but would by so doing ignore the extra structure of a derived sequence.

3.2 A first lower bound for $\langle\mathscr{A},\mathscr{A}\rangle$

We require information about the coefficients $c(\mathscr{F})$ and $c_k(\mathscr{F})$ which appear in the formulae (3.32) and (3.55) for $\langle\mathscr{A},\mathscr{A}\rangle$ and $<\mathscr{A}^{(k)},\mathscr{A}^{(k)}>$. When $|\mathscr{A}|=n$ we require $k<n$ in the definition (2.5) of $\mathscr{A}^{(k)}$.

In this section we treat the two cases together and it is convenient to write $\mathscr{X}_0$, c_0 etc. instead of $\mathscr{X}$, c in some of our formulae.

Lemma 3.8 *Let $0\le k<n$. Then $\mathscr{X}_k(\mathscr{F})$ and $c_k(\mathscr{F})$ are zero or non-zero together; moreover we have*

$$c_k(\mathscr{F})\ge b_k(\mathscr{F})\ge 2\mathscr{X}_k(\mathscr{F}). \tag{3.56}$$

This result should be compared with the written out expansions (2.104) and (3.33) of $t(\mathscr{A})$ and $\langle\mathscr{A},\mathscr{A}\rangle$ when $n=2$, and the remarks following (3.33).

The inequality on the left of (3.56) follows from the definitions (3.53), (3.54) and the fact that $a_k(\mathscr{E},\mathscr{F})$ is an integer.

Let $\mathscr{X}_k(\mathscr{F})=0$. By definition (3.51), the deficiency $\delta(\mathscr{F})>k$ whence $\mathscr{X}_k(\mathscr{G})=0$ throughout (3.52). Hence $a_k(\mathscr{E},\mathscr{F})=0$ for every $\mathscr{E}\subseteq\mathscr{F}$ and $c_k(\mathscr{F})=0$. Let $\mathscr{X}_k(\mathscr{F})=1$. Then $a_k(\mathscr{F},\mathscr{F})=\pm1$ and $b_k(\mathscr{F})\ge 1$. This proves the first part of our assertion. Next we have for all $\mathscr{F}$,

$$\sum_{\mathscr{E}\subseteq\mathscr{F}}(-1)^{|\mathscr{E}|}a_k(\mathscr{E},\mathscr{F})=\sum_{\mathscr{G}\subseteq\mathscr{F}}\mathscr{X}_k(\mathscr{G})\sum_{\mathscr{E}\subseteq\mathscr{G}}(-1)^{|\mathscr{G}|+|\mathscr{E}|}=0 \tag{3.57}$$

because the inner sum is zero unless $\mathscr{G}$ is empty, when $\mathscr{X}_k(\mathscr{G})=0$. From (3.53), (3.54) and (3.57), $b_k(\mathscr{F})$ and $c_k(\mathscr{F})$ are *even integers*, and (3.56) follows. This completes the proof.

So far we have made no use of $b_k(\mathscr{F})$ and it may be that it has no arithmetical significance. It was introduced because of the following result which is vital.

Lemma 3.9 *Let $\mathscr{F}\subseteq\mathscr{F}'$. Then*

$$b_k(\mathscr{F})\le b_k(\mathscr{F}'). \tag{3.58}$$

The corresponding inequality for $c_k(\mathscr{F})$ is false. For example let

$n = 3$, $k = 0$ and $\mathcal{F} = \{(12),(13),(23)\}$, $\mathcal{F}' = \{(1),(12),(13),(23)\}$. Then $b(\mathcal{F}) = b(\mathcal{F}') = c(\mathcal{F}') = 6$, $c(\mathcal{F}) = 8$.

Proof of the lemma We may assume $\mathcal{F}'$ has just one extra element, the set $T \subseteq S_0$. Let us write $\mathcal{E}' = \mathcal{E} \cup \{T\}$ for every $\mathcal{E} \subseteq \mathcal{F}$. Then by (3.52)

$$\begin{aligned} a_k(\mathcal{E},\mathcal{F}') &= \sum_{\mathcal{G}}\{(-1)^{|\mathcal{G}|}\mathcal{X}_k(\mathcal{G}) : \mathcal{E} \subseteq \mathcal{G} \subseteq \mathcal{F}\} \\ &\quad + \sum_{\mathcal{G}}\{(-1)^{|\mathcal{G}|+1}\mathcal{X}_k(\mathcal{G}') : \mathcal{E} \subseteq \mathcal{G} \subseteq \mathcal{F}\} \\ a_k(\mathcal{E}',\mathcal{F}') &= \sum_{\mathcal{G}}\{(-1)^{|\mathcal{G}|+1}\mathcal{X}_k(\mathcal{G}') : \mathcal{E} \subseteq \mathcal{G} \subseteq \mathcal{F}\} \end{aligned} \tag{3.59}$$

whence

$$a_k(\mathcal{E},\mathcal{F}') - a_k(\mathcal{E}',\mathcal{F}') = a_k(\mathcal{E},\mathcal{F}), \tag{3.60}$$

and

$$|a_k(\mathcal{E},\mathcal{F})| \leq |a_k(\mathcal{E},\mathcal{F}')| + |a_k(\mathcal{E}',\mathcal{F}')|\,. \tag{3.61}$$

The result follows from (3.53) and (3.61). This completes the proof.

We look for a class of families for which $c_k(\mathcal{F})$ is large.

Definition 3.10 *We say that $\mathcal{F}_1$ is k-minimal if the deficiency $\delta(\mathcal{F}_1) \leq k$, but $\delta(\mathcal{G}) > k$ for every $\mathcal{G} \subset \mathcal{F}_1$.*

Every family $\mathcal{F}$ such that $\delta(\mathcal{F}) \leq k$ has at least one k-minimal subfamily $\mathcal{F}_1$ (which may have deficiency $< k$). We sometimes refer to 0-minimal families as *minimally complete*.

Definition 3.11 *Let $\delta(\mathcal{F}) \leq k$. We define $\kappa(\mathcal{F},k)$ to be the maximum of the cardinalities of all the k-minimal subfamilies $\mathcal{F}_1$ of $\mathcal{F}$.*

Lemma 3.12 *Let $\mathcal{X}_k(\mathcal{F}) = 1$. Then*

$$c_k(\mathcal{F}) \geq b_k(\mathcal{F}) \geq 2^{\kappa(\mathcal{F},k)}. \tag{3.62}$$

Proof By hypothesis $\delta(\mathcal{F}) \leq k$ and so $\mathcal{F}$ has a k-minimal subfamily $\mathcal{F}_1$ of cardinality $\kappa(\mathcal{F},k)$. For every $\mathcal{E} \subseteq \mathcal{F}_1$, (3.52) implies that

$$a_k(\mathcal{E},\mathcal{F}_1) = (-1)^{|\mathcal{F}_1|} \tag{3.63}$$

whence by (3.53), (3.54),

$$b_k(\mathcal{F}_1) = c_k(\mathcal{F}_1) = 2^{|\mathcal{F}_1|} = 2^{\kappa(\mathcal{F},k)}. \tag{3.64}$$

The result follows from this and Lemma 3.9.

We may now identify the families for which $c_k(\mathscr{F}) = 2$.

Lemma 3.13 *We have* $c_k(\mathscr{F}) = 2$ *if and only if* $\mathscr{F}$ *has the form*

$$\mathscr{F} = \mathscr{D} \cup \{S_1, S_2, \ldots, S_h\} \tag{3.65}$$

where $\delta(\mathscr{D}) > k$ *and* $|S_i| \geq n - k$, $1 \leq i \leq h$. *In particular when* $k = 0$, $c(\mathscr{F}) = 2$ *if and only if* $\mathscr{F} = \mathscr{D} \cup \{S_0\}$ *where* $\mathscr{D}$ *is incomplete.*

Proof Let $\mathscr{F}$ have the form (3.65) and $\mathscr{E} \subseteq \mathscr{F}$. If $\mathscr{E} \subseteq \mathscr{G} \subseteq \mathscr{F}$ we can write $\mathscr{G} = \mathscr{H} \cup \mathscr{K}$ where

$$\begin{aligned} \mathscr{E} \cap \mathscr{D} \subseteq \; & \mathscr{H} \subseteq \mathscr{D} \\ \mathscr{E} \cap \{S_1, S_2, \ldots, S_h\} \subseteq \; & \mathscr{K} \subseteq \{S_1, S_2, \ldots, S_h\}. \end{aligned} \tag{3.66}$$

If $\mathscr{K}$ is non-empty then $\mathscr{X}_k(\mathscr{K}) = 1$ and so $\mathscr{X}_k(\mathscr{G}) = 1$ because $\mathscr{G} \supseteq \mathscr{K}$, $\delta(\mathscr{G}) \leq \delta(\mathscr{K})$. If $\mathscr{K}$ is empty then $\mathscr{X}_k(\mathscr{G}) = \mathscr{X}_k(\mathscr{H}) = 0$ because $\mathscr{H} \subseteq \mathscr{D}$, and $\mathscr{X}_k(\mathscr{D}) = 0$. Thus $\mathscr{X}_k(\mathscr{G}) = \mathscr{X}_k(\mathscr{K})$ and

$$a_k(\mathscr{E}, \mathscr{F}) = \sum_{\mathscr{H}} (-1)^{|\mathscr{H}|} \sum_{\mathscr{K}} (-1)^{|\mathscr{K}|} \mathscr{X}_k(\mathscr{K}) \tag{3.67}$$

with $\mathscr{H}$ and $\mathscr{K}$ as in (3.66). The first sum on the right is zero unless $\mathscr{E} \cap \mathscr{D} = \mathscr{D}$, that is $\mathscr{E} \supseteq \mathscr{D}$. Now let $\mathscr{E} = \mathscr{D} \cup \mathscr{L}$, where $\mathscr{L} \subseteq \{S_1, S_2, \ldots, S_h\}$. Then $\mathscr{H} = \mathscr{D}$, $\mathscr{L} \subseteq \mathscr{K} \subseteq \{S_1, S_2, \ldots, S_h\}$. By (3.67) we have

$$a_k(\mathscr{E}, \mathscr{F}) = (-1)^{|\mathscr{D}|} \sum_{\mathscr{K}} (-1)^{|\mathscr{K}|} \mathscr{X}_k(\mathscr{K}). \tag{3.68}$$

If $\mathscr{L}$ is non-empty then $\mathscr{X}_k(\mathscr{K}) = 1$ always and the sum on the right of (3.68) is zero unless $\mathscr{L} = \{S_1, S_2, \ldots, S_h\}$. In this case $\mathscr{E} = \mathscr{F}$. If $\mathscr{L}$ is empty, the sum is -1 because $\mathscr{X}_k(\mathscr{L}) = 0$ but $\mathscr{X}_k(\mathscr{K}) = 1$ for every $\mathscr{K} \supset \mathscr{L}$. In this case $\mathscr{E} = \mathscr{D}$. So there are two families $\mathscr{E} \subseteq \mathscr{F}$ for which $a_k(\mathscr{E}, \mathscr{F}) = \pm 1$, else it is zero. Thus $c_k(\mathscr{F}) = 2$.

Conversely let $c_k(\mathscr{F}) = 2$. Then by Lemmas 3.8, 3.12 we have $\mathscr{X}_k(\mathscr{F}) = 1$, $\kappa(\mathscr{F}, k) = 1$. Every k-minimal subfamily of $\mathscr{F}$ is therefore a singleton, say $\mathscr{F}_1 = \{S_i\}$, and $\delta(\mathscr{F}_1) \leq k$ requires $|S_i| \geq n - k$. We may suppose there are h such subfamilies, and let $\mathscr{D}$ comprise all the remaining sets in $\mathscr{F}$. We must have $\delta(\mathscr{D}) > k$ else $\mathscr{D}$ would have a k-minimal subfamily $\mathscr{D}_1$. Since $T \in \mathscr{D}_1$ implies $|T| < n - k$, $\delta(\mathscr{D}_1) \leq k$ requires $|\mathscr{D}_1| \geq 2$ and this contradicts the assertion that $\kappa(\mathscr{F}, k) = 1$. This proves the lemma.

Theorem 3.14 *Let* $0 \leq k \leq n$. *Then for all* $\mathscr{A}$,

$$\langle \mathscr{A}^{(k)}, \mathscr{A}^{(k)} \rangle \geq 2\mathbf{t}_k(\mathscr{A}) \tag{3.69}$$

with equality if and only if $\mathscr{P}(\mathscr{A}^{(k)})$ is a singleton, that is there is an element of $\mathscr{A}^{(k)}$ which divides all the others.

This theorem classifies certain extremal sequences. However the right-hand side of (3.69) is small and we shall have to look further for an applicable lower bound.

Proof The inequality follows directly from Theorems 2.21 and 3.7, and Lemma 3.8, so that we just have to consider the cases of equality. Since $\mathbf{t}_k(\mathscr{A}) = \mathbf{t}(\mathscr{A}^{(k)})$ we may assume $k = 0$, substituting $\mathscr{A}^{(k)}$ for $\mathscr{A}$ in the other cases. Let $a_i|a_j$ for some i and every j. Then $\mathscr{M}(\mathscr{A}) = a_i\mathbf{Z}^+$ and $\mathbf{t}(\mathscr{A}) = 1 - 1/a_i$, $\langle \mathscr{A}, \mathscr{A} \rangle = 2(1 - 1/a_i)$. The condition stated is sufficient for equality. Notice it involves $a_i = (a_1, a_2, \ldots, a_n) = d(S_0)$.

Suppose on the contrary that for every i we have $a_i > d(S_0)$, that is there exists $S^{(i)}$ for which

$$i \in S^{(i)} \subset S_0, \ d(S^{(i)}) \neq 1. \tag{3.70}$$

Put $\mathscr{F} = \{S^{(1)}, S^{(2)}, \ldots, S^{(n)}\}$. Then Span$\mathscr{F} = S_0$, that is $\mathscr{F}$ is complete and $\mathscr{X}(\mathscr{F}) = 1$. By (3.70), $\kappa(\mathscr{F}, 0) \geq 2$, and $c(\mathscr{F}) \geq 4$ by Lemma 3.12. By Theorem 3.6 and Lemma 3.8,

$$\langle \mathscr{A}, \mathscr{A} \rangle - 2\mathbf{t}(\mathscr{A}) \geq (c(\mathscr{F}) - 2) \prod_{i=1}^{n} \left(1 - \frac{1}{d(S^{(i)})}\right) \prod_{R \notin \mathscr{F}} \frac{1}{d(R)} \tag{3.71}$$

and the right-hand side is positive by (3.70). This completes the proof.

Theorem 3.15 *Let $\mathscr{A} = \{a_1, a_2, \ldots, a_n\}$ and $(a_1, a_2, \ldots, a_n) = 1$. Let $q = q(\mathscr{A})$ be the least integer with the property that whenever $1 \leq i_1 < i_2 < \cdots < i_q \leq n$ we have $(a_{i_1}, a_{i_2}, \ldots, a_{i_q}) = 1$. For $q \geq 2$ and $r \in \mathbf{Z}^+$ let $l(q, r)$ denote the least integer such that $l(q-1) \geq r$. Then for $0 \leq k < n$ we have*

$$\langle \mathscr{A}^{(k)}, \mathscr{A}^{(k)} \rangle \geq 2^{l(q,n-k)} \mathbf{t}_k(\mathscr{A}). \tag{3.72}$$

Proof Let $|S| \geq q$. If $i_1, i_2, \ldots, i_q \in S$ then $d(S)|(a_{i_1}, a_{i_2}, \ldots, a_{i_q})$ whence $d(S) = 1$. We may therefore restrict the sums over $\mathscr{F}$ in Theorems 2.21 and 3.7 to families $\mathscr{F}$ containing only sets S for which $|S| < q$. Let $\mathscr{F}$ be such a family and $\mathscr{X}_k(\mathscr{F}) = 1$. Any k-minimal subfamily $\mathscr{F}_1$ of $\mathscr{F}$ has cardinality at least $l(q, n-k)$ whence $\kappa(\mathscr{F}, k) \geq l(q, n-k)$ and by Lemma 3.12,

$$c_k(\mathscr{F}) \geq 2^{l(q,n-k)} \mathscr{X}_k(\mathscr{F}). \tag{3.73}$$

The result follows.

We remark that (3.72) is stronger, when $k \geq 1$, than the lower bound obtained by substituting $\mathscr{A}^{(k)}$ for $\mathscr{A}$ in the result for $k = 0$. We claim that

$$q(\mathscr{A}^{(k)}) > \binom{n}{k+1} - \binom{n - q(\mathscr{A}) + 1}{k+1}. \tag{3.74}$$

To see this notice that by the definition of $q(\mathscr{A})$ there are $q(\mathscr{A}) - 1$ elements of $\mathscr{A}$, say $a_1, a_2, \ldots, a_{q(\mathscr{A})-1}$ with common factor $d > 1$. Now let $i_0 < i_1 < \cdots < i_k$ and $[a_{i_0}, a_{i_1}, \ldots, a_{i_k}]$ be an element of $\mathscr{A}^{(k)}$ not divisible by d. Then $i_0 \geq q(\mathscr{A})$ whence

$$\operatorname{card}\{a : a \in \mathscr{A}^{(k)},\ d \nmid a\} \leq \binom{n - q(\mathscr{A}) + 1}{k+1}. \tag{3.75}$$

Since $q\left(\mathscr{A}^{(k)}\right)$ must exceed the number of elements of $\mathscr{A}^{(k)}$ divisible by d we obtain (3.74). It follows that if l' is the least integer such that

$$l' \left\{ \binom{n}{k+1} - \binom{n - q(\mathscr{A}) + 1}{k+1} \right\} \geq \binom{n}{k+1} \tag{3.76}$$

then $l' \geq l(q(\mathscr{A}^{(k)}), |A^{(k)}|)$. On the other hand it is not difficult to check that $l' \leq l(q(\mathscr{A}), n-k)$.

Theorem 3.15 does not apply in the case $(a_1, a_2, \ldots, a_n) > 1$. We rectify this as follows.

Lemma 3.16 *Let* $\mathscr{F}$ *be a family of sets, not containing the set* S_0 *and let* $\mathscr{F}' = \mathscr{F} \cup \{S_0\}$. *Then*

$$c_k(\mathscr{F}') = c_k(\mathscr{F}) + 2 - 2\mathscr{X}_k(\mathscr{F}). \tag{3.77}$$

Proof Let $\mathscr{E} \subseteq \mathscr{F}$ and $\mathscr{E}' = \mathscr{E} \cup \{S_0\}$. As in Lemma 3.9, (3.60) we have

$$a_k(\mathscr{E}, \mathscr{F}') - a_k(\mathscr{E}', \mathscr{F}') = a_k(\mathscr{E}, \mathscr{F}). \tag{3.78}$$

Now

$$\begin{aligned} a_k(\mathscr{E}', \mathscr{F}') &= \sum_{\mathscr{G}} \{(-1)^{|\mathscr{G}|} \mathscr{X}_k(\mathscr{G}) : \mathscr{E}' \subseteq \mathscr{G} \subseteq \mathscr{F}'\} \\ &= \sum_{\mathscr{G}} \{(-1)^{|\mathscr{G}|} : \mathscr{E}' \subseteq \mathscr{G} \subseteq \mathscr{F}'\} \end{aligned} \tag{3.79}$$

since $\mathscr{X}_k(\mathscr{G}) = 1$ throughout: $S_0 \in \mathscr{G}$. Hence

$$|a_k(\mathscr{E}', \mathscr{F}')| = \begin{cases} 1 & \text{if } \mathscr{E} = \mathscr{F}, \\ 0 & \text{else} \end{cases}. \tag{3.80}$$

From (3.78) and (3.80), we infer that

$$\sum_{\mathscr{E} \subset \mathscr{F}} a_k(\mathscr{E}, \mathscr{F})^2 = \sum_{\mathscr{E} \subset \mathscr{F}} \{a_k(\mathscr{E}, \mathscr{F}')^2 + a_k(\mathscr{E}', \mathscr{F}')^2\}, \tag{3.81}$$

whence

$$c_k(\mathscr{F}') = c_k(\mathscr{F}) + a_k(\mathscr{F},\mathscr{F}')^2 + a_k(\mathscr{F}',\mathscr{F}')^2 - a_k(\mathscr{F},\mathscr{F})^2. \tag{3.82}$$

We find that

$$\begin{aligned} |a_k(\mathscr{F},\mathscr{F})| &= \mathscr{X}_k(\mathscr{F}), \\ |a_k(\mathscr{F}',\mathscr{F}')| &= \mathscr{X}_k(\mathscr{F}') = 1, \\ |a_k(\mathscr{F},\mathscr{F}')| &= 1 - \mathscr{X}_k(\mathscr{F}). \end{aligned} \tag{3.83}$$

This implies (3.77) as required.

Theorem 3.17 *Let* $\mathscr{A} = \{a_1, a_2, \ldots, a_n\}$ *and* $(a_1, a_2, \ldots, a_n) = \Delta > 1$. *Let* $\mathscr{A}_1 = \{a_1/\Delta, a_2/\Delta, \ldots, a_n/\Delta\}$. *Then*

$$\langle \mathscr{A}^{(k)}, \mathscr{A}^{(k)} \rangle, = \langle \mathscr{A}_1^{(k)}, \mathscr{A}_1^{(k)} \rangle + 2\,(\Delta - 1)(1 - \mathfrak{t}_k(\mathscr{A}))\,. \tag{3.84}$$

In these circumstances we apply Theorem 3.15 to $\mathscr{A}_1$.

Proof Let the total decomposition sets of $\mathscr{A}$ and $\mathscr{A}_1$ be $\{d(S)\}$ and $\{d_1(S)\}$ respectively. We have $d_1(S) = d(S)$ if $S \neq S_0$, and $d_1(S_0) = 1$, $d(S_0) = \Delta$. Let $\sum^*$ denote summation over families $\mathscr{F}$ not containing S_0. We have

$$\langle \mathscr{A}_1^{(k)}, \mathscr{A}_1^{(k)} \rangle = \sum_{\mathscr{F}}{}^{*} c_k(\mathscr{F}) \prod_{S \in \mathscr{F}} \left(1 - \frac{1}{d_1(S)}\right) \prod_{R \notin \mathscr{F}} \frac{1}{d_1(R)} \tag{3.85}$$

because for the families which contain S_0 the product over S is zero. We remove the suffix 1 throughout the right-hand side: since S_0 is an R always this gives

$$\langle \mathscr{A}_1^{(k)}, \mathscr{A}_1^{(k)} \rangle = \Delta \sum_{\mathscr{F}}{}^{*} c_k(\mathscr{F}) \prod_{S \in \mathscr{F}} \left(1 - \frac{1}{d(S)}\right) \prod_{R \notin \mathscr{F}} \frac{1}{d(R)}. \tag{3.86}$$

Next we consider $\langle \mathscr{A}^{(k)}, \mathscr{A}^{(k)} \rangle$. With the convention that $\mathscr{F}' = \mathscr{F} \cup \{S_0\}$, this may be written in the form

$$\begin{aligned} \langle \mathscr{A}^{(k)}, \mathscr{A}^{(k)} \rangle &= \sum_{\mathscr{F}}{}^{*} c_k(\mathscr{F}) \prod_{S \in \mathscr{F}} \left(1 - \frac{1}{d(S)}\right) \prod_{R \notin \mathscr{F}} \frac{1}{d(R)} \\ &\quad + \sum_{\mathscr{F}}{}^{*} c_k(\mathscr{F}') \prod_{S \in \mathscr{F}'} \left(1 - \frac{1}{d(S)}\right) \prod_{R \notin \mathscr{F}'} \frac{1}{d(R)}. \end{aligned} \tag{3.87}$$

In the second term on the right, we replace $\mathscr{F}'$ by $\mathscr{F}$ in the specification of each product. This removes a factor $1 - 1/\Delta$ from the product over S, and introduces a factor $1/\Delta$ into the product over R. Thus

$$\langle \mathscr{A}^{(k)}, \mathscr{A}^{(k)} \rangle = \sum_{\mathscr{F}}{}^{*} c_k(\mathscr{F}) \prod_{S \in \mathscr{F}} \left(1 - \frac{1}{d(S)}\right) \prod_{R \notin \mathscr{F}} \frac{1}{d(R)}$$
$$+(\Delta - 1) \sum_{\mathscr{F}}{}^{*} c_k(\mathscr{F}') \prod_{S \in \mathscr{F}} \left(1 - \frac{1}{d(S)}\right) \prod_{R \notin \mathscr{F}} \frac{1}{d(R)}. \quad (3.88)$$

Now we subtract (3.86) from (3.88), and employ Lemma 3.16 to obtain

$$\langle \mathscr{A}^{(k)}, \mathscr{A}^{(k)} \rangle = \langle \mathscr{A}_1^{(k)}, \mathscr{A}_1^{(k)} \rangle$$
$$+2(\Delta - 1) \sum_{\mathscr{F}}{}^{*} (1 - \mathscr{X}_k(\mathscr{F})) \prod_{S \in \mathscr{F}} \left(1 - \frac{1}{d(S)}\right) \prod_{R \notin \mathscr{F}} \frac{1}{d(R)}. \quad (3.89)$$

The * is nugatory because $\mathscr{X}_k(\mathscr{F}) = 1$ for the missing families. This yields (3.84).

We conclude this section by working out the second example in (3.27), namely

$$\mathscr{A}_2 = \{p_1 p_2 \dots p_h : p_1 < p_2 < \dots < p_h \le y\}. \quad (3.90)$$

This is best envisaged as a derived sequence with $k = h - 1$ and

$$\mathscr{A} = \{p : p \le y\}. \quad (3.91)$$

Then $q(\mathscr{A}) = 2$, and $l = n - k = \pi(y) - h + 1$. Theorem 3.15 gives

$$\langle \mathscr{A}^{(k)}, \mathscr{A}^{(k)} \rangle \ge 2^{\pi(y) - h + 1} \mathbf{t}_k(\mathscr{A}). \quad (3.92)$$

We apply the formula (2.71) which implies that

$$\mathbf{t}_k(\mathscr{A}) \ge \prod_{p \le y} \left(1 - \frac{1}{p}\right) \sum_{g < h} \sum_{p_1 < p_2 < \dots < p_g \le y} (p_1 p_2 \dots p_g)^{-1}. \quad (3.93)$$

The inner sum on the right is at least

$$\frac{1}{g!} \left(\sum_{p \le y} \frac{1}{p}\right)^{g} - \frac{1}{(g-2)!} \left(\sum_{p \le y} \frac{1}{p}\right)^{g-2} \sum_{p \le y} \frac{1}{p^2}, \quad (3.94)$$

with no subtracted term if $g < 2$. We have $\sum p^{-2} < \frac{1}{2}$ whence, summing over g,

$$\mathbf{t}_k(\mathscr{A}) \ge \frac{1}{2} \prod_{p \le y} \left(1 - \frac{1}{p}\right) \sum_{g < h} \frac{1}{g!} \left(\sum_{p \le y} \frac{1}{p}\right)^{g}. \quad (3.95)$$

This is all we need. Note that we have a superior result to Hall (1990a): at that time there were no inequalities for derived sequences.

3.3 Upper bounds for $\langle \mathscr{A}, \mathscr{A} \rangle$

In this section we consider how large $\langle \mathscr{A}, \mathscr{A} \rangle$ can be, and we recall the assertion following (3.34) that for $n > 3$ the inequality $\langle \mathscr{A}, \mathscr{A} \rangle \le 2^n \mathbf{t}(\mathscr{A})$ is false.

Definition 3.18 *We define*

$$C(n) = \max_{\mathscr{F}} c(\mathscr{F}) \tag{3.96}$$

where the maximum is taken over all families $\mathscr{F}$ of subsets $S \subseteq \{1, 2, \ldots, n\}$.

Theorem 3.19 *For $n \le 3$ we have $C(n) = 2^n$. For $n \ge 4$ we have*

$$C_n \le C(n) \le \binom{2n}{n} \tag{3.97}$$

where the sequence C_n is given by

$$\begin{aligned} C_{2m} &= \frac{1}{4}\binom{2m}{m}^2 + \frac{1}{2}\binom{2m}{m} + 2^{2m-1} \\ C_{2m+1} &= \binom{2m}{m}^2 + 2^{2m-1}. \end{aligned} \tag{3.98}$$

Thus $C_n > 2^n$ for $n \ge 4$. Stirling's formula gives

$$C_n \sim \frac{4^n}{4\pi n}, \quad \binom{2n}{n} \sim \frac{4^n}{\sqrt{\pi n}} \tag{3.99}$$

so there is a substantial gap in (3.97). No value of $C(n)$, $n \ge 4$, is known. We have $20 \le C(4) \le 70$, and the reader will notice that there are 2^{15} candidate families in (3.96) when $n = 4$.

We have not attempted to carry the work in this section over to derived sequences ($k \ge 1$), primarily because the $C(n)$ problem itself has not been solved. We require some preparation for the proof of Theorem 3.19.

Definition 3.20 *Let $\mathscr{E}$ and $\mathscr{F}$ be families of subsets of $\{1, 2, \ldots, n\}$. We say that $\mathscr{E}$ is $\mathscr{F}$-faithful if*

$$\mathscr{E} = \{S : S \in \mathscr{F}, S \subseteq \operatorname{Span}\mathscr{E}\}. \tag{3.100}$$

Evidently $\mathscr{E} \subseteq \mathscr{F}$, moreover there is just one $\mathscr{F}$-faithful family $\mathscr{E}$ of given span. Thus there are, altogether 2^m $\mathscr{F}$-faithful families, where $m = |\operatorname{Span}\mathscr{F}| = n - \delta(\mathscr{F})$.

Lemma 3.21 *Let $a(\mathscr{E}, \mathscr{F})$ be as in (3.29). Then if $a(\mathscr{E}, \mathscr{F}) \neq 0$ we have $\delta(\mathscr{F}) = 0$ and $\mathscr{E}$ is $\mathscr{F}$-faithful.*

Proof of lemma That $\mathscr{F}$ must be complete ($\delta(\mathscr{F}) = 0$) is familiar from Lemma 3.8. We recall from (3.48) that

$$(-1)^{|\mathscr{E}|} a(\mathscr{E}, \mathscr{F}) = \sum_{H} \{(-1)^{|H|} : H \in \mathscr{H}(\mathscr{E}, \mathscr{F})\} \tag{3.101}$$

where $\mathscr{H}(\mathscr{E}, \mathscr{F})$ comprises the sets H which intersect every set belonging to $\mathscr{F} \setminus \mathscr{E}$ and no set belonging to $\mathscr{E}$. Since $a(\mathscr{E}, \mathscr{F}) \neq 0$ we know that $\mathscr{H}(\mathscr{E}, \mathscr{F})$ is not the empty family.

Now let us suppose $\mathscr{E}$ is not $\mathscr{F}$-faithful. Then there exists a set S satisfying both

$$S \in \mathscr{F} \setminus \mathscr{E}, \; S \subseteq \operatorname{Span}\mathscr{E}. \tag{3.102}$$

Any $H \in \mathscr{H}(\mathscr{E}, \mathscr{F})$ must intersect S, whence it intersects Span$\mathscr{E}$ and therefore intersects at least one member of $\mathscr{E}$. This is a contradiction and so $\mathscr{H}(\mathscr{E}, \mathscr{F})$ is empty. Since this is not the case, $\mathscr{E}$ is $\mathscr{F}$-faithful as required. This proves the lemma.

The idea of $\mathscr{F}$-faithful families can be useful in the evaluation of $c(\mathscr{F})$. For example let $\mathscr{F} = \{S : |S| = k\}$. We claim that

$$c(\mathscr{F}) = \binom{n-1}{k-1}^2 + \sum_{l=k}^{n} \binom{n}{l}. \tag{3.103}$$

Let us prove this. Choose l integers x_i such that $1 \leq x_1 < x_2 < \cdots < x_l \leq n$ and suppose $\mathscr{E} \subseteq \mathscr{F}$, Span$\mathscr{E} = \{x_1, x_2, \ldots, x_l\}$. Let $\mathscr{E}$ be $\mathscr{F}$-faithful, and consider the case $l < k$. This is impossible unless $l = 0$ and $\mathscr{E}$ is the empty family. $\mathscr{H}(\emptyset, \mathscr{F})$ comprises the sets H which intersect every member of $\mathscr{F}$, that is $|H| > n - k$, whence from (3.101), we have

$$a(\emptyset, \mathscr{F}) = (-1)^{n-k+1} \binom{n-1}{k-1}. \tag{3.104}$$

This accounts for the first term on the right of (3.103). Next, consider the case $l \geq k$. The $\mathscr{F}$-faithful family $\mathscr{E}$ with Span$\mathscr{E}$ as above is

$$\mathscr{E} = \{S : |S| = k, \, S \subseteq \{x_1, x_2, \ldots, x_l\}\}, \tag{3.105}$$

and $\mathscr{H}(\mathscr{E}, \mathscr{F})$ has just one member, the complement of $\{x_1, x_2, \ldots, x_l\}$. In this case $a(\mathscr{E}, \mathscr{F}) = \pm 1$ and the second term on the right of (3.103) is the number of choices for Span$\mathscr{E}$.

Proof of Theorem 3.19 For $n \leq 3$ this is by inspection, although the case $n = 3$ is tiresome. We assume from now on that $n \geq 4$ and we choose $k = [\frac{n+1}{2}]$ in (3.103) to obtain the lower bound $C(n) \geq C_n$. It remains to prove the upper bound.

Let H_1, H_2, H_3 be sets such that $H_1 \subseteq H_2 \subseteq H_3 \subseteq \{1,2,\dots,n\}$ and let $H_1, H_3 \in \mathscr{H}(\mathscr{E},\mathscr{F})$, for some family $\mathscr{F}$ and $\mathscr{F}$-faithful $\mathscr{E}$. Then $H_2 \in \mathscr{H}(\mathscr{E},\mathscr{F})$. Let us split all the subsets of $\{1,2,\dots,n\}$ into Sperner chains $\mathscr{C}_1, \mathscr{C}_2, \dots, \mathscr{C}_\nu$. We may write

$$(-1)^{|\mathscr{E}|}a(\mathscr{E},\mathscr{F}) = \sum_{\mu=1}^{\nu}\sum\{(-1)^{|H|} : H \in \mathscr{H}(\mathscr{E},\mathscr{F}) \cap \mathscr{C}_\mu\} \tag{3.106}$$

and we see from the argument above that $\mathscr{H}(\mathscr{E},\mathscr{F}) \cap \mathscr{C}_\mu$ is a subchain of $\mathscr{C}_\mu$, that is comprises sets $H_1, H_2, \dots, H_r$ say, with $H_i \subset H_{i+1}$, $|H_i| + 1 = |H_{i+1}|$, $1 \leq i < r$. Hence the inner sum in (3.106) has absolute value ≤ 1 and therefore

$$|a(\mathscr{E},\mathscr{F})| \leq \sum_{\mu=1}^{\nu} \varepsilon_\mu(\mathscr{E}) \tag{3.107}$$

where $\varepsilon_\mu(\mathscr{E}) = 1$ if $\mathscr{H}(\mathscr{E},\mathscr{F})$ intersects $\mathscr{C}_\mu$, else $\varepsilon_\mu(\mathscr{E}) = 0$. Notice that we have

$$\sum_{\mathscr{E}} \varepsilon_\mu(\mathscr{E}) \leq |\mathscr{C}_\mu| \tag{3.108}$$

because a given set H belongs to $\mathscr{H}(\mathscr{E},\mathscr{F})$ for at most one $\mathscr{E}$. For $g = 1, 2, 3, \dots$ put

$$y_g = \operatorname{card}\{\mu : |\mathscr{C}_\mu| \geq g\}. \tag{3.109}$$

Let $\mathscr{E}_1, \mathscr{E}_2, \dots, \mathscr{E}_h$ be distinct subfamilies of $\mathscr{F}$. By (3.107)–(3.109) we have

$$\begin{aligned}
\sum_{i=1}^{h} |a(\mathscr{E}_i,\mathscr{F})| &\leq \sum_{\mu=1}^{\nu}\sum_{i=1}^{h} \varepsilon_\mu(\mathscr{E}_i) \\
&\leq \sum_{\mu=1}^{\nu} \min(h, |\mathscr{C}_\mu|) \\
&\leq h\operatorname{card}\{\mu : |\mathscr{C}_\mu| \geq h\} + \sum_{g<h} g \operatorname{card}\{\mu : |\mathscr{C}_\mu| = g\} \\
&\leq hy_h + \sum_{g<h} g(y_g - y_{g+1}) \\
&\leq y_1 + y_2 + \cdots + y_h.
\end{aligned} \tag{3.110}$$

Lemma 3.22 *Let* $\{x_i\}_{i=1}^N$, $\{y_i\}_{i=1}^N$ *be non-negative, non-increasing sequences such that*

$$\begin{aligned} x_1 &\leq y_1 \\ x_1 + x_2 &\leq y_1 + y_2 \\ x_1 + x_2 + x_3 &\leq y_1 + y_2 + y_3 \\ &\vdots \\ x_1 + x_2 + \cdots + x_N &\leq y_1 + y_2 + \cdots + y_N. \end{aligned} \tag{3.111}$$

Then

$$x_1^2 + x_2^2 + \cdots + x_N^2 \leq y_1^2 + y_2^2 + \cdots + y_N^2. \tag{3.112}$$

Proof of lemma We use a simple variational argument. Let

$$X = \sup\{x_1^2 + x_2^2 + \cdots + x_N^2\} \tag{3.113}$$

subject to the constraints (3.111). X is attained, and we claim that this requires $x_i = y_i$ $(i \leq N)$. We suppose the contrary, that is

$$x_1^2 + x_2^2 + \cdots + x_N^2 = X, \; x_i \neq y_i \text{ for some } i. \tag{3.114}$$

Let g be the least i for which $x_i \neq y_i$. By (3.111) we must have $x_g < y_g$ and $x_1+x_2+\cdots+x_g < y_1+y_2+\cdots+y_g$. If $g \geq 2$ notice that $y_g \leq y_{g-1} = x_{g-1}$ whence $x_g < x_{g-1}$.

There exists h, $g < h \leq N$ such that $x_1+x_2+\cdots+x_h = y_1+y_2+\cdots+y_h$: if this were not the case we could increase x_g a little, leaving the other x_i fixed, without upsetting either the monotonicity of the sequence x_i or the constraints (3.111). By (3.114) this would give a sum of squares $> X$ which is impossible. Let k denote the first such h: we notice that this implies $x_k > y_k$. Also, by (3.111) $x_1+x_2+\cdots+x_{k+1} \leq y_1+y_2+\cdots+y_{k+1}$ so that $x_{k+1} \leq y_{k+1} \leq y_k$, and hence $x_{k+1} < x_k$. (If $k = N$ we have $0 < x_k$.) We introduce a small variation. There exists $\varepsilon > 0$ such that if we change x_g to $x_g + \varepsilon$ and x_k to $x_k - \varepsilon$ all the required inequalities still hold. The sum on the left of (3.114) becomes

$$X + 2\varepsilon(x_g - x_k) + 2\varepsilon^2 > X \tag{3.115}$$

and this is a contradiction. Hence $x_i = y_i$ for all i and $X = y_1^2 + y_2^2 + \cdots + y_N^2$. This proves the lemma.

To complete the proof of Theorem 3.19 we apply this as follows. Let us label the subfamilies $\mathscr{E}$ of $\mathscr{F}$ in such a way that

$$|a(\mathscr{E}_1, \mathscr{F})| \geq |a(\mathscr{E}_2, \mathscr{F})| \geq |a(\mathscr{E}_3, \mathscr{F})| \geq \cdots \tag{3.116}$$

and put $x_g = |a(\mathscr{E}_g, \mathscr{F})|$, keeping y_g as in (3.109). By (3.110), the hypotheses of Lemma 3.22 hold and we conclude that

$$c(\mathscr{F}) \leq \sum y_g^2. \tag{3.117}$$

However the sequence $\{y_g\}$ is the symmetric decreasing rearrangement (Hardy, Littlewood and Pólya (1934)) of the sequence of binomial coefficients $\binom{n}{g}$, and the right-hand side of (3.117) is therefore

$$\binom{n}{0}^2 + \binom{n}{1}^2 + \cdots + \binom{n}{n}^2 = \binom{2n}{n}. \tag{3.118}$$

This completes the proof.

Theorem 3.23 *We have both*

$$\sup\{\langle \mathscr{A}, \mathscr{A} \rangle : |\mathscr{A}| = n\} = C(n) \tag{3.119}$$

$$\sup\{\langle \mathscr{A}, \mathscr{A} \rangle \mathbf{t}(\mathscr{A})^{-1} : |\mathscr{A}| = n\} = C(n); \tag{3.120}$$

in (3.120) $\mathscr{A}$ is restricted to be non-trivial. The supremum in (3.119) is unattained. That in (3.120) is attained if and only if $n \leq 3$.

Proof The inequality

$$\langle \mathscr{A}, \mathscr{A} \rangle \leq C(n)\mathbf{t}(\mathscr{A}) \tag{3.121}$$

follows from Theorems 2.20, 3.6, Lemma 3.8 and the definition (3.96) of $C(n)$. Since $\mathbf{t}(\mathscr{A}) < 1$ it will be sufficient to show that

$$\sup \langle \mathscr{A}, \mathscr{A} \rangle \geq C(n) \tag{3.122}$$

to evaluate both suprema (3.119)–(3.120), moreover it is a consequence of (3.120) and $\mathbf{t}(\mathscr{A}) < 1$ that the supremum in (3.119) is unattained.

Choose a family $\mathscr{F}$ such that $c(\mathscr{F}) = C(n)$. By Theorem 3.6

$$\langle \mathscr{A}, \mathscr{A} \rangle \geq C(n) \prod_{S \in \mathscr{F}} \left(1 - \frac{1}{d(S)}\right) \prod_{R \notin \mathscr{F}} \frac{1}{d(R)}. \tag{3.123}$$

Let $|\mathscr{F}| = f$ and $p_1, p_2, \ldots p_f$ be distinct, large primes, say $z < p_1 < p_2 < \cdots < p_f$. We assign the values $d(S) = p_i$ in any order for $S \in \mathscr{F}$ and put $d(R) = 1$ if $R \notin \mathscr{F}$. By Theorem 0.21, this is the total decomposition set of some sequence $\mathscr{A}$ and by (3.123)

$$\langle \mathscr{A}, \mathscr{A} \rangle > C(n) \left(1 - \frac{1}{z}\right)^f. \tag{3.124}$$

Since z is arbitrary, (3.122) follows. It remains to determine under what circumstances we can have equality in (3.121). Suppose we do, and put

$$\mathscr{F}_0(\mathscr{A}) = \{S : d(S) \neq 1\}. \tag{3.125}$$

The non-zero terms in the expansions of $t(\mathscr{A})$ and $\langle \mathscr{A}, \mathscr{A} \rangle$ provided by Theorems 2.20, 3.6 arise from the complete subfamilies of $\mathscr{F}_0(\mathscr{A})$. If $\mathscr{A}$ is non-trivial, that is $a_i > 1$ for all i, then for each i there exists $S^{(i)}$ such that $i \in S^{(i)}$, $d(S^{(i)}) \neq 1$. Thus $\mathscr{F}_0(\mathscr{A})$ is complete if and only if $\mathscr{A}$ is non-trivial.

We must have $c(\mathscr{F}) = C(n)$ for every complete $\mathscr{F} \subseteq \mathscr{F}_0(\mathscr{A})$. Let $\mathscr{F}$ be minimally complete so that by (3.64) $c(\mathscr{F}) = 2^{|\mathscr{F}|}$. But any minimally complete family has cardinality $\leq n$ whence $C(n) \leq 2^n$ and, by Theorem 3.19, $n \leq 3$. In the case $n = 3$, (3.20) shows that the supremum is indeed attained.

It would plainly be of interest not only to know the value of $C(n)$ but to have some idea of the form $\mathscr{A}$ must have in order that

$$\langle \mathscr{A}, \mathscr{A} \rangle \sim C(n)\mathbf{t}(\mathscr{A}). \tag{3.126}$$

3.4 Primitive $\mathscr{A}$

In §3.2 we gave a lower bound for $\langle \mathscr{A}^{(k)}, \mathscr{A}^{(k)} \rangle$ valid for $0 \leq k < |\mathscr{A}|$. When we are concerned with derived sequences directly, or as in the example $\mathscr{A}_2$ from (3.27) worked out at the end of §3.2, we find it advantageous to treat a given sequence as a derivative, the condition that $\mathscr{A}$ should be primitive is inappropriate because $\mathscr{A}^{(k)}$ and $(\mathscr{P}(\mathscr{A}))^{(k)}$ may be unequal whenever $k \geq 1$. Thus Theorem 3.15 carries no condition of this sort: of course if $\mathscr{A}$ is extravagantly unprimitive or bloated with repeated elements there must be a price to pay, indeed in these and similar cases $q(\mathscr{A})$ will be large.

We recall the example

$$\mathscr{A}_1 = \{a+1, a+2, \ldots, 2a\} \tag{3.127}$$

from (3.27). This is a primitive sequence with a large q – because at least half the elements are even. For such sequences Theorem 3.15 is weak, and we now present a second lower bound for $\langle \mathscr{A}, \mathscr{A} \rangle$ motivated by (3.127). We require $\mathscr{A}$ to be primitive and we restrict our attention to the case $k = 0$.

We begin by setting

$$\mathscr{F}_0(\mathscr{A}) = \{S : d(S) \neq 1\} \tag{3.128}$$

as in (3.125), recalling from §3.3 that $\mathscr{F}_0(\mathscr{A})$ is a complete family if and only if $\mathscr{A}$ is non-trivial: this must be the case if $\mathscr{A}$ is primitive and $|\mathscr{A}| \geq 2$ as we may assume. For subfamilies $\mathscr{E}$ of $\mathscr{F}_0(\mathscr{A})$ we define

$$v(\mathscr{E}) = \prod_{T \in \mathscr{E}} \left(\frac{d(T)+1}{d(T)-1} \right) \tag{3.129}$$

and

$$a(\mathscr{E}, \mathscr{F}_0; \mathscr{A}) = \sum_{\mathscr{E} \subseteq \mathscr{G} \subseteq \mathscr{F}_0} (-1)^{|\mathscr{G}|} \mathscr{X}(\mathscr{G}) \prod_{T \in \mathscr{G}} \left(1 - \frac{1}{d(T)} \right). \tag{3.130}$$

Theorem 3.24 *For every* $\mathscr{A}$

$$\langle \mathscr{A}, \mathscr{A} \rangle = \sum_{\mathscr{E} \subseteq \mathscr{F}_0(\mathscr{A})} v(\mathscr{E}) a(\mathscr{E}, \mathscr{F}_0; \mathscr{A})^2. \tag{3.131}$$

This holds whether $\mathscr{A}$ is primitive or not. Notice that $a(\mathscr{E}, \mathscr{F}_0; \mathscr{A})$ is to some extent similar to $a(\mathscr{E}, \mathscr{F}_0)$, and the right-hand side of (3.131) is similar to $c(\mathscr{F}_0)$: however $a(\mathscr{E}, \mathscr{F})$ and $c(\mathscr{F})$, defined in (3.29), (3.31), were purely combinatorial in structure. Thus we could (to some extent) compute these coefficients without worrying about the total decomposition set of a particular sequence $\mathscr{A}$. This is no longer the case and Theorem 3.24 is, in this sense, of *mixed type.*

Proof For $S \in \mathscr{F}_0$ put

$$y(S) = (d(S)-1)^{-1}, \quad z(S) = 1 - d(S)^{-1} \tag{3.132}$$

so that $y \in (0, 1]$, $z \in [\frac{1}{2}, 1)$ and $z(S) = (1 + y(S))^{-1}$. By Theorem 3.6 we have

$$\langle \mathscr{A}, \mathscr{A} \rangle = \sum_{\mathscr{F} \subseteq \mathscr{F}_0} c(\mathscr{F}) \prod_{S \in \mathscr{F}} (1 - d(S)^{-1}) \prod_{R \notin \mathscr{F}} d(R)^{-1} \tag{3.133}$$

because for the remaining families the product over S vanishes. We can rewrite (3.133) in the form

$$\langle \mathscr{A}, \mathscr{A} \rangle = Z \prod_{S \in \mathscr{F}_0} (1 - d(S)^{-1}) \prod_{R \notin \mathscr{F}_0} d(R)^{-1} \tag{3.134}$$

where

$$Z = \sum_{\mathscr{F} \subseteq \mathscr{F}_0} c(\mathscr{F}) \prod_{T \in \mathscr{F}_0 \backslash \mathscr{F}} y(T). \tag{3.135}$$

From (3.29) and (3.31) we have

$$c(\mathscr{F}) = \sum_{\mathscr{G}_1 \subseteq \mathscr{F}} \sum_{\mathscr{G}_2 \subseteq \mathscr{F}} (-1)^{|\mathscr{G}_1| + |\mathscr{G}_2|} \mathscr{X}(\mathscr{G}_1) \mathscr{X}(\mathscr{G}_2) \sum_{\mathscr{E} \subseteq \mathscr{G}_1 \cap \mathscr{G}_2} 1 \tag{3.136}$$

and we insert this into (3.135), to obtain that

$$Z = \sum_{\mathscr{G}_1 \subseteq \mathscr{F}_0} \sum_{\mathscr{G}_2 \subseteq \mathscr{F}_0} (-1)^{|\mathscr{G}_1|+|\mathscr{G}_2|} \mathscr{X}(\mathscr{G}_1)\mathscr{X}(\mathscr{G}_2) \times 2^{|\mathscr{G}_1 \cap \mathscr{G}_2|} \sum_{\mathscr{G}_1 \cup \mathscr{G}_2 \subseteq \mathscr{F} \subseteq \mathscr{F}_0} \prod_{T \in \mathscr{F}_0 \setminus \mathscr{F}} y(T). \tag{3.137}$$

The innermost sum on the right of (3.137) is

$$\prod_{T \in \mathscr{F}_0 \setminus (\mathscr{G}_1 \cup \mathscr{G}_2)} (1 + y(T)) = \prod_{S \in \mathscr{F}_0} (1 - d(S)^{-1})^{-1} \prod_{T \in \mathscr{G}_1 \cap \mathscr{G}_2} (1 + y(T)) \prod_{T \in \mathscr{G}_1} z(T) \prod_{T \in \mathscr{G}_2} z(T) \tag{3.138}$$

and we notice from the definition (3.129) of $v(\mathscr{E})$ that

$$2^{|\mathscr{G}_1 \cap \mathscr{G}_2|} \prod_{T \in \mathscr{G}_1 \cap \mathscr{G}_2} (1 + y(T)) = \sum_{\mathscr{E} \subseteq \mathscr{G}_1 \cap \mathscr{G}_2} v(\mathscr{E}). \tag{3.139}$$

We assemble (3.137)–(3.139) to deduce that

$$\begin{aligned} Z \prod_{S \in \mathscr{F}_0} (1 - d(S)^{-1}) &= \sum_{\mathscr{G}_1, \mathscr{G}_2 \subseteq \mathscr{F}_0} (-1)^{|\mathscr{G}_1|+|\mathscr{G}_2|} \mathscr{X}(\mathscr{G}_1)\mathscr{X}(\mathscr{G}_2) \prod_{T \in \mathscr{G}_1} z(T) \prod_{T \in \mathscr{G}_2} z(T) \sum_{\mathscr{E} \subseteq \mathscr{G}_1 \cap \mathscr{G}_2} v(\mathscr{E}) \\ &= \sum_{\mathscr{E} \subseteq \mathscr{F}_0} v(\mathscr{E}) \left\{ \sum_{\mathscr{E} \subseteq \mathscr{G} \subseteq \mathscr{F}_0} (-1)^{|\mathscr{G}|} \mathscr{X}(\mathscr{G}) \prod_{T \in \mathscr{G}} z(T) \right\}^2. \end{aligned} \tag{3.140}$$

The inner sum here is $a(\mathscr{E}, \mathscr{F}_0; \mathscr{A})$, and the left-hand side of (3.139) is $\langle \mathscr{A}, \mathscr{A} \rangle$ by (3.134) because the product on the extreme right of (3.134) equals 1. This proves Theorem 3.24.

We look for families $\mathscr{E} \subseteq \mathscr{F}_0$ for which $a(\mathscr{E}, \mathscr{F}_0; \mathscr{A})$ is large.

Definition 3.25 *We say that $\mathscr{E}$ is maximally incomplete with respect to $\mathscr{F}$ if $\mathscr{E} \subset \mathscr{F}$, $\mathscr{X}(\mathscr{E}) = 0$ and $\mathscr{X}(\mathscr{G}) = 1$ for every $\mathscr{G}$ such that $\mathscr{E} \subset \mathscr{G} \subseteq \mathscr{F}$.*

The definition assumes $\mathscr{X}(\mathscr{F}) = 1$ which is in order since we are concerned with $\mathscr{F}_0$. Let $\mathscr{E}$ be maximally incomplete w.r.t. $\mathscr{F}_0$. Then (3.130) gives

$$\begin{aligned} (-1)^{|\mathscr{E}|} a(\mathscr{E}, \mathscr{F}_0; \mathscr{A}) &= \sum_{\mathscr{E} \subset \mathscr{G} \subseteq \mathscr{F}_0} (-1)^{|\mathscr{G}|-|\mathscr{E}|} \prod_{T \in \mathscr{G}} z(T) \\ &= \prod_{T \in \mathscr{E}} z(T) \left\{ \prod_{T \in \mathscr{F}_0 \setminus \mathscr{E}} (1 - z(T)) - 1 \right\}. \end{aligned} \tag{3.141}$$

Since $z(T) \geq \frac{1}{2}$ for every T this implies

$$|a(\mathscr{E}, \mathscr{F}_0; \mathscr{A})| \geq \frac{1}{2} \prod_{T \in \mathscr{E}} z(T) \tag{3.142}$$

for all maximally incomplete $\mathscr{E}$, whence for such $\mathscr{E}$ we have

$$v(\mathscr{E}) a(\mathscr{E}, \mathscr{F}_0; \mathscr{A})^2 \geq \frac{1}{4} \prod_{T \in \mathscr{E}} \left(1 - \frac{1}{d(T)^2}\right). \tag{3.143}$$

We combine this with Theorem 3.24 to give

Theorem 3.26 *For every non-trivial $\mathscr{A}$ we have*

$$\langle \mathscr{A}, \mathscr{A} \rangle \geq \frac{1}{4} \sum_{\mathscr{E} \subseteq \mathscr{F}_0(\mathscr{A})}^{*} \prod_{T \in \mathscr{E}} \left(1 - \frac{1}{d(T)^2}\right), \tag{3.144}$$

where $\sum^*$ denotes summation over families $\mathscr{E}$ which are maximally incomplete w.r.t. $\mathscr{F}_0$.

Theorem 3.27 *Let $\mathscr{A}$ be a primitive sequence of length $n \geq 2$. Then*

$$\langle \mathscr{A}, \mathscr{A} \rangle \geq \frac{1}{4} n \prod_{T \in \mathscr{F}_0(\mathscr{A})} \left(1 - \frac{1}{d(T)^2}\right). \tag{3.145}$$

Proof Let $i \neq j$. Then $a_i \nmid a_j$ if and only if $a_i/(a_i, a_j) > 1$, equivalently

$$\exists S \in \mathscr{F}_0(\mathscr{A}), \quad i \in S, \ j \notin S. \tag{3.146}$$

For any family $\mathscr{F}$, let us write

$$\mathscr{F}^{(j)} = \{S : S \in \mathscr{F}, j \notin S\}. \tag{3.147}$$

Then by (3.146) and (3.147) we have

$$\operatorname{Span} \mathscr{F}_0^{(j)}(\mathscr{A}) = \{i : a_i \nmid a_j\} \tag{3.148}$$

whence $\mathscr{A}$ is primitive if and only if

$$\operatorname{Span} \mathscr{F}_0^{(j)}(\mathscr{A}) = S_0 \setminus \{j\} \text{ for every } j. \tag{3.149}$$

It follows from (3.147) and (3.149) that if $\mathscr{A}$ is primitive then $\mathscr{F}_0^{(j)}(\mathscr{A})$ is maximally incomplete w.r.t. $\mathscr{F}_0$ for every j, that is there are at least n such families. Thus (3.145) follows from (3.144). This completes the proof.

Definition 3.28 *We denote by $e(\mathscr{A})$ the largest exponent e such that for some prime p and some i, $p^e|a_i$. In the usual notation*

$$e(\mathscr{A}) = \max_p \max_i v_p(a_i). \tag{3.150}$$

Theorem 3.29 *Let $\mathscr{A}$ be a primitive sequence of length $n \geq 2$ and $e(\mathscr{A})$ be as defined above. Then*

$$\langle \mathscr{A}, \mathscr{A} \rangle \geq \frac{1}{4} n \left(\frac{6}{\pi^2} \right)^{e(\mathscr{A})}. \tag{3.151}$$

Proof Recall that

$$[a_1, a_2, \ldots, a_n] = \prod \{d(S) : \text{ all } S\}. \tag{3.152}$$

The maximum power of p dividing the left-hand side is p^e and so for all p,

$$\text{card}\{S : p|d(S)\} \leq e(\mathscr{A}). \tag{3.153}$$

For each $T \in \mathscr{F}_0(\mathscr{A})$ put

$$P^+(T) = \max\{p : p \text{ prime}, p|d(T)\}. \tag{3.154}$$

Then

$$\begin{aligned} \prod_{T \in \mathscr{F}_0(\mathscr{A})} (1 - d(T)^{-2}) &\geq \prod_{T \in \mathscr{F}_0(\mathscr{A})} (1 - P^+(T)^{-2}) \\ &\geq \prod_p (1 - p^{-2})^{e(\mathscr{A})} \end{aligned} \tag{3.155}$$

by (3.153). We combine (3.145) and (3.155) to obtain (3.151). This completes the proof.

We work out, as an example, the case given in (3.127). We have $|\mathscr{A}_1| = a$ and

$$e(\mathscr{A}_1) \leq 1 + \frac{\log a}{\log 2}. \tag{3.156}$$

If $a = 1$ then $\mathscr{A}_1 = \{2\}$ and $\langle \mathscr{A}_1, \mathscr{A}_1 \rangle = 1$. If $a \geq 2$ we apply Theorem 3.29 which yields

$$< \mathscr{A}_1, \mathscr{A}_1 > \geq \frac{3}{2\pi^2} a \left(\frac{6}{\pi^2} \right)^{(\log a / \log 2)} > \frac{1}{7} a^{\beta} \tag{3.157}$$

with β as in (3.28), which therefore holds for all a. This method applies to $\mathscr{A} = \{a+1, a+2, \ldots, a+b\}$ provided b is not too large compared to a.

If $\mathscr{A}$ is a primitive sequence of squarefree numbers (exceeding 1) then

$$\langle \mathscr{A}, \mathscr{A} \rangle > \frac{n}{7}. \tag{3.158}$$

It seems likely that the condition that the elements should be squarefree is uneccessary.

Conjecture 3.30

$$\inf\{\langle \mathscr{A}, \mathscr{A} \rangle : \mathscr{A} \text{ primitive}, |\mathscr{A}| = n\} \gg n. \tag{3.159}$$

This would be best possible if true. We notice that, with p and q distinct primes,

$$\mathscr{A}_3 = \{p^{n-1}, p^{n-2}q, p^{n-3}q^2, \ldots, pq^{n-2}, q^{n-1}\} \tag{3.160}$$

has $\langle \mathscr{A}_3, \mathscr{A}_3 \rangle \asymp n$. This is a very interesting test case since none of our theorems cope with it, both $e(\mathscr{A}_3)$ and $q(\mathscr{A}_3)$ being large.

3.5 Perfect sequences

We consider, albeit somewhat briefly, the case when $\mathscr{P}(\mathscr{A})$ is infinite; that is $\mathscr{A}$ is not only infinite but contains an infinite primitive subsequence. We assume that $\mathscr{A}$ is Besicovitch.

Definition 3.31 *We say that $\mathscr{A}$ is perfect if $\mathscr{P}(\mathscr{A})$ is infinite and*

$$E(x, \mathscr{A}) = M(x) - \mathbf{d}\mathscr{M}(\mathscr{A})x = O(1) \tag{3.161}$$

where $M(x) = M(x, \mathscr{A})$ denotes the counting function of $\mathscr{M}(\mathscr{A})$.

Condition (3.161) is strong and it is natural to expect at first sight that we should be able to categorize the perfect sequences relatively easily and proceed to consider the rate of growth of $E(x, \mathscr{A})$. In fact we are only able to achieve the first of these objectives in some restricted cases.

The first of these, where a complete solution of the (initial) problem is possible, is that in which $\mathscr{A}$ is Behrend. This is easy because the error term $E(x, \mathscr{A})$ is negative and, moreover, cumulative: every integer omitted by $\mathscr{M}(\mathscr{A})$ decreases it. Hence (3.161) holds if and only if $\mathbf{N} \setminus \mathscr{M}(\mathscr{A})$ is finite.

Definition 3.32 *We say that $\mathscr{A}$ is a $\check{\mathscr{P}}$-sequence or of type $\check{\mathscr{P}}$ if*

(i) *$\mathscr{A}$ omits a finite set of primes $P = P(\mathscr{A})$,*

(ii) *for each $p \in P$, there exists a positive integer $\beta = \beta(p)$ such that $p^{\beta+1} \in \mathscr{A}$*

(iii) *$\mathscr{A}$ may contain some of the divisors of the number*

$$m(\mathscr{A}) = \prod\{p^{\beta(p)+1} : p \in P(\mathscr{A})\}.$$

For example, we might have

$$\mathscr{P}^{\vee} = (\mathscr{P} \setminus \{5, 7\}) \cup \{125, 35, 49\}$$

where $\mathscr{P}$ denotes the sequence of primes, in which case

$$\mathbf{N} \setminus \mathscr{M}(\mathscr{A}) = \{5, 7, 25\}.$$

We leave it as an exercise for the reader to prove the following result.

Theorem 3.33 *The Behrend sequence $\mathscr{A}$ is perfect if and only if it is of type $\mathscr{P}^{\vee}$.*

Let $\{b_1, b_2, \ldots, b_k\}$ be primitive, and $\mathscr{B}_i$ be Behrend for each $i \le k$. The sequence

$$\mathscr{A} = b_1\mathscr{B}_1 \cup b_2\mathscr{B}_2 \cup \cdots \cup b_k\mathscr{B}_k \tag{3.162}$$

is perfect if every $\mathscr{B}_i$ is of type $\mathscr{P}^{\vee}$. We have $\mathbf{t}(\mathscr{A}) = \mathbf{t}(\{b_1, b_2, \ldots, b_k\})$, and so (3.162) provides an example of a perfect sequence $\mathscr{A}$ for each prescribed value of $\mathbf{t}(\mathscr{A})$, provided this value is realized by some finite sequence. We do not know which rational numbers are values of $\mathbf{t}(\mathscr{B})$ for finite $\mathscr{B}$: if r is either a rational number outside this set, or is irrational, we do not have any example of a perfect sequence $\mathscr{A}$ such that $\mathbf{t}(\mathscr{A}) = r$. Notice that it is not necessary for $\mathscr{A}$ to be perfect that every $\mathscr{B}_i$ in (3.162) should be a $\mathscr{P}^{\vee}$-sequence. A counter example is

$$\mathscr{A} = 2\mathscr{P} \cup 3(\mathscr{P} \setminus \{2\}).$$

Here $\mathscr{M}(\mathscr{A})$ comprises all the integers divisible by 2 or 3 except the two integers themselves, and $\mathscr{P} \setminus \{2\}$ is not of type $\mathscr{P}^{\vee}$.

The second category of sequences to be considered here contains none which are perfect.

Theorem 3.34 *Let $\mathscr{A}$ contain infinite subsequence $\mathscr{A}'$ whose elements are pairwise coprime, and let $t(\mathscr{A}) > 0$. Then $\mathscr{A}$ is not perfect.*

Proof Let $\mathscr{A}' = \{a_1', a_2', a_3', \ldots\}$, and k be any positive integer. By the Chinese remainder theorem, we may find an integer n such that

$$n \equiv -j \pmod{a_j'}, \quad 1 \le j \le k$$

whence $n+1, n+2, \ldots, n+k \in \mathscr{M}(\mathscr{A})$ and

$$M(n+k) - M(n) = k. \tag{3.163}$$

By hypothesis $\mathscr{A}$ is not Behrend, and (3.163) contradicts (3.161) as k is unbounded. Hence $\mathscr{A}$ is not perfect.

We notice that an apparently weaker condition on $\mathscr{A}$ suffices for the above proof, namely that for every k we can find k elements of $\mathscr{A}$ which are pairwise coprime. In fact this implies the existence of $\mathscr{A}'$ as stated. For each prime p put

$$\mathscr{A}_p := \{a_i \in \mathscr{A} \, : \, a_i \equiv 0 \pmod p\}.$$

Then the 'weaker' condition implies that $\mathscr{A}$ is not equal to any finite union of the $\mathscr{A}_p$. We may therefore construct $\mathscr{A}'$ recursively as follows. Let $a_1' = a_1$, and suppose $a_1', a_2', \ldots, a_n'$ have been determined. Let $P(n)$ denote the set of prime factors of $a_1'a_2'\ldots a_n'$, and

$$a_{n+1}' = \min\left\{a_i \, : \, a_i \in \mathscr{A} \setminus \cup\{\mathscr{A}_p \, : \, p \in P(n)\}\right\}.$$

Theorem 3.34 is very simple and we expect that its hypotheses can be weakened. In view of Theorem 3.33 the condition $\mathbf{t}(\mathscr{A}) > 0$ is appropriate, and we may look for a relatively mild condition on $\mathscr{A}$ which implies that $\mathscr{M}(\mathscr{A})$ contains runs of consecutive elements of unbounded length, or less, namely that for every c there exist n and k such that

$$M(n+k) - M(n) > \mathbf{d}\mathscr{M}(\mathscr{A})k + c. \tag{3.164}$$

No verifiable condition of this sort, genuinely weaker than the one given above, has been found to date.

It is reasonable to suppose that perfect sequences are rare: there are none in the third category of sequences considered in this section. These are sparse, 'almost finite' sequences in which the a_i satisfy a rapid growth condition, viz, *for infinitely many n,*

$$a_{n+1} > \text{l.c.m.}[a_1, a_2, \ldots, a_n]. \tag{3.165}$$

When (3.165) holds, there is no interference between the multiples of $\{a_{n+1}, a_{n+2}, \ldots\}$ and the oscillations within the first period of $E(x, \mathscr{A}_n)$ where $\mathscr{A}_n = \{a_1, a_2, \ldots, a_n\}$. We may therefore apply the results of the earlier parts of this chapter to the finite sequences $\mathscr{A}_n$. Since we have so far been unable to settle the conjecture pertaining to (3.159) we have to impose a side condition on $\mathscr{A}$ restricting the exponents of the prime factors of the a_i. We write

$$v(a) = \max_p v_p(a) = \max\{\alpha \, : \, \exists p, \, p^\alpha | a\}. \tag{3.166}$$

Theorem 3.35 *Let $\mathscr{A}$ be a primitive sequence such that (3.165) holds infinitely often, and in addition*

$$\left(\frac{\pi^2}{6}\right)^{v(a_i)} = o(i), \quad i \to \infty. \tag{3.167}$$

Then $\mathscr{A}$ is not perfect.

Proof Let n be restricted to the sequence for which (3.165) holds, and consider $\mathscr{A}_n$. Plainly

$$\mathbf{d}\mathscr{M}(\mathscr{A}_n) < \mathbf{d}\mathscr{M}(\mathscr{A}). \tag{3.168}$$

We apply Theorem 3.29: in view of (3.167) we obtain from (3.151) that

$$\langle \mathscr{A}_n, \mathscr{A}_n \rangle \geq \xi(n) \tag{3.169}$$

where $\xi(n) \to \infty$. It follows from (3.9) and (3.169) that

$$\max_x \left|\overline{E}(x, \mathscr{A}_n)\right| \geq \left(\frac{\xi(n)}{12}\right)^{\frac{1}{2}}. \tag{3.170}$$

We recall that $\overline{E}(x, \mathscr{A}_n)$ is an odd function in the sense described in (3.7); it is of course periodic with period $[a_1, a_2, \dots, a_n]$. Hence there exists $x_n \leq [a_1, a_2, \dots, a_n]$ such that

$$\overline{E}(x_n, \mathscr{A}_n) \geq \left(\frac{\xi(n)}{12}\right)^{\frac{1}{2}}. \tag{3.171}$$

Let $M_n(x)$ denote the counting function of $\mathscr{M}(\mathscr{A}_n)$. By the definition (3.5) of $\overline{E}$, (3.171) implies

$$M_n(x_n) \leq [x_n] - \mathbf{t}(\mathscr{A}_n)x_n - \left(\frac{\xi(n)}{12}\right)^{\frac{1}{2}} \leq \mathbf{d}\mathscr{M}(\mathscr{A}_n)x_n - \left(\frac{\xi(n)}{12}\right)^{\frac{1}{2}}.$$

By (3.165), we have $M_n(x_n) = M(x_n)$, so in view of (3.168) this yields

$$M(x_n) - \mathbf{d}\mathscr{M}(\mathscr{A})x_n \leq -\left(\frac{\xi(n)}{12}\right)^{\frac{1}{2}} \tag{3.172}$$

whence $\mathscr{A}$ is not perfect. This completes the proof.

Notice that the strategy is to obtain an inequality opposite to (3.164).

4

Probabilistic group theory

4.1 Introduction

In this chapter we describe a method which can sometimes be used to evaluate the logarithmic density $\boldsymbol{\delta}\mathcal{M}(\mathcal{A})$ of the set of multiples of a given sequence $\mathcal{A}$. We may find that $\mathbf{d}\mathcal{M}(\mathcal{A})$ exists, that is $\mathcal{A}$ is Besicovitch, or that $\boldsymbol{\delta}\mathcal{M}(\mathcal{A}) = 1$, that is $\mathcal{A}$ is Behrend. In particular the method has been used to show that certain sequences $\mathcal{A}$ are Behrend when no other way of achieving this is known. An example will be the sequences $\mathcal{A}^*(t)$ defined in (1.81)–(1.83) which were used in Chapter 1 to establish that Theorem 1.7 is sharp. The content of Theorem 4.23 is that these sequences are Behrend.

The method is certainly rather specialized but, when it works, the information obtained is usually very strong and often best possible. It depends on a theorem of Erdös and Rényi (1965) and later developments of this; but the idea of applying such theorems in analytic number theory, and in particular to divisor problems, is due to Erdös (1965). Erdös proved the following result.

Theorem 4.1 *Let $F(x,D)$ denote the number of integers $n \le x$ such that n has at least one divisor in every congruence class prime to D. Then if $\alpha < \log 2$ is fixed and*

$$D < (\log x)^{\alpha} \tag{4.1}$$

we have

$$F(x,D) = x + o(x). \tag{4.2}$$

This is best possible in the sense that the conclusion becomes false if $\alpha \ge \log 2$. By the Hardy–Ramanujan theorem, for all but $o(x)$ integers

$n \le x$ we have

$$(\log x)^{\log 2-\varepsilon} < \tau(n) < (\log x)^{\log 2+\varepsilon} \tag{4.3}$$

and when n is counted by $F(x,D)$ we have $\tau(n) \ge \phi(D) \gg D/\log\log D$. Hence if $\alpha > \log 2$ almost all n will have too few divisors to fill the congruence classes. When $\alpha = \log 2$ we need more than (4.3) and we use the Erdös–Kac theorem (Erdös and Kac (1939), (1940), Elliott (1980), Theorem 12.3). As we shall need this later in the chapter we state it for convenience here.

Theorem 4.2 (Erdös–Kac) *Let $f(n)$ be a real-valued strongly additive function such that $|f(p)| \le 1$ always, and let*

$$A(x) = \sum_{p\le x} \frac{f(p)}{p}, \quad B(x) = +\sqrt{\sum_{p\le x} \frac{f(p)^2}{p}}, \tag{4.4}$$

moreover suppose $B(x) \to \infty$ as $x \to \infty$. Then

$$\operatorname{card}\{n : n \le x,\, f(n) - A(x) \le zB(x)\} = (G(z) + o(1))\, x \tag{4.5}$$

where

$$G(z) = \frac{1}{\sqrt{2\pi}} \int_{-\infty}^{z} e^{-u^2/2} du. \tag{4.6}$$

In particular, we have

$$\operatorname{card}\left\{n : n \le x,\, \frac{\omega(n) - \log\log x}{\sqrt{\log\log x}} \le z\right\} = (G(z) + o(1))\, x \tag{4.7}$$

and a similar result holds for $\Omega(n)$.

Of course Ω is not strongly additive but the deduction from (4.7) is easy: if $\xi(x) \to \infty$ then $\Omega(n) - \omega(n) < \xi(x)$ for all but $o(x)$ integers $n \le x$. Applying (4.7) to both ω and Ω, with $z = 0$, we may conclude that

$$\operatorname{card}\{n : n \le x,\, \tau(n) < (\log x)^{\log 2}\} \sim \frac{1}{2}x.$$

Since we may replace $(\log x)^{\log 2}$ above by $(\log x)^{\log 2} \exp\{o(\sqrt{\log\log x})\}$ we may conclude that

$$\limsup x^{-1} F(x,D) \le \frac{1}{2}, \quad D \sim (\log x)^{\log 2}.$$

In fact the limit exists and equals $\frac{1}{2}$. Erdös foresaw a refinement of his theorem, Theorem 4.10, which we state immediately after Theorem 4.8 below.

It may be helpful at this point to describe the general method as it would apply to show for example that a given $\mathscr{A}$ is Behrend. We begin by constructing an Abelian group $G = (G, +)$, (for consistency we shall always write the group operation as addition) together with a mapping $\rho : \mathbf{Z}^+ \to G$ with the following properties:

(i) for all $d, e \in \mathbf{Z}^+$ we have $\rho(de) = \rho(d) + \rho(e)$,
(ii) there exists a subset A of G such that $\rho(d) \in A$ implies $d \in \mathscr{A}$.

We then set out to prove that if we have a suitably large set of elements of G, say $g_1, g_2, \ldots, g_k$, then we can almost surely find a subset of them whose sum lies in A. There will be exceptional sets of k elements which possess no such subset (else we should have to take k much too large): for example the g_i may belong to a subgroup of G which does not intersect A. The number-theoretic part of our proof will be to establish that among the prime factors of almost all integers n we can find k distinct primes $p_1, p_2, \ldots, p_k$ such that the corresponding set of group elements $\rho(p_1), \rho(p_2), \ldots, \rho(p_k)$ are unexceptional in the sense described above. Evidently if $\rho(p_i) = g_i$, $1 \le i \le k$ and

$$g_{i_1} + g_{i_2} + \cdots + g_{i_h} \in A,$$

for some $i_1 < i_2 < \cdots < i_h \le k$ then by (i), (ii) above

$$p_{i_1} p_{i_2} \cdots p_{i_h} \in \mathscr{A}$$

and of course this product is a divisor of n.

The method requires G to be finite. In Theorem 4.1 it is the group of congruence classes prime to D under multiplication, but in many important applications the natural group which arises is $(\mathbf{R}/\mathbf{Z}, +)$. We have to impose a suitable finite group and make approximations, which unfortunately upsets property (i) above. In this sense Theorems 4.1 and 4.9 are *pure* and Theorems 4.16–4.23 *impure*.

We denote the order of G by N. In the following theorems a set of elements $g_1, g_2, \ldots, g_k$ will be called unexceptional, or simply *good*, if *every* element $g \in G$ has at least one representation

$$g_{i_1} + g_{i_2} + \cdots + g_{i_h} = g, \quad i_1 < i_2 < \cdots < i_h \le k$$

rather than just *some* elements $g \in A$. This requirement imposes the condition $2^k \ge N$ on k, simply in order that we can write down at least N formally distinct subsums of $g_1, g_2, \ldots, g_k$.

We now give the first theorem in probabilistic group theory, due to Erdös and Rényi (1965), of the type required for the work in this chapter and which Erdös used to prove Theorem 4.1. Rather than prove this, we state and prove a variant of it (the first of two variants presented in this chapter) which enables us to deal with a conjecture of Erdös which appeared in Astérisque (1979). We require the second variant to cope with the sequences $\mathscr{A}^*(t)$.

Theorem 4.3 (Erdös–Rényi) *Let G be an Abelian group of order N, and let k elements $g_1, g_2, \ldots, g_k$ be chosen from G, with repetitions allowed. For each $g \in G$ let $R(g) = R(g; g_1, g_2, \ldots, g_k)$ denote the number of representations of g in the form*

$$g_{i_1} + g_{i_2} + \cdots + g_{i_h} = g, \quad 1 \le i_1 < i_2 < \cdots < i_h \le k. \tag{4.8}$$

Then if $\delta > 0$ and

$$2^k \ge \frac{32}{\log 2}\delta^{-2} N \log N \tag{4.9}$$

we have $R(g) \ge 1$ for every g, for all but at most δN^k exceptional choices of the elements $g_1, g_2, \ldots, g_k$.

Erdös and Rényi used the terminology of probability theory in both the statement and the proof of their theorem; and from this point onward we follow their example. We say that the k elements $g_1, g_2, \ldots, g_k$ are chosen randomly and independently from G: the probability that g_i is any particular group element is $1/N$ and that $\{g_1, g_2, \ldots, g_k\}$ is any particular ordered k-tuple of elements is $1/N^k$. The conclusion becomes that (4.9) implies

$$\mathbf{Prob}(\min R(g) > 0) \ge 1 - \delta. \tag{4.10}$$

A useful alternative notation for sums like (4.8) is $\varepsilon_1 g_1 + \varepsilon_2 g_2 + \cdots + \varepsilon_k g_k$, where each $\varepsilon_i = 0$ or 1. This avoids the double suffices in (4.8) but strictly speaking is an abuse of notation: we mean $\varepsilon g = g$ if $\varepsilon = 1$, $\varepsilon g = 0_G$ if $\varepsilon = 0$.

It is natural to ask to what extent (4.9) is best possible with regard to the extra factors δ^{-2} and $\log N$ going beyond the obviously neccessary condition $2^k \ge N$. We notice that the Erdös-Rényi theorem imposes no condition on the structure of G (other than that it is Abelian), hence a (rather theoretical) approach to this question would be to attempt to determine the most unfavourable structure for G.

The *best* structure would appear to be a direct sum of cyclic groups of order 2. R.J.Miech (1967) observed that in this case, G may be regarded as a vector space over $\mathbf{Z}_2$: the elements g which are represented, that is

$R(g) > 0$, comprise a subgroup of G on which $R(g)$ is constant. Using these facts it may be shown that in this case the condition $2^k \geq \delta^{-1}N$ is sufficient to imply (4.10).

Erdös and Hall (1978) considered this problem, and imposed a condition on the structure of G which is satisfied by the groups with which we have to deal in this chapter. This is

Condition A *For each fixed $l \in \mathbf{Z}^+$, the number of elements of G of order l is $o(N)$.*

Theorem 4.4 (Erdös–Hall) *Let G satisfy condition A, and*

$$k = \frac{\log N}{\log 2} + O(1), \quad \lambda = \frac{2^k}{N}.$$

Then for each fixed $r \geq 0$ we have

$$\text{card}\{g \in G : R(g) = r\} \sim Ne^{-\lambda}\frac{\lambda^r}{r!}$$

with probability $\to 1$ as $N \to \infty$.

This shows that if, almost surely, every element is to be represented, then we must have $\lambda \to \infty$. It also suggests but does not prove that we need $\lambda > \log N$, just as in (4.9). If condition A is made quantitative, rather more can be said. For example if G is cyclic Erdös and Hall proved that $\lambda \gg \log\log N$ is necessary for every element to be represented with probability tending to 1.

In this chapter we have to consider two groups: the congruence classes prime to D under multiplication, and $(\mathbf{R}/\mathbf{Z}, +)$ which we approximate by a cyclic group of large order. In the latter case, condition A is trivial. In the former, we note that the number of elements of order l does not exceed $l^{\omega(D)+1}$. (The extra 1 occurs because the powers of 2 from 8 onwards do not have primitive roots.) We shall not apply Theorem 4.4 directly in what follows but draw the reader's attention to the behaviour of $R(g)$ as a Poisson variable: this is the best picture of what is happening that we have. We give some further information on the abstract theory in §4.5.

4.2 The Erdös–Rényi theorem: first variant

In this section we state and prove a variant of Theorem 4.3 which will enable us to settle a conjecture of Erdös (1979).

Theorem 4.5 *Let G be an Abelian group of even order N, and H be a subgroup of G of index 2. Let k elements $g_1, g_2, \ldots, g_k$ be chosen randomly and independently from G. Then the probability that $R(g; g_1, g_2, \ldots, g_k) \geq 1$ for every $g \in G$, conditional on the event that precisely r of $g_1, g_2, \ldots, g_k$ belong to H, is at least $1 - \delta$, provided $\delta > 0$, $r < k$ and*

$$2^k \geq \frac{128}{\log 2}\delta^{-2} N \log N. \tag{4.11}$$

This result is due to Erdös and Hall (1976a) (in which the right-hand side of (4.11) is claimed to be smaller; this error does not affect their application).

The condition $r < k$ is necessary: if $r = k$, $\varepsilon_1 g_1 + \varepsilon_2 g_2 + \cdots + \varepsilon_k g_k \in H$ always.

We emphasize at this point that in Theorem 4.3 there is no condition on the structure of G, and in Theorem 4.5 no condition other than N being even as stated, which is necessary and sufficient for the existence of a subgroup of index 2.

Proof of Theorem 4.5 We shall write **Prob**$(\ldots)$ and **Exp**$(\ldots)$ for the probability and expectation of the event in brackets; **Prob**$(\ldots|A)$ and **Exp**$(\ldots|A)$ are conditional on the event A. We call a set of k elements $g_1, g_2, \ldots, g_k$ good if $R(g; g_1, g_2, \ldots, g_k) \geq 1$ for every g: it is a (k, r)-set if exactly r of the elements belong to H. $\hat{G} = (\hat{G}, \chi)$ is the group of characters χ acting on G, of which χ_0 is the principal character, ($\chi_0(g) = 1$ for every $g \in G$) and χ_1 is such that $\chi_1(g) = 1$ if $g \in H$, $= -1$ else. $\overline{\chi}$ is the complex conjugate of χ, and the properties of these characters that we need are:

(i) each χ is a homomorphism from G into the unit circle, that is $|\chi(g)| = 1$ always, and $\chi(g + g') = \chi(g)\chi(g')$ for every $g, g' \in G$;

(ii) there are N distinct characters, and if we sum over all these we obtain

$$\frac{1}{N}\sum_{\chi} \overline{\chi}(g)\chi(g') = \begin{cases} 1 & \text{if } g = g' \\ 0 & \text{else;} \end{cases} \tag{4.12}$$

(iii) there holds the corresponding orthogonality relation

$$\frac{1}{N}\sum_{g} \overline{\chi}(g)\chi'(g) = \begin{cases} 1 & \text{if } \chi = \chi' \\ 0 & \text{else.} \end{cases} \tag{4.13}$$

In fact $\hat{G}$ is isomorphic to G apart from the change of group operation, but we shall not use this fact here. Notice that $\overline{\chi}(g) = \chi(-g)$.

Lemma 4.6 *Let l elements $g_1, g_2, \ldots, g_l$ be chosen randomly and independently from G, and*

$$M_0 = \operatorname{card}\{g : R(g; g_1, g_2, \ldots, g_l) = 0\}. \tag{4.14}$$

Then for any $s < l$ and $\eta > 0$ we have

$$\mathbf{Prob}_{(s)}(M_0 \geq \eta N^2 2^{-l}) < \eta^{-1} \tag{4.15}$$

where $\mathbf{Prob}_{(s)}$ denotes probability conditional on $\{g_1, g_2, \ldots, g_l\}$ being an (l, s)-set.

Proof of lemma Put $R(g) = R(g; g_1, g_2, \ldots, g_l)$. From (4.12) we have

$$R(g) = \frac{1}{N} \sum_{\chi} \overline{\chi}(g) \prod_{i=1}^{l} (1 + \chi(g_i)) \tag{4.16}$$

and we notice that the principal character χ_0 contributes $2^l/N$ to the right-hand of (4.16). This is the average value of $R(g)$ as g varies and we denote it by λ. If we subtract λ from both sides of (4.16) and employ the second orthogonality relation (4.13) we obtain

$$\sum_{g \in G} (R(g) - \lambda)^2 = \frac{1}{N} \sum_{\chi \neq \chi_0} \prod_{i=1}^{l} |1 + \chi(g_i)|^2. \tag{4.17}$$

For a given set of group elements $g_1, g_2, \ldots, g_l$ write

$$t = \sum_{i=1}^{l} \chi_1(g_i) \tag{4.18}$$

so that $t = 2s - l$ for (l, s)-sets. We consider the function of z, defined for $z \in \mathbf{C} \setminus \{0\}$ by

$$F(z) = \sum z^t \sum_{g \in G} (R(g; g_1, g_2, \ldots, g_l) - \lambda)^2 \tag{4.19}$$

where the outer sum is over all N^l sets of l elements and t is given by (4.18). We employ (4.17) to obtain

$$F(z) = \frac{1}{N} \sum_{\chi \neq \chi_0} \sum \prod_{i=1}^{l} z^{\chi_1(g_i)} |1 + \chi(g_i)|^2 \tag{4.20}$$

$$= \frac{1}{N} \sum_{\chi \neq \chi_0} \left\{ \sum_{g \in G} z^{\chi_1(g)} |1 + \chi(g)|^2 \right\}^l. \tag{4.21}$$

Notice that g has taken the place of g_i. We evaluate the inner sum. If $\chi \neq \chi_1$ it acts on H as a non-principal group character, and the inner sum is $N(z+z^{-1})$. If $\chi = \chi_1$ the sum is $2Nz$. Therefore

$$F(z) = \left(1 - \frac{2}{N}\right) N^l (z+z^{-1})^l + N^{l-1}(2z)^l. \tag{4.22}$$

Let us write $\sum_{(s)}$ for summation over all (l,s)-sets. Equating coefficients of z^{2s-l} in (4.19) and (4.22) yields, *provided* $s < l$,

$$\sum_{(s)} \sum_{g \in G} (R(g) - \lambda)^2 = \left(1 - \frac{2}{N}\right) \binom{l}{s} N^l. \tag{4.23}$$

Let us write $\mathbf{Exp}_{(s)}$ for expectation conditional on the event that $\{g_1, g_2, \ldots, g_l\}$ is an (l,s)-set. There are $\binom{l}{s}\left(\frac{N}{2}\right)^l$ such sets, and so (4.23) implies, for $s < l$,

$$\mathbf{Exp}_{(s)} \sum_{g \in G} (R(g) - \lambda)^2 = \left(1 - \frac{2}{N}\right) 2^l. \tag{4.24}$$

By (4.14), the sum on the left is at least $M_0 \lambda^2$ and we recall that $\lambda = 2^l/N$. Hence

$$\mathbf{Exp}_{(s)} M_0 \leq \left(1 - \frac{2}{N}\right) N^2 2^{-l}, \tag{4.25}$$

so that (4.15) follows by Markoff's inequality. This proves the lemma.

Let us write $k_0 = l + j$ where

$$j = \left[\frac{1}{\log 2} \log\left(\frac{\log N}{\log 2}\right)\right],$$
$$l = \frac{\log N}{\log 2} + u, \tag{4.26}$$

and u is a function of δ to be determined: it is positive, and such that $l \in \mathbf{Z}$. We begin by choosing just l random elements from G, appealing to Lemma 4.6 for an estimate of the probability that a significant proportion of the elements g already have $R(g; g_1, g_2, \ldots, g_l) > 0$. We then add j extra elements one at a time, claiming at each stage that the proportion of elements g represented is increased, with probability nearly 1. For this step we employ the following result.

Lemma 4.7 *Let H be a (fixed) subgroup of G of index 2 and B be any subset of G of cardinality M. Let an element g' be chosen, either randomly from H, or randomly from the complement of H (in either case the probability that g' is any particular group element is $2/N$). Let $M' = M'(g')$*

denote the number of elements g such that both $g \in B$, $g - g' \in B$. Then

$$\mathbf{Exp} M' \le \frac{2M^2}{N}. \tag{4.27}$$

Proof of lemma Let $S(g) = 0, 1$ or 2 according as neither, one or the other, or both of g and $g - g'$ lie in B. For group characters χ set

$$P(\chi, B) = \sum_{b \in B} \chi(b) \tag{4.28}$$

and let χ_1 be the non-principal character which is identically 1 on H, that is, principal as a character acting on H. We have

$$\begin{aligned} S(g) &= \frac{1}{N} \sum_{\chi} \overline{\chi}(g) P(\chi, B) \left(1 + \chi(g')\right) \\ \sum_{g \in G} S(g) &= P(\chi_0, B) \left(1 + \chi_0(g')\right) = 2M, \\ \sum_{g \in G} S(g)^2 &= \frac{1}{N} \sum_{\chi} |P(\chi, B)|^2 \, |1 + \chi(g')|^2 . \end{aligned} \tag{4.29}$$

Since we also have

$$\frac{1}{N} \sum_{\chi} |P(\chi, B)|^2 = M \tag{4.30}$$

we may subtract the second line of (4.29) from the third to obtain

$$M'(g') = \sum_{g \in G} \binom{S(g)}{2} = \frac{1}{N} \sum_{\chi} |P(\chi, B)|^2 \operatorname{Re} \chi(g'), \tag{4.31}$$

where Re denotes the real part. Suppose g' is to be chosen randomly from H. Then

$$\mathbf{Exp} \operatorname{Re} \chi(g') = \begin{cases} 1 \text{ if } \chi = \chi_0 & \text{or } \chi_1 \\ 0 & \text{else.} \end{cases} \tag{4.32}$$

On the other hand if g' is to be chosen from the complement of H we find that

$$\mathbf{Exp} \operatorname{Re} \chi(g') = \begin{cases} 1 & \text{if } \chi = \chi_0 \\ -1 & \text{if } \chi = \chi_1 \\ 0 & \text{else.} \end{cases} \tag{4.33}$$

In either case we have

$$\mathbf{Exp} M' \le \frac{1}{N} (|P(\chi_0, B)|^2 + |P(\chi_1, B)|^2) \tag{4.34}$$

and this implies (4.27) because $|P(\chi, B)| \leq M$ for any character. This proves the lemma.

To fix our ideas, let us assume that our final (k,r)-set is to be obtained by choosing the $k-r$ elements from the complement of H first: so when l elements have been chosen we must have an (l,s)-set, with $s = (l-k+r)^+$. Note that $r < k$ implies $s < l$. M_0 is as in (4.14) and A_0 is the event

$$M_0 < \eta N 2^{-u} = \eta N^2 2^{-l}. \tag{4.35}$$

Let A' denote the negation of the event A. Lemma 4.6 implies

$$\mathbf{Prob}_{(s)}(A_0') < \eta^{-1}. \tag{4.36}$$

Assume that A_0 occurs and choose a further element g_{l+1} from H or $G \setminus H$ as the case may be. For $0 \leq i \leq j$ put

$$\begin{aligned} s_i &= (l+i-k+r)^+ \\ B_i &= \{g : R(g; g_1, g_2, \ldots, g_{l+i}) = 0\} \\ M_i &= \operatorname{card} B_i. \end{aligned} \tag{4.37}$$

For $i \geq 1$, b_i is the set of elements g such that both g and $g - g_{l+i}$ belong to B_{i-1}, and Lemma 4.7 yields

$$\mathbf{Exp} M_i \leq \frac{2M_{i-1}^2}{N}. \tag{4.38}$$

So we have

$$\mathbf{Exp}_{(s_1)}(M_1 | A_0) < 2\eta^2 N 2^{-2u} \tag{4.39}$$

and we let A_1 be the event

$$M_1 < 4\eta^3 N 2^{-2u}. \tag{4.40}$$

We apply Markoff's inequality to (4.39) to obtain

$$\mathbf{Prob}_{(s_1)}(A_1' | A_0) < (2\eta)^{-1}. \tag{4.41}$$

Assume both A_0 and A_1 occur, denoting this event by A_0A_1. We pick g_{l+2} from H or $G \setminus H$, and apply Lemma 4.7 again to obtain

$$\mathbf{Exp}_{(s_2)}(M_2 | A_0A_1) < 32\eta^6 N 2^{-4u}. \tag{4.42}$$

Markoff gives

$$\mathbf{Prob}_{(s_2)}(A_2' | A_0A_1) < (4\eta)^{-1} \tag{4.43}$$

where A_2 is the event

$$M_2 < 128\eta^7 N 2^{-4u}. \tag{4.44}$$

We continue in this way. Thus A_i is the event

$$M_i < 2^{m_i}\eta^{2^{i+1}-1}N2^{-2^i u} \tag{4.45}$$

where $m_i = 3(2^i - 1) - i$. (The recurrence relation is $m_i = 2m_{i-1} + i + 1$, $m_0 = 0$.) Let us set

$$\eta = \frac{2}{\delta} \tag{4.46}$$

and fix u in such a way that l in (4.26) is an integer and $M_j < 1$. Let

$$\theta = \frac{1}{\log 2}\log\left(\frac{\log N}{\log 2}\right) - j \tag{4.47}$$

so that θ is the fractional part of the first quantity on the right, whence $0 \leq \theta < 1$. From (4.45) and the formula for m_j, we require

$$u \geq \frac{3(2^j - 1) - j}{2^j} + \frac{2^{j+1} - 1}{2^j \log 2}\log\frac{2}{\delta} + \frac{\log N}{2^j \log 2}$$

and, bearing in mind that we may have to increase u to make l an integer, we see that we can choose a suitable u satisfying

$$u < 4 + \frac{2}{\log 2}\log\frac{2}{\delta} + 2^\theta. \tag{4.48}$$

We put (4.26), (4.47) and (4.48) together to obtain

$$k_0 < \frac{\log N}{\log 2} + \frac{1}{\log 2}\log\left(\frac{\log N}{\log 2}\right) + \frac{2}{\log 2}\log\frac{2}{\delta} + 4 + 2^\theta - \theta$$

or, since $2^\theta - \theta \leq 1$,

$$2^{k_0} < \frac{128}{\log 2}\delta^{-2} N \log N. \tag{4.49}$$

The event A_j now includes $M_j < 1$, that is

$$R(g; g_1, g_2, \ldots, g_{k_0}) \geq 1 \text{ for every } g \in G. \tag{4.50}$$

If k satisfies (4.11) then $k > k_0$ is also sufficient. We have

$$\begin{aligned} \mathbf{Prob}_{(r)}(A'_j) \leq\ & \mathbf{Prob}_{(r)}(A'_j | A_0 A_1 \ldots A_{j-1}) \\ & + \mathbf{Prob}_{(s_{j-1})}(A'_{j-1} | A_0 A_1 \ldots A_{j-2}) \\ & \vdots \\ & + \mathbf{Prob}_{(s_1)}(A'_1 | A_0) \\ & + \mathbf{Prob}_{(s)}(A'_0) \end{aligned} \tag{4.51}$$

and by (4.36), (4.41), (4.43) etc., the right-hand side does not exceed

$$\frac{1}{\eta} + \frac{1}{2\eta} + \frac{1}{4\eta} + \cdots + \frac{1}{2^j\eta} < \frac{2}{\eta} = \delta.$$

Thus A_j occurs with probability at least $1 - \delta$. This completes the proof of Theorem 4.5.

We are now in a position to deal with one of the conjectures contained in Erdös (1979). Set

$$\mathscr{A}(D, \alpha) = \{d : d \equiv 1 (\mathrm{mod}\ D),\ 1 < d \leq \exp D^\alpha\}, \tag{4.52}$$

and

$$A(D, \alpha) = \mathbf{d}\mathscr{M}\left(\mathscr{A}(D, \alpha)\right). \tag{4.53}$$

Trivially $A(D, \alpha) < D^{-1}(D^\alpha + 1) \to 0$ as $D \to \infty$ if $\alpha < 1$. Erdös stated that he could prove that

$$A(D, 1) \to 0 \text{ as } D \to \infty. \tag{4.54}$$

He conjectured the existence of a number α_0, exceeding 1 and such that both

$$\lim_{D\to\infty} A(D, \alpha) = 0, \quad \text{if} \quad \alpha < \alpha_0 \tag{4.55}$$

$$\lim_{D\to\infty} A(D, \alpha) = 1, \quad \text{if} \quad \alpha > \alpha_0. \tag{4.56}$$

We shall prove that (4.55) and (4.56) hold, with $\alpha_0 = (\log 2)^{-1} = 1.442695\ldots$, obtaining slightly stronger results in both cases, in which $\alpha \to \alpha_0 - 0$, $\alpha_0 + 0$ (respectively) as $D \to \infty$.

Theorem 4.8 *Let $A(D, \alpha)$ be as defined above, and*

$$\alpha < \frac{1}{\log 2} - c\sqrt{\frac{\log\log\log D}{\log D}} \tag{4.57}$$

where $c > (\frac{2}{\log 2})^{\frac{1}{2}}$ is fixed. Then $A(D, \alpha) \to 0$ as $D \to \infty$.

Theorem 4.9 *Let $A^*(D, \alpha)$ denote the density of the integers n such that for every a prime to D, n has a divisor $d \equiv a (\mathrm{mod}\ D)$ in the interval*

$$\exp\exp\left(\sqrt{\log D}\right) < d \leq \exp D^\alpha. \tag{4.58}$$

Then if

$$\alpha = \frac{1}{\log 2} + \xi(D)(\log D)^{-1/2} \tag{4.59}$$

where $\xi(D) \to \infty$ as $D \to \infty$, we have $A^*(D, \alpha) \to 1$.

This is a suitable point to state the refinement of Erdös' Theorem 4.1 mentioned earlier.

Theorem 4.10 (Erdös and Hall (1976)) *Let $F(x, D)$ be as in Theorem 4.1 and*

$$D = 2^{\log\log x + (X + o(1))\sqrt{\log\log x}}. \tag{4.60}$$

Then

$$F(x, D) = (x + o(x))\frac{1}{\sqrt{2\pi}}\int_X^\infty e^{-u^2/2}du \tag{4.61}$$

This result had been conjectured by Erdös (1965) and the idea has been a leitmotif in his work in this area. The parameter α in (4.1) has a *threshold* at $\log 2$, by which we mean that the arithmetical function $F(x, D)$ satisfies different asymptotic formulae on either side of the threshold, in this case

$$\begin{aligned} F(x, D) \quad &\sim x \quad \alpha < \log 2 \\ &= o(x), \quad \alpha > \log 2). \end{aligned}$$

Erdös' idea is to allow α to converge to the threshold and to introduce a further parameter, say X, and a probability distribution function $\Xi(X)$ to give an asymptotic formula on the threshold. $\Xi(X)$ provides a smooth transition or homotopy.

Conjecture 4.11 *There exists a probability distribution function $\Xi(X)$ such that if $A(D, \alpha)$ is as defined above and $D \to \infty$, $\alpha \to 1/\log 2$ together in such a way that*

$$\alpha = \frac{1}{\log 2} + (X + o(1))(\log D)^{-1/2} \tag{4.62}$$

then

$$\lim A(D, \alpha) = \Xi(X), \quad X \in \mathbf{R}. \tag{4.63}$$

Theorems 4.9 and 4.10 share the property that apparently Theorem 4.3 provides insufficient input from probabilistic group theory to prove them, because of a shortfall in our knowledge about the distribution of primes (mod D), specifically about the possible Siegel zero (mod D). Theorem 4.4 bypasses the difficulty such a zero raises.

Proof of Theorem 4.8

Lemma 4.12 (Erdös (1946)) *Let $\varepsilon > 0$ be fixed, and $t_o \to \infty$. Then for almost all integers n, and every t, $t_o < t \leq n$, we have*

$$|\Omega(n, t) - \log\log t| < \sqrt{(2 + \varepsilon)\log_2 t \log_4 t}. \tag{4.64}$$

Erdös indicated that this could be derived from the law of the iterated logarithm in probability theory. There is a number theoretic proof in Hall and Tenenbaum (1988), Chapter 1 together with references to complete probabilistic proofs, e.g. Kubilius (1964).

We shall apply Lemma 4.12 with $t_0 = D$. $\mathscr{G}$ is the sequence of integers n for which (4.64) holds, and

$$\mathscr{B}' = \mathscr{B}(\mathscr{A}(D,\alpha)) \cap \mathscr{G}. \tag{4.65}$$

Since $\mathscr{A}(D,\alpha)$ is finite, its set of multiples $\mathscr{B}$ is a finite union of arithmetic progressions and as such has asymptotic density equal to its logarithmic density. So it will be sufficient to show that $\delta\mathscr{B}' = o(1)$ or

$$\sum_{n\in\mathscr{B}'} n^{-\sigma} = o\left((\sigma-1)^{-1}\right) \tag{4.66}$$

uniformly for $\sigma > 1$; in what follows limit processes are as $D \to \infty$. We set

$$f(t) = \sqrt{(2+\varepsilon)\log_2 t \log_4 t} \tag{4.67}$$

where ε is to be fixed later. We have, for $n \in \mathscr{G}$ and $y \leq 1$,

$$\begin{aligned} \tau(n,\mathscr{A}) &\leq \sum_{d|n,\, d\in\mathscr{A}} y^{\Omega(n,d)-\log\log d - f(d)} \\ &\leq Z \sum_{d|n,\, d\in\mathscr{A}} y^{\Omega(n,d)}(\log d)^{-\log y} \end{aligned} \tag{4.68}$$

where

$$Z = \exp\{(-\log y)\sqrt{(2+2\varepsilon)\alpha\log D\log\log\log D}, \tag{4.69}$$

assuming as we may that D is so large that $(2+\varepsilon)\log\log\log D^{\alpha} \leq (2+2\varepsilon)\log\log\log D$. This implies that

$$\begin{aligned} \sum_{n\in\mathscr{B}'} n^{-\sigma} &\leq \sum_{n\in\mathscr{G}} n^{-\sigma}\tau(n,\mathscr{A}) \\ &\leq Z\sum_{n=1}^{\infty} n^{-\sigma} \sum_{d|n,\, d\in\mathscr{A}} y^{\Omega(n,d)}(\log d)^{-\log y} \\ &\leq Z\sum_{d\in\mathscr{A}} d^{-\sigma} y^{\Omega(d)}(\log d)^{-\log y} \sum_{m=1}^{\infty} m^{-\sigma} y^{\Omega(m,d)} \\ &\leq Z\zeta(\sigma)\sum_{d\in\mathscr{A}} d^{-\sigma} y^{\Omega(d)}(\log d)^{-\log y} g(y,\sigma,d) \end{aligned} \tag{4.70}$$

where

$$g(y,\sigma,d) = \prod_{p\le d}\left(1-\frac{y}{p^\sigma}\right)^{-1}\left(1-\frac{1}{p^\sigma}\right) \ll \exp\left\{(y-1)\sum_{p\le d}p^{-\sigma}\right\}. \tag{4.71}$$

By the mean value theorem,

$$\sum_{p\le d}p^{-\sigma} \ge \sum_{p\le d}p^{-1}-(\sigma-1)\sum_{p\le d}p^{-1}\log p \ge \log\log d-(\sigma-1)\log d+O(1), \tag{4.72}$$

uniformly for all d and $1<\sigma\le 2$. We insert this into (4.71) to obtain

$$g(y,\sigma,d)\ll d^{\sigma-1}(\log d)^{y-1} \tag{4.73}$$

and we substitute this into (4.70) to obtain

$$\sum_{n\in\mathscr{B}'}n^{-\sigma} \ll \frac{Z}{\sigma-1}\sum_{d\in\mathscr{A}}d^{-1}y^{\Omega(d)}(\log d)^{y-1-\log y} \ll \frac{Z}{\sigma-1}D^{\alpha(y-1-\log y)}\sum_{d\in\mathscr{A}}d^{-1}y^{\Omega(d)} \tag{4.74}$$

provided $y-1-\log y\ge 0$, because for all $d\in\mathscr{A}$ we have $\log d\le D^\alpha$. We apply an estimate of Shiu (1980) to the sum on the right. This yields

$$\sum_{d\in\mathscr{A}}d^{-1}y^{\Omega(d)}\ll D^{\alpha y-1} \tag{4.75}$$

so that we now have

$$\sum_{n\in\mathscr{B}'}n^{-\sigma}\ll Z(\sigma-1)^{-1}D^{\alpha(2y-1-\log y)-1}. \tag{4.76}$$

We choose $y=\frac{1}{2}$ to minimize the exponent of D (checking that $y-1-\log y\ge 0$) and we see that (4.66) will follow from

$$ZD^{\alpha\log 2-1}=o(1),\quad D\to\infty. \tag{4.77}$$

Let us put $\alpha=(\log 2)^{-1}-\eta$, and refer to (4.69) for the definition of Z. For (4.77), it will be sufficient to have

$$\eta\log D-\sqrt{\left(\frac{2+2\varepsilon}{\log 2}\right)\log D\log\log\log D}\to+\infty \tag{4.78}$$

and this holds, by (4.57), if we choose $\varepsilon<\varepsilon_0(c)$. This completes the proof of Theorem 4.8.

Proof of Theorem 4.9 Let $D \geq 2$ and $I(D)$ denote the interval $(u, v]$ where

$$u = \exp\exp(\sqrt{\log D}), \quad v = \frac{1}{2}\exp\left(\frac{D^\alpha}{2\alpha\log D}\right). \tag{4.79}$$

By hypothesis $\alpha > 1/\log 2$. We have

$$Dv^{2\alpha\log D} < \exp D^\alpha \tag{4.80}$$

and if D is sufficiently large, as we may assume, $v > u > D$. We construct divisors of n in every congruence class prime to D using only prime factors of n lying in $I(D)$ and possibly one extra prime factor less than D. We show that we are able to do this for integers n belonging to a sequence of asymptotic density $\to 1$ as $D \to \infty$. We begin with the following assumptions about n.

(i) the prime factors $\in I(D)$ are distinct,
(ii) n possesses a prime factor $q < D$, $q \nmid D$.
(iii) $\alpha\log D - \xi_1(D)\sqrt{\log D} < \omega(n; u, v) \leq 2\alpha\log D$.

where $\xi_1(D) \to \infty$ as $D \to \infty$. The integers n which do not satisfy either (i) or (ii) have density

$$< \frac{D}{\varphi(D)}\prod_{p<D}\left(1 - \frac{1}{p}\right) + \sum_{p>u}\frac{1}{p^2} = o(1) \text{ as } D \to \infty,$$

and those failing to satisfy (iii) have density $\ll \xi_1(D)^{-2} = o(1)$ by Turán's familiar variance argument. Notice that a divisor d of n constructed in the manner described above satisfies

$$d < Dv^{\omega(n;u,v)} < \exp D^\alpha \tag{4.81}$$

by (4.80). Next, we strengthen condition (i) above to

(i′) the prime factors $\in I(D)$ are incongruent (mod D).

For each l prime to D define

$$Z(l) = \sum\{p^{-1} : u < p \leq v, p \equiv l (\text{mod } D)\}. \tag{4.82}$$

We deduce from (4.91) below that

$$Z(l) \ll \frac{1}{\varphi(D)}\log\left(\frac{\log v}{\log u}\right)$$

and the integers n which fail condition (i′) have density

$$\leq \sum_{(l,d)=1} Z(l)^2 \ll \frac{1}{\varphi(D)} \log^2\left(\frac{\log v}{\log u}\right) = o(1),$$

noticing that because $u > D$ the primes in $I(D)$ do not divide D.

For a particular n, (we assume henceforth that conditions (i′), (ii), (iii) are satisfied), put $\omega(n; u, v) = t$ and let $p_1, p_2, \ldots, p_t$ be the prime factors in $I(D)$. Put $p_i \equiv l_i(\text{mod} D)$, $1 \leq i \leq t$ so that the congruence classes l_i are prime to D and distinct. We say that $\{l_1, l_2, \ldots, l_t\}$ is a *good set* of congruence classes if for every l prime to D we can find exponents $\varepsilon_1, \varepsilon_2, \ldots, \varepsilon_t$ such that $\varepsilon_i = 0$ or 1, $1 \leq i \leq t$ and

$$l_1^{\varepsilon_1} l_2^{\varepsilon_2} \cdots l_t^{\varepsilon_t} \equiv l(\text{mod } D). \tag{4.83}$$

Else we call $\{l_1, l_2, \ldots, l_t\}$ a *bad set*. Let us show that if it is a good set we can construct divisors of n of the shape required. We begin with the case $l \not\equiv 1$ (mod D), when we simply solve (4.83) for the ε_i which of course cannot be all zero, and put

$$d = p_1^{\varepsilon_1} p_2^{\varepsilon_2} \cdots p_t^{\varepsilon_t}. \tag{4.84}$$

We have

$$d | n, \ d \equiv l(\text{mod } D), \quad u < d < \exp D^{\alpha} \tag{4.85}$$

using (4.81). When $l \equiv 1$ (mod D) we have to take care to ensure $d > 1$. By assumption (ii) above, n has a prime factor q less than, and prime to D. Clearly $q \not\equiv 1$ (mod D) so that if l' is such that $ql' \equiv 1$ (mod D) we have $l' \not\equiv 1$ (mod D) and we replace l by l' in (4.83) and put

$$d = q p_1^{\varepsilon_1} p_2^{\varepsilon_2} \cdots p_t^{\varepsilon_t}. \tag{4.86}$$

Since $l' \not\equiv 1$ (mod D) the ε_i are not all zero and so $d > u$ as before; in fact d satisfies (4.85) as required.

It remains to show that the density of the integers n which satisfy conditions (i′), (ii) and (iii) above but are associated with bad sets $\{l_1, l_2, \ldots, l_t\}$ is $o(1)$ as $D \to \infty$. Let $\{l_1, l_2, \ldots, l_t\}$ be a bad set of congruence classes (mod D) and $p_1, p_2, \ldots p_t$ be primes in $I(D)$ such that $p_i \equiv l_i(\text{modD})$, $1 \leq i \leq t$. The density of the integers n with precisely these prime factors in $I(D)$ is

$$(p_1 p_2 \cdots p_t)^{-1} \prod_{u<p\leq v} \left(1 - \frac{1}{p}\right) \tag{4.87}$$

and so the density of the integers n associated with this particular bad set is

$$\prod_{u<p\leq v}\left(1-\frac{1}{p}\right)\prod_{i=1}^{t}Z(l_i) \tag{4.88}$$

with $Z(l)$ as in (4.82).

Now let $\sum^{(t)}$ denote summation over all bad sets of t distinct congruence classes (mod D), counting different orders of the classes multiply. Since each bad set of t distinct classes has $t!$ (plainly bad) permutations, and we are not concerned about the order of the primes in (4.87), the density of the exceptional integers n associated with these sets is

$$\prod_{u<p\leq v}\left(1-\frac{1}{p}\right)\frac{1}{t!}\sum^{(t)}\prod_{i=1}^{t}Z(l_i). \tag{4.89}$$

We should explain at this point that we have counted the bad sets in this way because we are going to apply Theorem 4.4. In the probabilistic model adopted in this book, we choose ordered sets of not necessarily distinct group elements $\{g_1, g_2, \ldots, g_k\}$: if N denotes the order of the group then the number of choices of such a set is N^k and the probability that a particular set is chosen is $1/N^k$. It would be possible to work with a model in which unordered sets of distinct group elements were chosen: the formulae would be slightly more complicated and more importantly we should be concerned at every turn with independence. It seems both easier and clearer to proceed in the present manner, accepting a little clumsiness at this stage of the argument.

We want an upper bound for the sum

$$\prod_{u<p\leq v}\left(1-\frac{1}{p}\right)\sum_{t}{}^{*}\frac{1}{t!}\sum^{(t)}\prod_{i=1}^{t}Z(l_i) \tag{4.90}$$

in which * denotes that $t=\omega(n;u,v)$ is restricted by condition (iii) above. We therefore require an approximation to $Z(l)$ and it is neccessary to take account of the influence of a Siegel zero (mod D) if such exists: in our present state of knowledge of these zeros this may not be negligible and cannot be treated as an error term. We refer the reader to Prachar (1957), Chapter IV and in particular to Satz 7.3 (p.136), from which we deduce that for $(l,d)=1$ we have

$$Z(l)=\frac{A}{\varphi(D)}(1-\chi_1(l)B)+O\left(\frac{1}{\varphi(D)\log u}\right) \tag{4.91}$$

where

$$A = \log\left(\frac{\log v}{\log u}\right), \quad AB = \frac{1}{\beta}\int_u^v \frac{w^{\beta-2}dw}{\log w}. \tag{4.92}$$

Here $\beta \in (0,1)$ is the (supposed) Siegel zero of the Dirichlet L-function $L(s,\chi_1)$. The Dirichlet character χ_1 must be real and non-principal, i.e. $\chi_1^2 = \chi_0$, $\chi_1 \neq \chi_0$; if β does not exist we put $B = 0$. Since $\chi_1(l) = \pm 1$ taking each value on half the congruence classes prime to D, the effect of the Siegel zero is to increase half the quantities $Z(l)$ at the expense of the other half. These perturbations may not be small and we have to show that their effects cancel.

We apply Theorem 4.5. Let $\widehat{G}$ be the group of congruence classes prime to D under multiplication and $\widehat{H}$ the subgroup of index 2 on which the Dirichlet character χ_1 in (4.91) takes the value $+1$. Now G is isomorphic to $\widehat{G}$, with the group operation $+$, H corresponding to $\widehat{H}$ under the isomorphism. We say that $\{l_1, l_2, \ldots, l_t\}$ is a (t,r)-set if exactly r of the congruence classes l_i belong to $\widehat{H}$, and $\sum^{(t,r)}$ denotes summation over bad (t,r)-sets. The l_i are ordered and we do not demand from now on that they should be distinct. We deduce from (4.91) that the sum (4.90) is

$$\ll \frac{\log u}{\log v}\sum_t{}^{*}\frac{\varphi(D)^{-t}}{t!}\sum_{r\leq t}\sum{}^{(t,r)}U^r V^{t-r} \tag{4.93}$$

where, A and B being as in (4.92),

$$U = A(1-B)+\theta, \quad V = A(1+B)+\theta, \quad \theta \ll \frac{1}{\log u}. \tag{4.94}$$

Put $N = \varphi(D)$, $k = t$, $\delta = \delta(D)$ in Theorem 4.5. Here $\delta(D) = o(1)$ will be specified later. Provided condition (4.11) is satisfied, that is

$$2^t \geq \frac{128}{\log 2}\delta(D)^{-2}\varphi(D)\log\varphi(D) \tag{4.95}$$

we may deduce that the number of bad (t,r)-sets is

$$\leq \begin{cases} \delta(D)\binom{t}{r}\left(\frac{\varphi(D)}{2}\right)^t & \text{if } r < t, \\ \left(\frac{\varphi(D)}{2}\right)^t & \text{if } r = t, \end{cases} \tag{4.96}$$

whence the sum over r in (4.93) is

$$\begin{aligned} &< \delta(D)\left(\frac{\varphi(D)}{2}\right)^t (U+V)^t + \left(\frac{\varphi(D)}{2}\right)^t U^t \\ &< \varphi(D)^t\{\delta(D)(A+\theta)^t + 2^{-t}(A - AB + \theta)^t\}, \end{aligned} \tag{4.97}$$

and the whole sum is

$$\begin{aligned} &\ll \frac{\log u}{\log v}\left\{\delta(D)e^{A+\theta} + e^{(A+\theta)/2}\right\} \\ &\ll \delta(D) + \left(\frac{\log u}{\log v}\right)^{\frac{1}{2}} = o(1), \end{aligned} \tag{4.98}$$

as $D \to \infty$, using (4.94) and then (4.92). We have to check that (4.95) holds, with (say) $\delta(D) = 1/\log D$. Plainly it would be sufficient to have

$$t \geq 8 + \frac{3}{\log 2}\log\log D + \frac{\log D}{\log 2} \tag{4.99}$$

whereas by condition (iii) at the beginning of our proof we have $t > \alpha \log D - \xi_1(D)\sqrt{\log D}$, where we must choose $\xi_1(D) \to \infty$. We put $\xi_1(D) = \frac{1}{2}\xi(D)$ with $\xi(D)$ as in (4.59), (the definition of α) and (4.99) holds for sufficiently large D. Thus (4.98) holds, and the density of the integers n without a divisor of the type specified is $o(1)$ as $D \to \infty$, that is $A^*(D,\alpha) \to 1$. This completes the proof of Theorem 4.9.

The argument presented above to prove Theorem 4.9 may be extended to give a lower bound for $A(D,\alpha)$ when, as in (4.62), we have

$$\alpha = \frac{1}{\log 2} + (X + o(1))(\log D)^{-1/2}, \quad X \in \mathbf{R}. \tag{4.100}$$

We require u to be smaller, such that $\log\log u = o\left(\sqrt{\log D}\right)$; a suitable choice is

$$u = \exp\exp((\log D)^{1/3}), \tag{4.101}$$

with v as in (4.79). We assume that the integer n satisfies conditions (i′) and (ii) as before, the exceptional n having density $o(1)$, and Theorem 4.5 shows that the integers n which also satisfy, (see 4.99),

$$\omega(n;u,v) \geq 8 + \frac{3}{\log 2}\log\log D + \frac{\log D}{\log 2} \tag{4.102}$$

but do not have divisors (of the construction described in the proof of Theorem 4.9) in every congruence class prime to D, have density $o(1)$. We strengthen condition (iii) to:

(iii′) $\log(\frac{\log v}{\log u}) + Y\sqrt{\log(\frac{\log v}{\log u})} < \omega(n;u,v) \leq 2\alpha \log D$,

which implies

$$\omega(n;u,v) \geq \alpha \log D + Y\alpha^{1/2}(\log D)^{1/2} + O((\log D)^{1/3}), \tag{4.103}$$

and we can choose $Y = Y(X,D)$ in such a way that (4.100) and (4.103) imply (4.102), moreover

$$Y = \frac{-X}{\sqrt{\log 2}} + o(1), \quad D \to \infty, \tag{4.104}$$

By the Erdös–Kac Theorem, (see §4.1, also Elliott (1980) Theorem 12.3) the integers n such that (iii′) holds have density

$$\sim \frac{1}{\sqrt{2\pi}} \int_{Y}^{\infty} e^{-w^2/2} dw$$

whence

$$\liminf_{D\to\infty} A(D,\alpha) \geq \frac{1}{\sqrt{2\pi}} \int_{-X/\sqrt{\log 2}}^{\infty} e^{-w^2/2} dw. \tag{4.105}$$

This is a partial result in the direction of our conjecture (4.63). The upper bound needs a new idea since the first step would be to remove the iterated logarithm on the right of (4.57), which arises from Lemma 4.12. (This latter result is of course, in itself, sharp.)

4.3 The Erdös–Rényi theorem: second variant

We now state and prove the second variant of Theorem 4.3, which we need for our subsequent proof that $\mathscr{A}^*(t)$ is a Behrend sequence.

Theorem 4.13 *Let* $\alpha \in (0, \frac{1}{2}]$ *and* $\delta > 0$ *be given and let*

$$\mu(\alpha) = -\alpha \log \alpha - (1-\alpha)\log(1-\alpha), \tag{4.106}$$

$$\kappa(\alpha) = \frac{1}{2} - \frac{\log(\alpha(1-\alpha))}{\log 2} \tag{4.107}$$

$$\xi(\alpha,\delta) = 2\log\frac{1}{\delta} + \frac{1}{2}\log\frac{1}{\mu(\alpha)} + 7. \tag{4.108}$$

Let $N \geq \max(N_0(\alpha), \delta^{-10})$ *where* $N_0(\alpha)$ *is sufficiently large, and* $G = (G,+)$ *be an Abelian group of order* N. *Let* k *be an integer such that*

$$k \geq \frac{1}{\mu(\alpha)}\{\log N + \kappa(\alpha)\log\log N + \xi(\alpha,\delta)\}, \tag{4.109}$$

and let h *be any integer satisfying*

$$\min(h, k-h) \geq \alpha k. \tag{4.110}$$

Then if k *elements* $g_1, g_2, \ldots, g_k$ *are chosen randomly and independently from* G, *with probability exceeding* $1-\delta$ *every element* $g \in G$ *may be written as a sum of precisely* h *of these* k *elements.*

It is important that h be specified in advance of the choice of $g_1, g_2, \ldots, g_k$. When $\alpha = \frac{1}{2}$, k must be even and $h = \frac{k}{2}$. We then have $\mu = \log 2$, $\kappa = \frac{5}{2}$, so that (4.109) is not much stronger than the corresponding condition (4.8) in Theorem 4.3, and we do get some extra information.

The factor $k^{-1}\min(h, k-h)$ is crucial. We can only write down $\binom{k}{h}$ sub-sums of h elements and this must be at least N. It is a straightforward calculation with Stirling's formula, or more swiftly, with the inequality

$$\binom{k}{h} \leq \min_{0<x<1} x^{-h}(1-x)^{h-k} = \frac{k^k}{h^h(k-h)^{k-h}} \tag{4.111}$$

to see that the factor $\mu(\alpha)^{-1}$ in (4.109) is best possible. We need two lemmas corresponding to Lemmas 4.6, 4.7.

Lemma 4.14 *Let l elements $g_1, g_2, \ldots, g_l$ be chosen randomly and independently from G, and*

$$M_0 = \text{card}\{g : R^{(j)}(g; g_1, g_2, \ldots, g_l) = 0\}. \tag{4.112}$$

Then for any $j \leq l$ and $\eta > 0$ we have

$$\textbf{Prob}\left(M_0 \geq \frac{\eta N^2}{\binom{l}{j}}\right) < \eta^{-1}. \tag{4.113}$$

Proof of lemma Put $R^{(j)}(g) = R^{(j)}(g; g_1, g_2, \ldots, g_l)$. For any complex number z we have, from (4.12),

$$\sum_{j=0}^{l} R^{(j)}(g)z^j = \frac{1}{N}\sum_{\chi} \overline{\chi}(g)\prod_{i=1}^{l}(1+\chi(g_i)z) \tag{4.114}$$

whence

$$\sum_{h=0}^{l}(R^{(j)}(g) - \lambda^{(j)})z^j = \frac{1}{N}\sum_{\chi\neq\chi_0} \overline{\chi}(g)\prod_{i=1}^{l}(1+\chi(g_i)z) \tag{4.115}$$

where

$$\lambda^{(j)} = \frac{1}{N}\binom{l}{j} \tag{4.116}$$

is the average value of $R^{(j)}(g)$ as g varies. We put $z = re^{i\theta}$ and apply

Parseval's formula, which yields

$$\sum_{j=0}^{l}(R^{(j)}(g)-\lambda^{(j)})^2 r^{2j}$$
$$=\frac{1}{N^2}\sum_{\chi_1\neq\chi_0}\bar{\chi}_1(g)\sum_{\chi_2\neq\chi_0}\chi_2(g)\frac{1}{2\pi}\int_0^{2\pi}\prod_{i=1}^{l}(1+\chi_1(g_i)z)(1+\bar{\chi}_2(g_i)\bar{z})d\theta \tag{4.117}$$

and sum over g to obtain

$$\sum_{g\in G}\sum_{j=0}^{l}(R^{(j)}(g)-\lambda^{(j)})^2 r^{2j}=\frac{1}{N}\sum_{\chi\neq\chi_0}\frac{1}{2\pi}\int_0^{2\pi}\prod_{i=1}^{l}|1+\chi(g_i)z|^2 d\theta. \tag{4.118}$$

For non-principal characters χ we have, for each i,

$$\mathbf{Exp}\,|1+\chi(g_i)z|^2=1+r^2, \tag{4.119}$$

and because the g_i are independent, (4.118) and (4.119) imply that

$$\mathbf{Exp}\sum_{g\in G}\sum_{j=0}^{l}(R^{(j)}(g)-\lambda^{(j)})^2 r^{2j}=\left(1-\frac{1}{N}\right)(1+r^2)^l. \tag{4.120}$$

We equate coefficients of r^{2j} in (4.120) to obtain, for each $j\leq l$,

$$\mathbf{Exp}\sum_{g\in G}(R^{(j)}(g)-\lambda^{(j)})^2=\left(1-\frac{1}{N}\right)\binom{l}{j}. \tag{4.121}$$

The sum on the left-hand side is at least $M_0N^{-2}\binom{l}{j}^2$, so that we have

$$\mathbf{Exp}M_0<N^2\binom{l}{j}^{-1} \tag{4.122}$$

and Markoff's inequality gives (4.113). This proves the lemma.

Lemma 4.15 *Let B be an arbitrary subset of G of cardinality M. Let f and f' be chosen randomly and independently from G, and let $M'=M'(B;f,f')$ denote the number of elements $g\in G$ such that both $g-f, g-f'$ belong to B. Then*

$$\mathbf{Exp}M'=\frac{M^2}{N}. \tag{4.123}$$

Proof of lemma We define, for each group character χ,

$$P(\chi,B)=\sum_{b\in B}\chi(b), \tag{4.124}$$

and for $g, g, f' \in G$,

$$S(g; f, f') = \frac{1}{N} \sum_{\chi} \overline{\chi}(g) \left(\chi(f) + \chi(f')\right) P(\chi, B) \qquad (4.125)$$

so that $S = 2, 1$ or 0 according as both, exactly one, or neither of $g - f, g - f' \in B$. Hence

$$M' = \sum_{g \in G} \binom{S(g; f, f')}{2}. \qquad (4.126)$$

Now

$$\sum_{g \in G} S(g; f, f') = 2M \qquad (4.127)$$

and

$$\begin{aligned} \sum_{g \in G} S(g; f, f')^2 &= \frac{1}{N} \sum_{\chi} |\chi(f) + \chi(f')|^2 \, |P(\chi, B)|^2 \\ &= \frac{4M^2}{N} + \frac{1}{N} \sum_{\chi \neq \chi_0} |\chi(f) + \chi(f')|^2 \, |P(\chi, B)|^2. \qquad (4.128) \end{aligned}$$

For non-principal characters,

$$\mathbf{Exp} 2\mathrm{Re}\chi(f)\overline{\chi}(f') = 0 \qquad (4.129)$$

so that, from (4.128) and (4.129) we have

$$\begin{aligned} \mathbf{Exp} \sum_{g \in G} S(g; f, f')^2 &= \frac{4M^2}{N} + \frac{2}{N} \sum_{\chi \neq \chi_0} |P(\chi . B)|^2 \\ &= \frac{2M^2}{N} + \frac{2}{N} \sum_{\chi} |P(\chi, B)|^2 \\ &= \frac{2M^2}{N} + 2M. \qquad (4.130) \end{aligned}$$

It follows from (4.127) and (4.130) that

$$\mathbf{Exp} \sum_{g \in G} \binom{S(g; f, f')}{2} = \frac{M^2}{N} \qquad (4.131)$$

and together with (4.126), this gives the result stated.

Proof of Theorem 4.13 Our argument is based on that of Erdös and Rényi (1965) and this being the case it is similar to the proof of Theorem 4.5. We require an extra mechanism to ensure that the representations of each $g \in G$ as sub-sums of the given group elements $g_1, g_2, \ldots, g_k$ contain precisely h summands. For $g, g_1, g_2, \ldots, g_l, f_1, f'_1, f_2, f'_2, \ldots, f_v, f'_v \in G$ denote by $R_0^{(j)}(g; g_1, g_2, \ldots, g_l; f_1, f'_1, \ldots, f_v, f'_v)$ the number of solutions of the system of equations (in which every $\varepsilon_i, \varepsilon'_i = 0$ or 1),

$$\begin{aligned} g &= \varepsilon_1 g_1 + \cdots + \varepsilon_l g_l + \varepsilon_{l+1} f_1 + \varepsilon'_{l+1} f'_1 + \varepsilon_{l+2} f_2 + \varepsilon'_{l+2} f'_2 + \cdots \\ &\quad + \varepsilon_{l+v} f_v + \varepsilon'_{l+v} f'_v, \\ &\varepsilon_1 + \varepsilon_2 + \cdots + \varepsilon_l = j, \\ &\varepsilon_{l+1} + \varepsilon'_{l+1} = \varepsilon_{l+2} + \varepsilon'_{l+2} = \cdots = \varepsilon_{l+v} + \varepsilon'_{l+v} = 1. \end{aligned} \tag{4.132}$$

We have

$$\begin{aligned} &R_0^{(j)}(g; g_1, g_2, \ldots, g_l; f_1, f'_1, \ldots, f_v, f'_v) \\ &\qquad \leq R^{(j+v)}(g; g_1, g_2, \ldots, g_l, f_1, f'_1, \ldots, f_v, f'_v), \end{aligned}$$

and we put

$$l = k - 2w, \; j = h - w, \tag{4.133}$$

where

$$w = \left[\frac{\log \log N}{\log 2} \right] \tag{4.134}$$

and begin by choosing just l elements $g_1, g_2, \ldots, g_l$ randomly and independently from G: we shall then add w pairs of elements f_i, f'_i, one pair at a time in an iterative process. At the start, let

$$B_0 = \{g \in G : \; R^{(j)}(g; g_1, g_2, \ldots, g_l) = 0\}, \tag{4.135}$$

and put $M_0 = \text{card} B_0$. Put

$$\binom{l}{j} = Ne^u, \; \eta > 0 \tag{4.136}$$

and let A_0 denote the event

$$M_0 < \eta N e^{-u} \tag{4.137}$$

so that if A'_0 is the negation of A_0, we have from Lemma 4.14 that

$$\mathbf{Prob} A'_0 < \eta^{-1}. \tag{4.138}$$

We assume A_0 occurs, and pick the first pair of extra group elements f_1, f'_1 randomly and independently. Let

$$B_1 = \{g \in G : \; R_0^{(j)}(g; g_1, \ldots, g_l; f_1, f'_1) = 0\}, \tag{4.139}$$

and $M_1 = \operatorname{card} B_1$. For $g \in B_1$, we have both $g - f_1 \in B_0$, $g - f_1' \in B_0$, and we apply Lemma 4.15 which yields

$$\mathbf{Exp}(M_1|A_0) < \eta^2 N e^{-2u}. \tag{4.140}$$

Let A_1 denote the event

$$M_1 < 2\eta^3 N e^{-2u} \tag{4.141}$$

so that if A_1' is the negation of A_1, Markoff's inequality implies that

$$\mathbf{Prob}(A_1'|A_0) < (2\eta)^{-1}. \tag{4.142}$$

We assume that A_1 occurs, and pick f_2, f_2' randomly and independently from G. Let

$$B_2 = \{g : R_0^{(j)}(g; g_1, \ldots, g_l; f_1, f_1', f_2, f_2') = 0\}, \tag{4.143}$$

and $M_2 = \operatorname{card} B_2$. For $g \in B_2$ we have both $g - f_2$, $g - f_2' \in B_1$ and Lemma 4.15 yields

$$\mathbf{Exp}(M_2|A_0 A_1) < 4\eta^6 N e^{-4u}. \tag{4.144}$$

Throughout, the product of events denotes their joint occurrence, and A' denotes the negation of the event A. We let A_2 denote the event

$$M_2 < 16\eta^7 N e^{-4u} \tag{4.145}$$

so that Markoff's inequality applied to (4.144) yields

$$\mathbf{Prob}(A_2'|A_0 A_1) < (4\eta)^{-1}. \tag{4.146}$$

We continue this process. We put, for $v \leq w$,

$$B_v = \{g : R_0^{(j)}(g; g_1, \ldots, g_l; f_1, f_1', \ldots, f_v, f_v') = 0\}, \tag{4.147}$$

and $M_v = \operatorname{card} B_v$. We denote by A_v the event

$$M_v < 2^{m_v} \eta^{2^{v+1}-1} N e^{-2^v u} \tag{4.148}$$

where $m_v = 2(2^v - 1) - v$. (The recurrence relation is $m_v = 2m_{v-1} + v$, $m_0 = 0$.) We put

$$\eta = 2\delta^{-1}, \; u \geq 2\log\delta^{-1} + \log 16 + 2 \tag{4.149}$$

so that, by (4.134),

$$2^{w+1}\log 2 + 2^{w+1}\log\eta + \log N - 2^w u \leq 0 \tag{4.150}$$

and the event A_w implies $M_w < 1$. This means that B_w is empty, and for every g

$$R^{(j+w)}(g; g_1, g_2, \ldots, g_l, f_1, f_1', \ldots, f_w, f_w') > 0 \tag{4.151}$$

as required. As in the proof of Theorem 4.5, we have

$$\begin{aligned}\mathbf{Prob}(A_w') &\leq \mathbf{Prob}(A_w'|A_0A_1\ldots A_{w-1})\\ &\quad +\mathbf{Prob}(A_{w-1}'|A_0A_1\ldots A_{w-2})\\ &\quad \vdots\\ &\quad +\mathbf{Prob}(A_1'|A_0)\\ &\quad +\mathbf{Prob}(A_0')\\ &< \frac{2}{\eta} = \delta.\end{aligned} \tag{4.152}$$

It remains to calculate how large we need k to be, given the restriction $\min(h, k-h) \geq \alpha k$. Substituting from (4.149) into (4.136), we find that we need

$$\binom{l}{j} \geq 16e^2\delta^{-2}N. \tag{4.153}$$

For $k \geq 2$, $0 < h < k$, we have the (sharp) inequality

$$\binom{k}{h} \geq \frac{k^{k+\frac{1}{2}}}{h^h(k-h)^{k-h}\sqrt{8h(k-h)}} \tag{4.154}$$

which may be derived from Stirling's formula in the form: for $n \geq 1$,

$$n! = \sqrt{2\pi n}\left(\frac{n}{e}\right)^n \exp\left(\frac{\theta_n}{12n}\right), \quad \theta_n \in (0,1) \tag{4.155}$$

in all the cases in which $k \geq 4$, and which checks in the remaining cases. Hence

$$\binom{k}{h} \geq (2k)^{-1/2}\exp(\mu(\alpha)k) \tag{4.156}$$

with $\mu(\alpha)$ as in (4.106). Next, we have, on recalling that $l = k - 2w$, $j = h - w$,

$$\binom{l}{j} \geq \binom{k}{h}\frac{h^w(k-h)^w}{k^{2w}}\prod_{v<w}\left(1-\frac{v}{h}\right)\left(1-\frac{v}{k-h}\right). \tag{4.157}$$

By (4.110) we have $h(k-h)/k^2 \geq \alpha(1-\alpha)$. In view of the inequality $1 - x \geq e^{-2x}$, valid for $0 \leq x \leq \frac{1}{2}$, we also have

$$1 - \frac{kv}{h(k-h)} + \frac{v^2}{h(k-h)} \geq \exp\left(\frac{-2kv}{h(k-h)}\right)$$

if we assume that $2w \le \alpha(1-\alpha)k$. Thus (4.157) implies

$$\binom{l}{j} \ge \binom{k}{h} (\alpha(1-\alpha))^w \exp\left(\frac{-w^2}{\alpha(1-\alpha)k}\right). \tag{4.158}$$

Since $2^k \ge N$ and by (4.134), $w \log 2 \le \log\log N$, we shall ceratinly have $2w \le \alpha(1-\alpha)k$ for such $N > N_0(\alpha)$, indeed we may assume that for such N the exponential factor on the right of (4.158) is close to 1, say $> 1/\sqrt{2}$ for convenience. We combine (4.156) with (4.158) to obtain

$$\begin{aligned}\binom{l}{j} &\ge \frac{1}{2}k^{-1/2} (\alpha(1-\alpha))^w \exp(\mu(\alpha)k) \\ &\ge \frac{1}{2}k^{-1/2}(\log N)^{\frac{1}{2}-\kappa} \exp(\mu(\alpha)k) \end{aligned} \tag{4.159}$$

with $\kappa = \kappa(\alpha)$ as in (4.107), using $w \log 2 \le \log\log N$ again. The right-hand side of (4.159) is an increasing function of k – the condition $\mu(\alpha)k > \frac{1}{2}$ follows from (4.109) – and so we may replace k by the smaller number k_0 (which need not be an integer) satisfying $\mu(\alpha)k_0 = \log N + \kappa(\alpha)\log\log N + \xi(\alpha, \delta)$. Hence

$$\binom{l}{j} \ge \frac{1}{2}\left(\frac{\log N}{k_0}\right)^{\frac{1}{2}} Ne^{\xi(\alpha,\delta)}$$

so that (4.153) will follow from

$$\left(\frac{\log N}{k_0}\right)^{\frac{1}{2}} e^{\xi(\alpha,\delta)} \ge 32e^2\delta^{-2} \tag{4.160}$$

which, by the definition (4.108) of $\xi(\alpha, \delta)$ is equivalent to

$$\frac{\log N}{k_0} \ge \left(\frac{2}{e}\right)^{10} \mu(\alpha) \tag{4.161}$$

for which a sufficient condition is

$$20.5 \log N \ge \kappa(\alpha)\log\log N + \xi(\alpha, \delta). \tag{4.162}$$

By hypothesis $N \ge \max(N_0(\alpha), \delta^{-10})$, and this implies (4.162) if $N_0(\alpha)$ is sufficiently large. This completes the proof of Theorem 4.13.

4.4 The Behrend sequences $\mathscr{A}^*(t)$

The main result in this section is Theorem 4.17, from which it will be a short step to Theorem 4.23 that $\mathscr{A}^*(t)$ is Behrend. We begin by stating a result of Erdös and Hall (1974) which was a precursor of Theorem 4.17 and has a similar but technically easier proof.

Theorem 4.16 *Let c be a fixed real number and* $0 \le \theta < 1$. *Then the integers n having a divisor d such that*

$$0 < \|\log d - \theta\| < 2^{-\log\log n - c\sqrt{\log\log n}} \tag{4.163}$$

have asymptotic density

$$\frac{1}{\sqrt{2\pi}} \int_c^\infty e^{-y^2/2} dy; \tag{4.164}$$

moreover if there exists a (necessarily unique) integer $m(\theta)$ *such that* $\theta \equiv \log m(\theta) \,(\mathrm{mod}\, 1)$ *and we allow equality on the left in (4.163) then the density is increased to*

$$\frac{1}{m(\theta)\sqrt{2\pi}} \int_{-\infty}^c e^{-y^2/2} dy + \frac{1}{\sqrt{2\pi}} \int_c^\infty e^{-y^2/2} dy. \tag{4.165}$$

Furthermore we may replace c in (4.163) by $\xi(n)$, *or* $-\xi(n)$, *where* $\xi(n) \to \infty$ *as* $n \to \infty$. *In the former case the asymptotic densities in (4.164) and (4.165) become* 0 *and* $1/m(\theta)$ *respectively; in the latter case both densities become* 1.

This result should be compared with Theorem 4.10. The underlying group, namely $(\mathbf{R}/\mathbf{Z}, +)$ is infinite in Theorems 4.16 and 4.17 and we have to replace it by a finite cyclic group of suitable order. As explained in §4.1, this involves an approximation (which will be made clear in the proof of Theorem 4.17) for which there is a price: property (i) of the mapping ρ no longer holds exactly, and our application of probabilistic group theory is therefore *impure.* It will turn out not to present any serious difficulty.

Some explanation of Theorem 4.16 may be helpful in preparation for the statement of Theorem 4.17. Recall that by the Erdös–Kac theorem (1940), stated in §4.1, each of the sequences

$$\{n : \omega(n) \ge \log\log n + c\sqrt{\log\log n}\}, \tag{4.166}$$

$$\{n : \Omega(n) \ge \log\log n + c\sqrt{\log\log n}\}, \tag{4.167}$$

has asymptotic density given by (4.164). Since

$$2^{\omega(n)} \le \tau(n) \le 2^{\Omega(n)} \tag{4.168}$$

the same is therefore true of the sequence

$$\{n : \tau(n) \ge 2^{\log\log n + c\sqrt{\log\log n}}\}. \tag{4.169}$$

Thus the theorem is in accord with a very simple heuristic model in which the fractional part of $\log d$ is uniformly distributed on $[0, 1)$ when d runs through the divisors of a large random integer n. That is, these fractional parts are spaced $1/\tau(n)$ apart. This model appears again in the next chapter but the reader should beware of interpreting such arguments too literally. For example, Hall (1975a) proved that if

$$g(n) := \min\{\|\log(d/d')\| : d, d'|n,\ d \neq d'\} \tag{4.170}$$

then

$$g(n) = (\log n)^{-\log 3 + o(1)} p.p., \tag{4.171}$$

a result which should be compared with the Maier–Tenenbaum theorem ((1984), *Divisors*, Chapter 5) in which $\|\ \|$ in (4.170) is replaced by $|\ |$ and the right-hand side of (4.171) is multiplied by $\log n$. An equally simple, *but different*, heuristic model involving the distinct ratios of the divisors (cf. *Divisors* §3.3) predicts the behaviour of $g(n)$.

A corollary of Theorem 4.16 is that the sequence

$$\mathscr{A}^*\left(\frac{1}{2}\right) = \{a > 1 : \|\log a\| < (\log a)^{-\log 2} \exp(\xi(a)\sqrt{\log\log a})\} \tag{4.172}$$

is Behrend, provided $\xi(a) \to \infty$ as $a \to \infty$; moreover it is necessary for the conclusion that $\xi(a)$ be unbounded.

We proceed to Theorem 4.17. This depends on Theorem 4.13 in probabilistic group theory. In a parallel fashion, Theorem 4.16 depends on Theorem 4.3.

Theorem 4.17 *Let c be a fixed real number and $0 \le \theta < 1$, $t \in (0, \frac{1}{2}) \cup (\frac{1}{2}, 1)$. Then the integers n having a divisor d such that both*

$$0 < \|\log d - \theta\| < T^{-\log\log n - c\sqrt{\log\log n}} \tag{4.173}$$

and

$$\Omega(d) \begin{cases} \le t\Omega(n), & (\text{if } t < \frac{1}{2}), \\ \ge t\Omega(n), & (\text{if } t > \frac{1}{2}), \end{cases} \tag{4.174}$$

where $T = \exp(\mu(t))$ and $\mu(t) = -t\log t - (1-t)\log(1-t)$ as in (4.106), have asymptotic density

$$\frac{1}{\sqrt{2\pi}} \int_c^\infty e^{-y^2/2} dy. \tag{4.175}$$

Furthermore we may replace c in (4.173) by $\xi(n) \to \infty$ or $-\infty$ when the density becomes respectively 0 *or* 1, *and we may if we wish require that d, in addition to satisfying (4.173) and (4.174), is a multiple of any particular prime factor of n.*

The last part of the theorem provides a device, useful in some applications, to ensure that d is fairly large. For example we may choose $P^+(n)$ as the particular prime and since $P^+(n) > n^{o(1)}$ *p.p.* we then have

$$\log\log d > \log\log n - \xi_0(n)\, p.p., \tag{4.176}$$

where $\xi_0(n)$ is any function tending to infinity with n. We use this device later when we show that $\mathscr{A}^*(t)$, $(t < \frac{1}{2})$ is a Behrend sequence.

Proof of Theorem 4.17 The proofs for $t < \frac{1}{2}$, $t > \frac{1}{2}$ are similar and we treat these cases together, pointing out the slight differences as we meet them.

We split the integers $n \le x$ into four main classes, which we specify in turn as we encounter them. The zero class comprises those integers for which $\Omega(n) \ge \log\log x + (\log\log x)^{\frac{2}{3}}$: it contains $o(x)$ integers by the Hardy–Ramanujan theorem.

The first class comprises those integers $n \le x$ not in the zero class for which

$$\Omega(n) \ge \log\log x + c\sqrt{\log\log x} + (j_0(t)+1)(\log\log x)^{\frac{1}{3}}, \tag{4.177}$$

where $j_0(t)$ is an integer depending on t only, to be specified later. By the Erdös–Kac theorem, the first class has cardinality

$$x\frac{1}{\sqrt{2\pi}}\int_c^\infty e^{-y^2/2}dy + o(x) \tag{4.178}$$

and we shall prove that all but $o(x)$ of these integers have a divisor of the required kind. We split the first class into sub-classes C_j, $j_0(t) \le j \le j_1(x)$, assigning n to C_j if

$$\left[\frac{\Omega(n) - \log\log x - c\sqrt{\log\log x}}{(\log\log x)^{\frac{1}{3}}}\right] = j+1. \tag{4.179}$$

Here

$$j_1(x) = [(\log\log x)^{1/3} - c(\log\log x)^{1/6}]. \tag{4.180}$$

We put $I(x) = (u, v]$ where

$$u = \exp\left((\log\log x)^{1/3}\right), \quad v = x^{1/(\log\log x)^2} \tag{4.181}$$

and we have

$$\sum_{n\le x}(\Omega(n) - \Omega(n; u, v)) \sim 5x\log\log\log x. \tag{4.182}$$

We assume henceforth that first class n satisfy

$$\Omega(n) - \Omega(n; u, v) < (\log\log x)^{1/3}; \tag{4.183}$$

by (4.182) the number of exceptional n is $o(x)$. We denote by C_j' the set of integers n in C_j for which (4.183) is satisfied. For $n \in C_j'$ we deduce from (4.179) and (4.183) that $\Omega(n; u, v) \geq k(j)$ where

$$k(j) = [\log\log x + c\sqrt{\log\log x} + j(\log\log x)^{\frac{1}{3}}]. \tag{4.184}$$

Let the $k(j)$ smallest prime factors of n which lie in $I(x)$ be denoted by $p_1, p_2, \ldots, p_{k(j)}$. We consider divisors d of n which are products of exactly $h(j)$ of these primes, where $h(j)$ is specified below. We are going to apply Theorem 4.13 and this involves a parameter α, at our disposal, and requires in condition (4.110) that $\min(h(j), k(j) - h(j)) \geq \alpha k(j)$. We consider the cases $t < \frac{1}{2}$, $t > \frac{1}{2}$ separately. If $t < \frac{1}{2}$ we put

$$h(j) = [\alpha k(j)] + 1, \ \alpha = t. \tag{4.185}$$

We shall then have $\Omega(d) = h(j) \leq t\Omega(n)$ if $k(j) + t^{-1} \leq \Omega(n)$. By (4.179) and (4.184) a sufficient condition for this is $t^{-1} < (\log\log x)^{1/3}$ or $x > x_1(t)$. We may assume this holds, so that any divisor d of the kind described above satisfies (4.174). If $t > \frac{1}{2}$ we put

$$h(j) = [(1-\alpha)k(j)], \ \alpha = 1 - t - \eta \tag{4.186}$$

where $\eta = 3(\log\log x)^{-2/3}$. From (4.179) and (4.184) we have $\Omega(n) - k(j) < 2(\log\log x)^{1/3} + 1$, and $\eta k(j) > 2(\log\log x)^{1/3} + 2$ provided $x > x_2(c, t)$. Hence $\eta k(j) \geq t(\Omega(n) - k(j)) + 1$ which implies $[(t+\eta)k(j)] \geq t\Omega(n)$. Thus any divisor d with $\Omega(d) = h(j)$ satisfies (4.174) when $t > \frac{1}{2}$. We assume henceforth that $x > \max(x_1, x_2)$.

For $t < \frac{1}{2}$, $\mu(\alpha) = \mu(t)$, and we shall need a relation between $\mu(\alpha)$ and $\mu(t)$ when $t > \frac{1}{2}$. Recall from (4.106) that $\mu(t) = -t\log t - (1-t)\log(1-t)$, so that $\mu'(1-t) = -\log(1-t) + \log t < -\log(1-t)$ because $t < 1$, and $\mu' \leq -\log(1-t)$ in the interval to the left of $1-t$. It follows from the mean value theorem and (4.186) that there exists $\eta_3(t)$ such that $\eta < \eta_3(t)$ implies $\mu(\alpha) > \mu(1-t) + \eta\log(1-t)$. Hence for $x > x_3(t)$ and all t, we have

$$\mu(\alpha) > \mu(t) + 3\log(1-t)(\log\log x)^{-2/3}. \tag{4.187}$$

Next we set, for each j, $j_0 \leq j \leq j_1$,

$$N_j = [\exp\{\mu(\alpha)\left(\log\log x + c\sqrt{\log\log x} + (j-1)(\log\log x)^{1/3}\right)\}] \tag{4.188}$$

and we define G_j to be the cyclic group of congruence classes (mod

N_j) under addition. We are going to apply Theorem 4.13 with $G = G_j$, $N = N_j$, $k = k(j)$, $h = h(j)$ and $\delta = (\log\log x)^{-1}$, and we note at this point that there exists an $x_4 = x_4(c,t)$ such that for $x > x_4$, the conditions $N_j \geq \max(N_0(\alpha), \delta^{-10})$, (4.109) and (4.110) are all satisfied. We assume henceforth that $x > \max(x_1, x_2, x_3, x_4)$.

We define the mapping $\rho_j : \mathbf{Z}^+ \to G_j$ by the relation

$$[N_j \log m] \equiv \rho_j(m) \pmod{N_j}; \tag{4.189}$$

this does not satisfy $\rho(mm') = \rho(m) + \rho(m')$ for all m, m' but does so approximately, and this is sufficient for our purpose. Indeed, if $g(\theta) \in G$, $g(\theta) \equiv [N_j\theta] \pmod{N_j}$ and

$$\rho(p_{i_1}) + \rho(p_{i_2}) + \cdots + \rho(p_{i_h}) \equiv g(\theta) \pmod{N_j} \tag{4.190}$$

then

$$\|\log(p_{i_1}p_{i_2}\dots p_{i_h}) - \theta\| < \frac{h(j)}{N_j}. \tag{4.191}$$

We are now in a position to specify the integer $j_0(t)$ which appears in the definition (4.177) of the first class. From (4.184), (4.187) and (4.188) we have, for sufficiently large j,

$$\frac{k(j)}{N_j} < T^{-\log\log x - c\sqrt{\log\log x}} \tag{4.192}$$

where $T = \exp(\mu(t))$ as in the statement of the theorem. We define $j_0(t)$ to be the smallest integer such that $j \geq j_0(t)$ implies (4.192).

We have to refine the class C_j' further. We say that $n \in C_j''$ if $n \in C_j'$ and the congruence classes $\rho(p) \pmod{N_j}$ are distinct, where p runs through all the $\Omega(n; u, v)$ prime factors of n belonging to $I(x)$. Of course this also implies that these prime factors are distinct.

Lemma 4.18 *For each fixed* $\beta > 0$ *and every subinterval* E *of* $[0,1)$ *we have*

$$\sum_{\substack{u<p\leq v \\ \log p \in E \pmod 1}} p^{-1} = |E|\left(\log\left(\frac{\log v}{\log u}\right) + O\left(\frac{1}{\log u}\right)\right) + O_\beta(e^{-\beta\sqrt{\log u}}), \tag{4.193}$$

where $|E|$ *is the length of* E.

This is a straightforward deduction from the Prime Number Theorem in the form

$$\pi(z) = li z + O_\beta(ze^{-2\beta\sqrt{\log z}}) \tag{4.194}$$

and we omit the details.

For each l, $0 \le l < N_j$, let $P_j(l)$ denote the set of primes $p \in I(x)$ such that $\rho(p) = l$, that is

$$[N_j \log p] \equiv l \,(\mathrm{mod}\ N_j). \tag{4.195}$$

An equivalent condition is

$$\log p \in E_j(l) \,(\mathrm{mod}\ 1), \quad E_j(l) = \left[\frac{l}{N_j}, \frac{l+1}{N_j}\right). \tag{4.196}$$

We deduce from Lemma 4.18 that

$$\sum_{p \in P_j(l)} p^{-1} \ll N_j^{-1} \log\log x \tag{4.197}$$

uniformly for $j_0(t) \le j \le j_1(x)$, whence

$$\begin{aligned} \mathrm{card}\{C_j' \setminus C_j''\} &\le x \sum_l \Big(\sum_{p \in P_j(l)} p^{-1} \Big)^2 \\ &\ll x N_j^{-1} (\log\log x)^2, \end{aligned} \tag{4.198}$$

and

$$\sum_{j_0(t) \le j \le j_1(x)} \mathrm{card}\{C_j' \setminus C_j''\} \ll x N_{j_0(t)}^{-1} (\log\log x)^3 = o(x). \tag{4.199}$$

Let $n \in C_j''$ and write $\rho_j(p_i) = g_i$ for $1 \le i \le k(j)$: the g_i are therefore distinct congruence classes (mod N_j). We say that n is *good* if $\{g_1, g_2, \dots, g_{k(j)}\}$ is a *good set of congruence classes* in the sense that every congrunce class g (and so in particular $g(\theta)$) has at least one representation

$$g \equiv g_{i_1} + g_{i_2} + \cdots + g_{i_{h(j)}} \,(\mathrm{mod}\ N_j) \tag{4.200}$$

in which $i_1 < i_2 < \cdots < i_{h(j)}$. A good integer n has a divisor d satisfying the requirements of the theorem and we have to estimate the number of members of C_j'' which are not good, let us say *bad*. Accordingly let $g_1, g_2, \dots, g_m$ be distinct congruence classes (mod N_j) where $m \ge k(j)$, such that no sub-set of $k(j)$ of these classes is good in the sense described above, and moreover

$$m < \log\log x + (\log\log x)^{\frac{2}{3}}. \tag{4.201}$$

By Theorem 4.13, the number of choices of $g_1, g_2, \dots, g_m$ is

$$< \delta N_j^m. \tag{4.202}$$

If n is a bad integer associated with this set of congruence classes, we have $n = p_1 p_2 \cdots p_m q$ where $p_i \in I(x)$, $\log p_i \in E_j(g_i)$ (mod 1) for $1 \le i \le m$

and q has no prime factor in $I(x)$. If n is to be in the first class (4.201) holds.

Lemma 4.19 *Let I be a set of primes of which none exceeds x. Then*

$$\mathit{card}\{q : q \le x, p|q \Rightarrow p \notin I\} \ll x \prod_{p \in I} \left(1 - \frac{1}{p}\right). \tag{4.203}$$

For a proof see van Lint and Richert (1965), Hall (1974a) or Chapter 3 of Halberstam and Richert (1974).

The number of integers n as above is

$$\text{card}\{q : q \le \frac{x}{p_1 \cdots p_m}, p|q \Rightarrow p \notin I(x)\} \tag{4.204}$$

and we note that by the definition (4.181) of v and (4.201) we have $v \le x/v^m$ so that Lemma 4.19 is applicable and the cardinality in (4.204) is

$$\ll \frac{x}{p_1 p_2 \cdots p_m} \prod_{p \in I(x)} \left(1 - \frac{1}{p}\right). \tag{4.205}$$

Hence the number of bad n corresponding to the set $g_1, g_2, \dots, g_m$ is

$$\ll \quad x \left(\frac{\log u}{\log v}\right) \prod_{i=1}^{m} \left\{ \sum_{\substack{u<p\le v \\ \log p \in E_j(g_i)}} p^{-1} \right\} \tag{4.206}$$

$$\ll \quad x \left(\frac{\log u}{\log v}\right) \left(N_j^{-1} \log \left(\frac{\log v}{\log u}\right)\right)^m \tag{4.207}$$

by (4.193) and (4.201). Hence the number of bad integers in C_j'' such that $\Omega(n; u, v) = m$ is, combining (4.202) and (4.207),

$$\ll \delta x \left(\frac{\log u}{\log v}\right) \frac{1}{m!} \left(\log \left(\frac{\log v}{\log u}\right)\right)^m \tag{4.208}$$

and we sum over m, and then over j to deduce that there are

$$\ll \delta x (\log\log x)^{\frac{2}{3}} = o(x) \tag{4.209}$$

exceptional integers in the first class. Our treatment of this class is complete.

The second class comprises the integers $n \le x$ such that (4.177) is false, but

$$\Omega(n) \ge \log\log x + c\sqrt{\log\log x} - (\log\log x)^{\frac{1}{3}}. \tag{4.210}$$

Since $\Omega(n) = \log\log x + (c + o(1))\sqrt{\log\log x}$ for all these integers, the cardinality of the class is $o(x)$ by the Erdös–Kac theorem.

The third class comprises the integers $n \le x$ for which (4.210) is false, and we show that the number of these integers having a divisor of the type specified in the theorem is $o(x)$. We retain u as in (4.181) and we begin by disposing of the integers with such a divisor d satisfying the extra condition $P^+(d) \le u$. For this purpose we replace condition (4.173) by

$$0 < \|\log d - \theta\| < (\log x)^{-\mu(t)/2} \tag{4.211}$$

which is weaker except for small integers n whose number is negligible. We are going to show that

$$\sum_{P^+(d)\le u}^{*} d^{-1} = \sum_{\substack{P^+(d)\le u\\ d>d_0}}^{*} d^{-1} + \sum_{\substack{P^+(d)\le u\\ d\le d_0}}^{*} d^{-1} = o(1), \tag{4.212}$$

where we have spilt the sum at $d_0 = \exp((\log\log x)^7)$, the * denoting that (4.211) holds. The first sum on the right does not exceed

$$\begin{aligned}\sum_{\substack{P^+(d)\le u\\ d>d_o}} d^{-1} &\le (\log d_0)^{-1} \sum_{P^+(d)\le u} d^{-1}\log d\\ &\le (\log d_0)^{-1} \prod_{p\le u}\left(1-\frac{1}{p}\right)^{-1} \sum_{p\le u}\frac{\log p}{p-1}\\ &\ll (\log\log x)^{-7}(\log u)^2 = o(1)\end{aligned} \tag{4.213}$$

because $\log u = (\log\log x)^3$. Notice we have not used (4.211): obviously we must do so in estimating the remaining sum in (4.212). Let $m = 0, 1, 2, 3, \ldots$; the sum of the reciprocals of the d's in the range

$$m - (\log x)^{-\mu(t)/2} < \log d - \theta < m + (\log x)^{-\mu(t)/2} \tag{4.214}$$

is

$$\ll (\log x)^{-\mu(t)/2} + e^{-m}, \tag{4.215}$$

if indeed there exists an integer d satisfying (4.214): recall that in Theorem 4.17 we do not allow $\log d = \theta$ in the case $\theta \in \log \mathbf{Z}^+ (\mathrm{mod}\, 1)$. Let $d(x,\theta)$ denote the smallest integer d for which (4.211) holds, and let $d(x,\theta)$ satisfy (4.214) when $m = m(x,\theta)$. We have $d(x,\theta) \to \infty$, $m(x,\theta) \to \infty$ as $x \to \infty$, and we sum (4.215) over the range $m(x,\theta) \le m \le \log d_0 + O(1)$

(if this is non-empty) to obtain

$$\sum_{\substack{P^+(d)\le u\\ d\le d_0}}^{*} d^{-1} \ll \frac{(\log\log x)^7}{(\log x)^{\mu(t)/2}} + d(x,\theta)^{-1} = o(1). \tag{4.216}$$

Notice that in this part of the sum (4.212) we have not used the condition $P^+(d) \le u$. Together with (4.213) this establishes (4.212).

Next we consider the integers int the third class with a divisor d satisfying the requirements of the theorem but none such with $P^+(d) \le u$. We refer to these integers as being in the fourth class. We want to show that the cardinality if this class is $o(x)$.

Let us exclude from the class integers n for which

$$P^+(n) \le w = x^{1/\log\log x}. \tag{4.217}$$

By Lemma 4.19, the number of such integers is

$$\ll x \prod_{w<p\le x}\left(1-\frac{1}{p}\right) \ll x(\log\log x)^{-1} = o(x). \tag{4.218}$$

For the remaining integers n in the fourth class we write $n = mp$, $p > w$, and we consider whether such a decomposition, in which m also belongs to the fourth class is possible or not. The number of cases when this is possible is

$$\le \sum_{m<x/w}^{*} \pi\left(\frac{x}{m}\right) \ll \frac{x\log\log x}{\log x}\sum^{*}\frac{1}{m} \tag{4.219}$$

where * denotes that m is in the fourth class. Let us write $m = qp_1p_2\cdots p_r$ where $P^+(q) \le u$ and $u < p_1 < p_2 < \cdots < p_r$. Of course $P^+(m) > u$ because $d \nmid q$, and we may assume that the prime factors of m which exceed u are distinct: the m's with a repeated prime factor $> u$ contribute

$$\ll \frac{x\log\log x}{\log x}\sum_{p>u}\frac{1}{p^2}\sum_{s\le x}\frac{1}{s} = o(x). \tag{4.220}$$

Next, m has a divisor d of the form

$$d = ap_1^{\varepsilon_1}p_2^{\varepsilon_2}\cdots p_r^{\varepsilon_r},\ a|q,\ \varepsilon_i = 0 \text{ or } 1\ \forall i, \tag{4.221}$$

which satisfies (4.173) and (4.174). We may restrict our attention to integers $n > \sqrt{x}$ so that (4.173) implies

$$0 < \|\log d - \theta\| < \frac{l}{2} \tag{4.222}$$

where

$$l \asymp_c T^{-\log\log x - c\sqrt{\log\log x}}. \qquad (4.223)$$

By definition of the fourth class, $P^+(d) > u$ and hence there exists j, $1 \le j \le r$, such that $\varepsilon_j = 1$. But then (4.222) restricts the fractional part of $\log p_j$ to lie within the union of M sub-intervals of $[0,1)$ each of length l. Here M is the number of divisors of $qp_1p_2\cdots p_{j-1}p_{j+1}\cdots p_r$ which satisfy (4.174), and plainly M does not exceed the number of divisors of n itself with this property, which we denote by $M(n,t)$. It follows that if $\sum'$ denotes the summation over the primes p_j as above then by Lemma 4.18 (in which we substitute x for v), we have

$$\sum{}' \frac{1}{p_j} \ll M(n,t) l \log\left(\frac{\log x}{\log u}\right). \qquad (4.224)$$

Lemma 4.20 *Let s be a positive integer and X, Y be real numbers such that $0 < X \le \frac{1}{2}s \le Y < s$. Then the number of s-tuples $\{\delta_1, \delta_2, \ldots, \delta_s\}$ in which each $\delta_i = 0$ or 1 and, respectively*

$$\delta_1 + \delta_2 + \cdots + \delta_s \le X \qquad (4.225)$$

$$\delta_1 + \delta_2 + \cdots + \delta_s \ge Y \qquad (4.226)$$

does not exceed

$$\exp\left\{s\mu\left(\frac{X}{s}\right)\right\}, \quad \exp\left\{s\mu\left(\frac{Y}{s}\right)\right\}, \qquad (4.227)$$

where μ is defined by (4.106).

Proof of lemma The two results are equivalent: if we replace δ_i by $1-\delta_i$ and X by $s-Y$ in (4.225) it becomes (4.226). We restrict our attention to (4.225).

Let $0 < z \le 1$. The number of s-tuples for which (4.225) is satisfied does not exceed

$$\sum_{\delta_1=0}^{1}\sum_{\delta_2=0}^{1}\cdots\sum_{\delta_s=0}^{1} z^{\delta_1+\delta_2+\cdots+\delta_s-X} = (1+z)^s z^{-X} \qquad (4.228)$$

and we put $z = X/(s-X)$ which does not exceed 1 by the hypothesis about X. This gives the result stated.

Lemma 4.21 *Let $M(n,t)$ denote the number of divisors of n such that (4.174) holds. Then*

$$M(n,t) \le \exp\{\Omega(n)\mu(t)\}, \tag{4.229}$$

where μ is defined by (4.106).

Proof of lemma Let n be squarefree, $\Omega(n) = s$, $n = p_1 p_2 \cdots p_s$ and a typical divisor be

$$d = p_1^{\delta_1} p_2^{\delta_2} \cdots p_s^{\delta_s}, \quad \text{each } \delta_i = 0 \text{ or } 1. \tag{4.230}$$

We apply Lemma 4.20 with $X = t\Omega(n)$ if $t < \frac{1}{2}$ and $Y = t\Omega(n)$ if $t > \frac{1}{2}$. This gives the result stated in this case.

In the general case let $n = p_1^{\alpha_1} p_2^{\alpha_2} \cdots p_k^{\alpha_k}$ where $\alpha_1 + \alpha_2 + \cdots + \alpha_k = s$, and let

$$n' = p_1^{(1)} p_1^{(2)} \cdots p_1^{(\alpha_1)} p_2^{(1)} p_2^{(2)} \cdots p_2^{(\alpha_2)} \cdots p_k^{(1)} p_k^{(2)} \cdots p_k^{(\alpha_k)} \tag{4.231}$$

be any squarefree number with s prime factors. We define a multiplicative, injective mapping ϕ from the set of divisors of n into the set of divisors of n' by setting

$$\phi(p_i^{\beta_i}) = p_i^{(1)} p_i^{(2)} \cdots p_i^{(\beta_i)}, \quad \beta_i \le \alpha_i \tag{4.232}$$

and we have $\Omega(\phi(d)) = \Omega(d)$ for every d. Hence $M(n,t) \le M(n',t) \le \exp(s\mu(t))$ because n' is squarefree. This completes the proof.

We return to (4.224) and we recall that for integers n in the fourth class, (4.209) is false, that is

$$\Omega(n) < \log\log x + c\sqrt{\log\log x} - (\log\log x)^{1/3}. \tag{4.233}$$

Hence (4.223) and (4.229) imply

$$M(n,t)l < \Lambda \tag{4.234}$$

where

$$\Lambda = \exp\{-(\log\log x)^{1/3}\}. \tag{4.235}$$

We may now estimate from above the sum on the right of (4.219). This does not exceed

$$\sum_{P^+(q)\le u} q^{-1} \sum_{r\le r_0} \sum_{j=1}^{r} \sum{}'' (p_1 p_2 \cdots p_{j-1} p_{j+1} \cdots p_r)^{-1} \sum{}' p_j^{-1} \tag{4.236}$$

where the sum $\sum''$ runs over all products of $r-1$ distinct primes $p_i \in (u, x]$, $\sum'$ is as in (4.224) and r_0 is the maximum value of r. By (4.233), $r_0 \ll \log\log x$ and so by (4.224) and (4.234) the sum in (4.236) is

$$\begin{aligned} &\ll (\log u) r_0^2 \Lambda \sum_{r \le r_0} \frac{1}{r!} \left(\log\left(\frac{\log x}{\log u}\right)\right)^r \\ &\ll r_0^2 \Lambda \log x. \end{aligned} \tag{4.237}$$

It follows from (4.219), (4.220), (4.235) and (4.237) that the number of integers n in the fourth class of the form mp, $p > w$, where m is also in the fourth class is

$$\ll \Lambda x (\log\log x)^3 + o(x) = o(x). \tag{4.238}$$

It remains to consider the integers n in the fourth class for which such a decomposition is impossible, that is if p is any prime factor of n exceeding w then $m = n/p$ is not in the fourth class.

Lemma 4.22 *There exists an absolute constant B such that if $y \ge 2$, $1 \le w < y$, and E is any sub-interval of* $[0, 1)$ *then*

$$\sum_{\substack{w<p\le y \\ \log p \in E (\mathrm{mod}\ 1)}} 1 \le B\,|E|\,y(\log y)^{-1} + O_\beta(y e^{-\beta\sqrt{\log w}}) \tag{4.239}$$

where $|E|$ is the length of E and β is arbitrary.

This is a straightforward deduction from the Prime Number Theorem (4.194).

If n is one of the above integers it has a divisor d, necessarily a multiple of p satisfying (4.174) and (4.222). This places the fractional part of $\log p$ within the union of not more than $M(n, t)$ sub-intervals of $[0, 1)$ each of length l (the position of these sub-intervals depending on $m = n/p$), and so the number of such n does not exceed

$$\sum_{m<x/w} \sum_{w<p\le x/m}{}^{(m)} 1 \tag{4.240}$$

where $^{(m)}$ denotes that $\log p$ lies in the relevant set (mod 1). We apply Lemma 4.22, with $w = x^{1/\log\log x}$, $y = x/m$, $|E| = l$. By (4.223) the error term is negligible, fixing $\beta = 1$ say, and so (4.240) becomes, using (4.234),

$$\begin{aligned} &\ll \sum_{m \le x/w} \Lambda \frac{x}{m} \left(\log \frac{x}{m}\right)^{-1} \\ &\ll \Lambda x (\log w)^{-1} \log x = o(x). \end{aligned} \tag{4.241}$$

This completes our treatment of the fourth class integers and our proof of Theorem 4.17 except for the last part which involves a small change in the treatment of the first class integers only, and which we leave to the reader.

Theorem 4.23 *Let* $0 < t < 1$, *and* $\xi(a) \to \infty$ *as* $a \to \infty$. *Then the sequence* $\mathscr{A}^*(t)$ *is a Behrend sequence, where* $\mathscr{A}^*(\frac{1}{2})$ *is defined by (4.172), and for* $t < \frac{1}{2}$, $t > \frac{1}{2}$ *respectively*

$$\begin{aligned}\mathscr{A}^*(t) &= \{a > 1 : \Omega(a) \le t \log\log a + \xi(a)\sqrt{\log\log a}, \\ &\qquad \|\log a\| < T^{-\log\log a + \xi(a)\sqrt{\log\log a}}\}, \qquad (4.242)\\ \mathscr{A}^*(t) &= \{a > 1 : \Omega(a) \ge t \log\log a - \xi(a)\sqrt{\log\log a}, \\ &\qquad \|\log a\| < T^{-\log\log a + \xi(a)\sqrt{\log\log a}}\}, \qquad (4.243)\end{aligned}$$

T being as in Theorem 4.17, that is $T = \exp\{-t\log t - (1-t)\log(1-t)\}$.

Proof of Theorem 4.23 The case $t = \frac{1}{2}$ is a corollary of Theorem 4.16 and is left to the reader. We consider the case $t > \frac{1}{2}$. Put

$$\xi_1(n) = \min_{d|n}\left(\frac{\log\log n - \log\log d}{\sqrt{\log\log n}} + \xi(d)\frac{\sqrt{\log\log d}}{\sqrt{\log\log n}}\right). \qquad (4.244)$$

Then $\xi_1(n) \to \infty$ as $n \to \infty$, for either we have $\log\log d \le \frac{1}{4}\log\log n$ when the first term on the right of (4.244) is at least $\frac{3}{4}\sqrt{\log\log n}$ or else the second term exceeds $\frac{1}{2}\xi(d)$ and $d \to \infty$. By (4.244) we have, for every divisor d of n,

$$\log\log d - \xi(d)\sqrt{\log\log d} \le \log\log n - \xi_1(n)\sqrt{\log\log n}. \qquad (4.245)$$

We apply Theorem 4.17 with $\theta = 0$ and $c = -\xi_1(n)$. For almost all n, there exists a divisor d satisfying both (4.173) and (4.174). We call this divisor a and by (4.173) and (4.245) a fulfils the second condition in (4.243). We may assume $\Omega(n) \ge \log\log n - \xi_1(n)\sqrt{\log\log n}$ by the Hardy–Ramanujan theorem and we see from (4.174) and (4.245) that a also fulfils the first condition in (4.243), whence $a \in \mathscr{A}^*(t)$. Thus almost all integers n have a divisor in $\mathscr{A}^*(t)$ and $\mathscr{A}^*(t)$ is a Behrend sequence.

It remains to deal with the case $t < \frac{1}{2}$, when (4.173) is again stronger than we need for the second part of (4.242) but (4.174) is not sufficient for the first part without some extra information. Let $\xi_1(n)$ be as in (4.244). We apply Theorem 4.17 with $\theta = 0$ and $c = -\xi_1(n)$ as before, so that for almost all n we can find a divisor d satisfying both (4.173), (4.174) which is, in addition, a multiple of the greatest prime factor $P^+(n)$ of n.

We choose $\xi_0(n) = \frac{1}{2}\sqrt{\log\log n}$ in (4.176) so that if n is large enough, we have

$$\log\log n < \log\log d + \sqrt{\log\log d} \; p.p.$$

whence

$$\log\log n < \log\log d + \frac{1}{2}\xi(d)\sqrt{\log\log d} \; p.p. \tag{4.246}$$

because $\xi(d) \to \infty$ as $d \to \infty$ by hypothesis. As in the case $t > \frac{1}{2}$, (4.173) and (4.245) imply that a $(= d)$ fulfils the second condition in (4.242). Next, by the Hardy–Ramanujan theorem we have $\Omega(n) \leq \log\log n + \xi_1(n)\sqrt{\log\log n}$ *p.p.* and we combine this with (4.245) to obtain

$$\Omega(n) \leq 2\log\log n - \log\log d + \xi(d)\sqrt{\log\log d} \; p.p.,$$

whence by (4.246)

$$t\Omega(n) \leq t\log\log d + 2t\xi(d)\sqrt{\log\log d} \; p.p. \tag{4.247}$$

We substitute (4.247) into (4.174), noticing that in the present case $t < \frac{1}{2}$, so that a $(= d)$ also fulfils the first condition in (4.242) *p.p.*, whence $\mathscr{A}^*(t)$ is a Behrend sequence as required.

4.5 A conjecture of Erdös and Rényi

In this final section we discuss the accuracy of the Poisson model for $R(g)$ provided by Theorem 4.4 when $\lambda = 2^k/N$ is large. We recall that one of the hypotheses of this theorem is that $\log\lambda = O(1)$. If we ignore this condition and allow λ to be large, then we are led to the conjecture that almost surely

$$\text{card}\{g \in G : R(g) = r\} = 0 \tag{4.248}$$

when λ is so large that

$$e^{-\lambda}\frac{\lambda^r}{r!} < \frac{1}{N}. \tag{4.249}$$

Thus we expect to require $\lambda > \log N$ for every element to be represented. If we put

$$\lambda = (1+\theta)\log N, \; r = s\log N \tag{4.250}$$

then (4.249) reduces, approximately, to

$$\theta > s\log(1+\theta) - s\log s + s. \tag{4.251}$$

For positive θ, (4.251) holds if $s < s_1(\theta)$ or $s > s_2(\theta)$ where

$$0 < s_1(\theta) < 1 + \theta < s_2(\theta) < \infty \tag{4.252}$$

and so we expect the values of $R(g)$ to cluster about their average λ, with approximately normal distribution of variance λ. It is not difficult to show that as θ grows, the ratio $s_2(\theta)/s_1(\theta) \to 1$, whence $R(g) \sim \lambda$ for all g.

The above argument is speculative. In their original paper, Erdös and Rényi proved (Theorem 1) that if

$$2^k \geq \varepsilon^{-2}\delta^{-1}N^2 \tag{4.253}$$

then with probability $\geq 1 - \delta$ we have for all g,

$$(1 - \varepsilon)\lambda \leq R(g) \leq (1 + \varepsilon)\lambda. \tag{4.254}$$

In fact this is a straightforward deduction from the formula

$$\mathbf{Exp} \sum_{g \in G} (R(g) - \lambda)^2 = \left(1 - \frac{1}{N}\right) 2^k \tag{4.255}$$

which follows from (4.17). Notice that k in (4.253) is very large, indeed $\lambda > N$. Nevertheless, Erdös and Rényi conjectured that *the exponent of* N *in* (4.253) *cannot be reduced if* (4.254) *is to hold, without extra conditions on the group structure.*

Although it disagrees with the heuristic model derived from Theorem 4.4 explained above, the conjecture seems reasonable. Various authors including Miech (1967), Bognár (1970), Hall (1972), and Hall and Sudbery (1972) studied this conjecture, in each case reducing the exponent of N in (4.253) at the expense of restrictions on G, more precisely on the orders of the elements. These arise because the high moments

$$\mu_m(G) = \mathbf{Exp} \sum_{g \in G} R(g)^m \tag{4.256}$$

depend on the group structure: Bognár (1970) showed that this is the case if $m \geq 4$.

Erdös and Hall (1976b) obtained the following result. This disproves the conjecture of Erdös and Rényi, but there is still a large gap between what we know to be true and the heuristic model.

Theorem 4.24 *Let A be fixed and*

$$\frac{1}{\varepsilon} < N^{A/\log\log N} \tag{4.257}$$

$$k\log 2 > \left(1 + B\frac{\log\log\log N}{\log\log N}\right)\log N \tag{4.258}$$

where $B = B(A)$ depends on A only. Then with probability $\to 1$ we have, for all g, that (4.254) holds.

Although there is some way to go before such results are applicable to, say, the discrepancy estimates in Chapter 6, the author believes that this is one of the promising avenues of further study.

5

Divisor density

5.1 Introduction

The idea of divisor density is due to Hall (1978).

Definition 5.1 *Let $\mathscr{A}$ be a (positive) integer sequence and $0 \le z \le 1$. We say that $\mathscr{A}$ possesses divisor density z, and write $\mathbf{D}\mathscr{A} = z$ in this circumstance, if*

$$\tau(n, \mathscr{A}) = \text{card}\{a : a \in \mathscr{A}, a|n\} = (z + o(1))\,\tau(n)\, p.p., \tag{5.1}$$

that is there exists a further sequence of asymptotic density 1 *on which the ratio $\tau(n, \mathscr{A})/\tau(n)$ converges to z.*

Thus a random divisor of a large random integer has probability z of belonging to $\mathscr{A}$. This only makes sense if $\tau(n)$ is large (indeed if z is irrational (5.1) requires $\tau(n) \to \infty$) and so we only require (5.1) to hold *p.p.*

We stress that the $o(1)$ term in (5.1) tends to zero *as n tends to infinity* (through a suitable sequence of asymptotic density 1), not as $\tau(n) \to \infty$. This latter condition would be stronger but too rigid for our purpose. We may of course assume if we wish that on the sequence of asymptotic density 1 referred to in our definition, we have $\tau(n) \to \infty$ or even

$$\tau(n) > (\log n)^{\log 2 - \varepsilon}, \tag{5.2}$$

and we leave as an exercise for the reader to check that if z is not a dyadic rational then the weaker of these assumptions is obligatory.

The definition is unusual for two main reasons: firstly the involvement of another, supplementary sequence of asymptotic density 1 and secondly because it is not at all clear from the start that any sequence $\mathscr{A}$ exists having divisor density z, $z \ne 0$ or 1. Indeed, Erdös was initially foxed and wrote in a letter dated January 1976 that it should be possible to prove that whenever it exists, divisor density must be equal to 0 or 1. Erdös observed that no arithmetic progression other than $\mathbf{Z}^+$ (or $\mathbf{Z}^+ \setminus [1, n_0]$) has divisor density and nor does the sequence of squarefree numbers. (The reader should verify these observations.) However, the sequence

$$\mathscr{A}_0(z) = \{a : \log a \leq z (\mathrm{mod}\ 1)\} \tag{5.3}$$

(i.e. the fractional part of $\log a$ lies in $[0, z]$) satisfies $\mathbf{D}\mathscr{A}_0(z) = z$: this follows from Theorem 5.14 below. This was the first example of such a sequence, but is not logically the most straightforward. We construct a fairly basic sequence $\mathscr{A}$ below such that $\mathbf{D}\mathscr{A} = z$, from first principles. (The reader may not care to take an example such as (5.3) on trust, fearing that the remainder of the chapter may be nugatory.) Once convinced by these examples Erdös immediately made a far-sighted conjecture about divisor density which we shall meet later.

Clearly if $\mathbf{D}\mathscr{A} = z > 0$ then $\mathscr{A}$ is a Behrend sequence. Even when $\mathbf{D}\mathscr{A} = 0$, it may be possible to show that there exists a sequence $\{z_n\}$ such that $z_n \to 0$, and

$$z_n > \frac{1}{\tau(n)}, \quad \tau(n, \mathscr{A}) \sim z_n \tau(n)\ p.p. \tag{5.4}$$

which again implies $\mathscr{A}$ is Behrend. Thus we might have $z_n = (\log n)^{-\beta}, n \geq 2, \beta < \log 2$. This idea will be taken up in the next chapter.

We note two straightforward consequences of our definition. If $\mathscr{A}_1$ and $\mathscr{A}_2$ are disjoint sequences such that $\mathbf{D}\mathscr{A}_1 = z_1$ and $\mathbf{D}\mathscr{A}_2 = z_2$ then $\mathbf{D}(\mathscr{A}_1 \cup \mathscr{A}_2) = z_1 + z_2$. If $\mathscr{A}_1'$ is the complement of $\mathscr{A}_1$ then $\mathbf{D}\mathscr{A}_1' = 1 - z_1$.

We proceed to the construction of a sequence $\mathscr{A}_1(z)$ such that $\mathbf{D}\mathscr{A}_1(z) = z$. This cannot be too easy in view of Erdös' remarks quoted above. We shall see later that there is quite a simple result (Theorem 5.2) about the structure of an integer sequence with divisor density which immediately excludes sequences of the wrong form; and this will help to explain the 'counterexamples' mentioned above.

Let $a \in \mathbf{Z}$, $b \in \mathbf{Z}^+$. We begin by showing that the sequence

$$\mathscr{A} = \{d : \Omega(d) \equiv a (\mathrm{mod}\ b)\} \tag{5.5}$$

has $\mathbf{D}\mathscr{A} = 1/b$. We have (with $e(\theta) = \exp(2\pi i \theta)$),

$$\tau(n, \mathscr{A}) = \frac{1}{b} \sum_{r=0}^{b-1} e\left(\frac{-ar}{b}\right) \sum_{d|n} e\left(\frac{r\Omega(d)}{b}\right). \tag{5.6}$$

The inner sum is multiplicative and when n is prime it has absolute value $2\,|\cos(\pi r/b)|$. Let $m(n)$ be the product of the prime factors p of n such that $p^2 \nmid n$. Then from (5.6) we have

$$\left|\tau(n,\mathscr{A})-\frac{1}{b}\tau(n)\right| \le \frac{1}{b}\tau\left(\frac{n}{m(n)}\right)\sum_{r=1}^{b-1}\left|2\cos\frac{\pi r}{b}\right|^{\omega(m)} \le \tau(n)\left(\cos\frac{\pi}{b}\right)^{\omega(m)}. \tag{5.7}$$

(We assume $b \ge 2$ else there is nothing to prove.) We have $\omega(m(n)) = \omega(n) - \omega(\frac{n}{m(n)})$ and since the average order of $\omega(\frac{n}{m(n)})$ is a constant, for $\xi(n) \to \infty$ as $n \to \infty$ we have $\omega(\frac{n}{m(n)}) < \xi(n)\, p.p.$, whence $\omega(m(n))$ has normal order $\log\log n$. Hence the right-hand side of (5.7) is $o(\tau(n))$, and $\mathbf{D}\mathscr{A} = 1/b$ as required. Now set, for $i = 1, 2, 3, \ldots$,

$$\begin{aligned}\mathscr{A}_i &= \{d : \Omega(d) \equiv 2^{i-1} (\bmod\ 2^i)\}\\ &= \{d : 2^{i-1} \| \Omega(d)\}.\end{aligned} \tag{5.8}$$

From the above, $\mathbf{D}\mathscr{A}_i = 2^{-i}$. Let the binary expansion of z be

$$z = .\varepsilon_1\varepsilon_2\varepsilon_3 \ldots \ (\varepsilon_i = 0 \text{ or } 1). \tag{5.9}$$

We claim that

$$\mathscr{A} = \bigcup\{\mathscr{A}_i : \varepsilon_i = 1\} \tag{5.10}$$

possesses divisor density $\mathbf{D}\mathscr{A} = z$. This requires a little care since it is not true in general that an infinite disjoint union of sequences possessing divisor density has divisor density equal to the sum of the individual densities; a gegenbeispiel is $\mathscr{B}_j = \{j\}$, $(j = 1, 2, 3, \ldots)$ in which $\mathbf{D}\mathscr{B}_j = 0$ for every j but $\cup\mathscr{B}_j = \mathbf{Z}^+$. We therefore give a detailed argument employing the selection principle.

For $\varepsilon > 0$ let $\mathscr{B}(\varepsilon)$ denote the sequence of integers n such that

$$\left|\frac{\tau(n,\mathscr{A})}{\tau(n)} - z\right| > \varepsilon \tag{5.11}$$

where $\mathscr{A}$ and z are given by (5.9) and (5.10). We claim that if $\mathscr{B}(\varepsilon)$ has zero asymptotic density for every ε then $\mathbf{D}\mathscr{A} = z$. To see this let $\chi(n, \varepsilon)$ be the characteristic function of $\mathscr{B}(\varepsilon)$ so that by hypothesis

$$\sum_{n\le x}\chi(n,\varepsilon) = o(x). \tag{5.12}$$

Let $j \in \mathbf{Z}^+$. By (5.12) there exists x_j such that $x > x_j$ implies

$$\sum_{n\le x}\chi(n, j^{-1}) < j^{-1}x. \tag{5.13}$$

Put $\chi(n) = \chi(n, j^{-1})$ for $x_j < x \le x_{j+1}$. For these values of x we have

$$\sum_{n \le x} \chi(n) \le \sum_{n \le x} \chi(n, j^{-1}) < j^{-1}x, \tag{5.14}$$

because $\chi(n, j^{-1})$ is an increasing function of j, whence

$$\sum_{n \le x} \chi(n) = o(x), \quad x \to \infty. \tag{5.15}$$

The sequence on which $\chi(n) = 0$ therefore has asymptotic density 1 and from the defintion of $\chi(n)$ and (5.11) we have $(\tau(n, \mathscr{A})/\tau(n)) \to z$ on this sequence, that is $\mathbf{D}\mathscr{A} = z$.

It remains to show that $\mathbf{d}\mathscr{B}(\varepsilon) = 0$ always. There exists h so that

$$.\varepsilon_1\varepsilon_2\varepsilon_3 \dots \varepsilon_h > z - \frac{1}{2}\varepsilon \tag{5.16}$$

and we truncate the union (5.10) at h. We can add divisor densities in a finite disjoint union whence $\tau(n, \mathscr{A}) > \left(z - \frac{1}{2}\varepsilon + o(1)\right) \tau(n)\, p.p.$ Next, put

$$\mathscr{A}_i^+ = \bigcup_{j > i} \mathscr{A}_j = \{d : \Omega(d) \equiv 0 (\text{mod } 2^i)\}. \tag{5.17}$$

Then $\mathbf{D}\mathscr{A}_i^+ = 2^{-i}$. Also there exists k such that

$$.\varepsilon_1\varepsilon_2\varepsilon_3 \dots \varepsilon_k 111 \dots < z + \frac{1}{2}\varepsilon \tag{5.18}$$

and since

$$\mathscr{A} \subseteq \bigcup_{i=1}^{k} \{\mathscr{A}_i : \varepsilon_i = 1\} \cup \mathscr{A}_k^+ \tag{5.19}$$

we have $\tau(n, \mathscr{A}) < (z + \frac{1}{2}\varepsilon + o(1))\, p.p.$ – together with the lower bound already obtained this establishes that $\mathbf{d}\mathscr{A}(\varepsilon) = 0$, whence $\mathbf{D}\mathscr{B} = z$. Thus for each $z \in [0, 1]$ there exists an integer sequence $\mathscr{A}$ which possesses divisor density z.

We now state our first theorem. This easy result, already referred to above, is very useful for showing that a given sequence cannot have divisor density; indeed even in such cases as the squarefree numbers it provides a negative test which is as quick as a direct argument. Moreover we gain some insight into the structure of sequences with divisor density. Surprisingly this theorem made its first appearance rather late in Hall and Tenenbaum (1986).

Theorem 5.2 *Let $\mathscr{A}$ be an integer sequence. In order that $\mathscr{A}$ should possess divisor density $\mathbf{D}\mathscr{A}$ equal to z, it is necessary that for each fixed integer q we have $\mathbf{D}\mathscr{A}(q) = z$, where*

$$\mathscr{A}(q) := \{m \in \mathbf{Z}^+ : mq \in \mathscr{A}\}. \tag{5.20}$$

As an example, let us dismiss the possibility that the sequence of squarefree numbers should possess divisor density, by taking q to be any square exceeding 1. $\mathscr{A}(q)$ is empty, so that z would equal zero, which is clearly false.

Proof We have $\mathscr{A}(q_1q_2) = \mathscr{A}(q_1)(q_2)$ for every q_1 and q_2, hence we may assume that q is prime. Then

$$\begin{aligned} \tau(n, \mathscr{A}(q)) &= \operatorname{card}\{d : d|n,\ dq \in \mathscr{A}\} \\ &= \operatorname{card}\{d : dq|nq,\ dq \in \mathscr{A}\} \\ &= \tau(nq, \mathscr{A}) - \operatorname{card}\{t : t|nq,\ t \in \mathscr{A},\ q \nmid t\} \\ &= \tau(nq, \mathscr{A}) - \tau(nq^{-\beta}, \mathscr{A}), \end{aligned} \tag{5.21}$$

where $q^\beta \| n$. Let us fix β. By hypothesis, $\mathbf{D}\mathscr{A} = z$ whence the right-hand side of (5.21) is

$$(z + o(1))\, \tau(nq) - (z + o(1))\, \tau(nq^{-\beta}) = (z + o(1))\, \tau(n)\ p.p.,$$

for the n such that $q^\beta \| n$, which have asymptotic density $q^{-\beta}(1 - q^{-1})$. Hence if $\beta \le \beta_0$ where $\beta_0 \to \infty$ sufficiently slowly, we have $\tau(n, \mathscr{A}(q)) = (z + o(1))\, \tau(n)$, that is $\mathbf{D}\mathscr{A}(q) = z$ as required.

There is no sufficiency version of this theorem in the literature, and it is difficult to envisage a situation in which it would be easier to establish that $\mathscr{A}(q)$ possesses divisor density than that $\mathscr{A}$ itself does so. However it would be of some interest from a theoretical standpoint to determine minimal conditions on a sequence $\mathscr{Q} \subseteq \mathbf{Z}^+ \setminus \{1\}$ such that if $\mathbf{D}\mathscr{A}(q) = z$ for every $q \in \mathscr{Q}$ then necessarily $\mathbf{D}\mathscr{A} = z$. By Theorem 5.2 we may restrict our attention to primitive $\mathscr{Q}$. (There is a mistake in Hall and Tenenbaum (1986) at this point, where we stated that it would be sufficient to consider the case when $\mathscr{Q}$ contains primes only.) An alternative proof of Theorem 5.2 follows from Theorem 5.9 below; Theorem 5.9 is also likely to be relevant to this problem about $\mathscr{Q}$.

Let

$$\sum_{d \in \mathscr{A}} d^{-1} = \infty. \tag{5.22}$$

Then

$$\sum_{n\le x}\tau(n,\mathscr{A})\sim x\sum_{\substack{d\le x\\ d\in\mathscr{A}}}d^{-1} \tag{5.23}$$

and the right-hand side of (5.23) is

$$(z+o(1))x\log x=(z+o(1))\sum_{n\le x}\tau(n) \tag{5.24}$$

if and only if $\mathscr{A}$ has logarithmic density $\boldsymbol{\delta}\mathscr{A}=z$. This might suggest some causal connection between divisor density and logarithmic densities, although of course we know from the examples already considered that $\boldsymbol{\delta}\mathscr{A}=z$ cannot, in general, imply $\mathbf{D}\mathscr{A}=z$. The truth is quite clear-cut.

Theorem 5.3 *Divisor and logarithmic density are independent. More precisely, given* $(z,w)\in[0,1]^2$ *there exists a sequence* $\mathscr{A}$ *such that* $\boldsymbol{\delta}\mathscr{A}=z$, $\mathbf{D}\mathscr{A}=w$, *or if we wish, either one of these equations holds and the other density is not defined.*

We shall deduce this from the second part (concerning $\mathscr{D}^{\pm}$) of the following theorem. We include the first, negative, part of this theorem as a precautionary note: it might at first sight be imagined that these sequences should possess divisor density.

Theorem 5.4 *Let* z *be a fixed real number. Then the sequence*

$$\mathscr{D}(z)=\{d\ge 3:\ \omega(d)\le\frac{1}{2}\log\log d+z\sqrt{\log\log d}\}$$

does not possess divisor density. On the other hand the sequence

$$\mathscr{D}^{+}(z)=\{d\ge 3:\ \omega(d)\le\frac{1}{2}\log\log d+\theta(d)\sqrt{\log\log d}\}$$

in which $\theta(d)\to\infty$, *has* $\mathbf{D}\mathscr{D}^{+}=1$. *If instead* $\theta(d)\to-\infty$, *the corresponding sequence being denoted by* $\mathscr{D}^{-}$, *then* $\mathbf{D}\mathscr{D}^{-}=0$.

We begin with some lemmas.

Lemma 5.5 *Let* $0\le\alpha<y<2$. *Then*

$$\sum_{n\le x}y^{\Omega(n)}\sum_{d|n}{}'\left(\frac{\log n}{\log d}\right)^{\alpha}\ll_{\alpha,y}x(\log x)^{2y-1}, \tag{5.25}$$

where tha dash indicates that $d=1$ *is omitted from the sum.*

Proof of lemma Denote the sum above by S. We invert summations, put $n = md$, and employ the inequalities

$$(\log md)^\alpha \ll (\log m)^\alpha + (\log d)^\alpha, \tag{5.26}$$
$$\sum_{m\le w} y^{\Omega(m)} \ll_y w(\log 2w)^{y-1}, \tag{5.27}$$

of which (5.27) is a special case of (0.38). We obtain

$$S \ll_y S(x,\alpha) + S(x,0) \tag{5.28}$$

where

$$\begin{aligned} S(x,\alpha) &= {\sum_{d\le x}}' \frac{y^{\Omega(d)}}{(\log d)^\alpha}\frac{x}{d}\left(\log\frac{2x}{d}\right)^{y-1+\alpha} \\ &\ll {\sum_{d\le x}}' \frac{y^{\Omega(d)}}{(\log d)^\alpha}\sum_{r\le x/d}(\log 2r)^{y-1+\alpha} \\ &\ll \sum_{r\le x}(\log 2r)^{y-1+\alpha}{\sum_{d\le x/r}}'\frac{y^{\Omega(d)}}{(\log d)^\alpha} \\ &\ll \sum_{r\le x}(\log 2r)^{y-1+\alpha}\frac{x}{r}\left(\log\frac{2x}{r}\right)^{y-1-\alpha}. \end{aligned} \tag{5.29}$$

At this point we note that from our hypotheses both exponents exceed -1, whence we obtain

$$S(x,\alpha) \ll_{\alpha,y} x(\log x)^{2y-1}$$

which we insert into (5.28). This proves the lemma.

Lemma 5.6 *Let $\xi(n) \to \infty$ as $n \to \infty$. Then we have*

$$\operatorname{card}\{d > 1 : d|n,\ \log\log d < \log\log n - \xi(n)\} = o(\tau(n)) \ p.p.$$

Proof Denote this quantity by $\Delta(n,\xi)$. For these divisors of n we have

$$\frac{\log n}{\log d} > e^{\xi(n)},$$

and we apply Lemma 5.5 with $y = \frac{1}{2}$ and $\alpha = \frac{1}{3}$, which implies that

$$\sum_{n\le x}\frac{\Delta(n,\xi)}{2^{\Omega(n)}}e^{\xi(n)/3} \ll x \tag{5.30}$$

whence for all but $o(x)$ integers $n \le x$ we have

$$\Delta(n,\xi) < e^{-\xi(n)/4}2^{\Omega(n)}.$$

We have $\tau(n) \geq 2^{\omega(n)}$, and for almost all n we also have $\Omega(n) - \omega(n) < \xi(n)/20\log 2$. Hence for $x + o(x)$ integers $n \leq x$,

$$\Delta(n, \xi) < e^{-\xi(n)/5}\tau(n).$$

This completes the proof.

Lemma 5.7 *The number of divisors d of n such that*

$$\omega(d) \leq \frac{1}{2}\omega(n) + t\sqrt{\omega(n)} \tag{5.31}$$

is

$$\left(G_0(2t) + o(1) + O\left(\frac{\Omega(n) - \omega(n)}{\sqrt{\omega(n)}}\right)\right)\tau(n) \tag{5.32}$$

where

$$G_0(u) = \frac{1}{\sqrt{2\pi}}\int_{-\infty}^{u} e^{-v^2/2}dv$$

and the $o(1)$ in (5.32) is a function of $\omega(n)$. This result is uniform in t.

Perhaps the quickest proof is to treat $\omega(d)$ as the result of Bernoulli trials on the squarefree kernel of n, and apply the Central Limit Theorem. We leave it to the reader to construct a proof to his taste.

Proof of Theorem 5.4 We begin by showing that for fixed z, the sequence $\mathscr{D}(z)$ does not possess divisor density. Let $n \geq n_0(z)$ and

$$\omega(n) = \log\log n + w\sqrt{\log\log n}. \tag{5.33}$$

Then if $d|n$ and $d \in \mathscr{D}(z)$ we have

$$\begin{aligned}\omega(d) &\leq \frac{1}{2}\log\log n + z\sqrt{\log\log n}\\ &\leq \frac{1}{2}(\omega(n) - w\sqrt{\omega(n)}) + z\sqrt{\omega(n)} + O\left((1+|z|)\left(|w| + \frac{w^2}{\sqrt{\omega(n)}}\right)\right),\end{aligned} \tag{5.34}$$

that is

$$\omega(d) \leq \frac{1}{2}\omega(n) + \left(z - \frac{1}{2}w + o(1)\right)\sqrt{\omega(n)}. \tag{5.35}$$

Alternatively let $d|n$ and

$$\omega(d) \leq \frac{1}{2}\log\log n + z\sqrt{\log\log n} - (\log\log n)^{\frac{1}{3}}.$$

This is again equivalent to (5.35), and either $d \in \mathscr{D}(z)$ or if not, then

$$\log\log d < \log\log n - \xi(n)$$

where $\xi(n) \to \infty$. By Lemma 5.6 the number of such exceptional d is $o(\tau(n))$ *p.p.*, and so negligible. For all the remaining divisors the condition $d \in \mathscr{D}(z)$ is equivalent to (5.35), and by Lemma 5.7 the number of these divisors is

$$(G_0(2z - w) + o(1))\,\tau(n)\,p.p., \tag{5.36}$$

since we clearly have $\Omega(n) - \omega(n) = o(\sqrt{\omega(n)})\,p.p.$ The variable w defined by (5.33) is normally distributed, whence from (5.36) $\mathbf{D}\mathscr{D}(z)$ does not exist.

We may replace z throughout this argument by a function $\theta(d)$ tending slowly towards $\pm\infty$. The z appearing in (5.36) becomes $\theta(n)$. We require that the error term in (5.34) is $o\left(\sqrt{\omega(n)}\right)$. We deduce that $\mathbf{D}\mathscr{D}^+ = 1$, $\mathbf{D}\mathscr{D}^- = 0$ as claimed. This completes the proof.

Proof of Theorem 5.3 We set

$$\mathscr{B} = \{b \geq 3 : \omega(b) \leq \frac{2}{3}\log\log b\}. \tag{5.37}$$

By the Hardy–Ramanujan theorem we have $\mathbf{d}\mathscr{B} = 0$, and by Theorem 5.4, $\mathbf{D}\mathscr{B} = 1$. Let $\mathscr{B}'$ be the complement of $\mathscr{B}$ so that we have $\boldsymbol{\delta}\mathscr{B} = 0$, $\boldsymbol{\delta}B' = 1$. We can find a sequence $\mathscr{A}_1$ such that $\boldsymbol{\delta}\mathscr{A}_1 = z$, or $\boldsymbol{\delta}\mathscr{A}_1$ does not exist, and a sequence $\mathscr{A}_2$ such that $\mathbf{D}\mathscr{A}_2 = w$, or $\mathbf{D}\mathscr{A}_2$ does not exist. We put

$$\mathscr{A} = (\mathscr{A}_1 \cap \mathscr{B}') \cup (\mathscr{A}_2 \cap \mathscr{B})$$

which has the properties required. This proves the theorem.

We have now shown that for each $z \in [0, 1]$ there exists a sequence $\mathscr{A}$ such that $\mathbf{D}\mathscr{A} = z$, moreover this definition of density is quite distinct from the densities with which we are more familiar. A sequence $\mathscr{A}$ possessing positive divisor density is necessarily Behrend; the converse is false (witness the sequence of primes). This provides a method (sometimes the only method known) for establishing that certain sequences are Behrend.

In the next section we shall give several necessary and/or sufficient conditions for $\mathbf{D}\mathscr{A}$ to exist.

5.2 Necessary and sufficient conditions

In the previous section we introduced the idea of divisor density and showed that it is non-vacuous. We now seek conditions which are either necessary or sufficient, and in one important case both, for a given integer sequence to have divisor density.

We recall that an essential part of the definition requires the existence of a supplementary sequence of asymptotic density 1 on which $\tau(n, \mathscr{A}) = (z + o(1))\,\tau(n)$. This is unavoidable, but the condition can be eased in a rather surprising and useful way. We shall find that it is sufficient to have $\tau(n, \mathscr{A}) = (z + o(1))\,\tau(n)$ on a sequence of logarithmic density 1.

From a practical point of view this is rather important. If we want to employ Dirichlet series, with or without complex variable methods, to establish that a given sequence has divisor density, the series employed are almost certain to have abscissae of convergence $\sigma_0 = 1$. We then need a Tauberian theorem to infer back from properties of the infinite series in the half-plane $\sigma > 1$ to properties of the finite set of integers not exceeding x. The most natural tool at this point will be the Hardy-Littlewood-Karamata theorem, which will leave a residual weight n^{-1} on the integer n, $n \leq x$. Therefore our conclusion will be about logarithmic density.

We may compare what happens when we consider sets of multiples rather than divisor density. Here the classical theorem is that of Davenport and Erdös (1937), (1951), (Theorem 0.2) that $\mathscr{M}(\mathscr{A})$ possesses logarithmic density. The pre-war proof of this theorem reproduced in Chapter 0 depends on the Hardy-Littlewood-Karamata theorem in just the manner described above. However in this case there is no possibility of a second step to asymptotic density since by Besicovitch's theorem (Theorem 0.1) $\mathbf{d}\mathscr{M}(\mathscr{A})$ may not exist.

These observations notwithstanding, we recall that the Davenport-Erdös theorem contains a second conclusion, that the lower asymptotic density $\underline{\mathbf{d}}\mathscr{M}(\mathscr{A})$ of a set of multiples is equal to the logarithmic density $\boldsymbol{\delta}\mathscr{M}(\mathscr{A})$. As we pointed out in Chapter 0, this means that $\boldsymbol{\delta}\mathscr{M}(\mathscr{A}) = 1$, that is $\mathscr{A}$ is Behrend, already implies $\mathbf{d}\mathscr{M}(\mathscr{A}) = 1$. It was this fact which originally suggested the truth of the next theorem, (although it is not actually used directly in the proof). We shall employ the notation *p.p.l.* to mean 'on a sequence of logarithmic density 1'.

Theorem 5.8 *Let $\mathscr{A}$ be an integer sequence such that*

$$\tau(n, \mathscr{A}) = (z + o(1))\,\tau(n)\ p.p.l. \tag{5.38}$$

Then $\mathscr{A}$ has divisor density $\mathbf{D}\mathscr{A} = z$.

Proof We deduce from (5.38) that

$$\sum_{n \le x} \frac{1}{n} \left(\frac{\tau(n, \mathscr{A})}{\tau(n)} - z \right)^2 = o(\log x) \tag{5.39}$$

and the essence of the proof is to deduce from (5.39) that

$$\sum_{n \le x} \left(\frac{\tau(n, \mathscr{A})}{\tau(n)} - z \right)^2 = o(x). \tag{5.40}$$

However the factor $\tau(n)^{-2}$ is technically rather inconvenient and we consider the simpler sums

$$L(x) := \sum_{n \le x} \frac{(\tau(n, \mathscr{A}) - z\tau(n))^2}{n 4^{\Omega(n)}} \tag{5.41}$$

and

$$S(x) := \sum_{n \le x} \frac{1}{4^{\Omega(n)}} (\tau(n, \mathscr{A}) - z\tau(n))^2 . \tag{5.42}$$

Since $\tau(n) \le 2^{\Omega(n)}$ (5.39) implies

$$L(x) = o(\log x). \tag{5.43}$$

Suppose that we can prove that

$$S(x) = o(x). \tag{5.44}$$

This implies (5.40). To see this, notice that since the average order of $\Omega(n) - \omega(n)$ is a constant, for any $\xi(x) \to \infty$ we have

$$\frac{2^{\Omega(n)}}{\tau(n)} \le 2^{\Omega(n) - \omega(n)} \le \xi(x) \tag{5.45}$$

for all but $o(x)$ integers $n \le x$. Hence the sum on the left of (5.40) does not exceed

$$\xi(x)^2 S(x) + o(x) = o(x) \tag{5.46}$$

provided $\xi(x) \to \infty$ sufficiently slowly. Of course (5.40) implies that $\mathbf{D}\mathscr{A} = z$.

We therefore set out to deduce (5.44) from (5.43), and, denoting the characteristic function of $\mathscr{A}$ by χ, we begin by defining

$$r(m) := \sum \{(\chi(d) - z)(\chi(d') - z) : [d, d'] = m\} \tag{5.47}$$

so that we have both

$$L(x) = \sum_{m \le x} \frac{r(m)}{m.4^{\Omega(m)}} \sum_{h \le x/m} \frac{1}{h.4^{\Omega(h)}} \tag{5.48}$$

and

$$S(x) = \sum_{m \le x} \frac{r(m)}{4^{\Omega(m)}} \sum_{h \le x/m} \frac{1}{4^{\Omega(h)}}. \tag{5.49}$$

For each fixed $y \in (0, 2)$ we have

$$\sum_{n \le x} y^{\Omega(n)} = \left(C(y) + O\left(\frac{1}{\log x}\right)\right) x(\log x)^{y-1} \tag{5.50}$$

where

$$C(y) = \frac{H(y)}{\Gamma(y)}, \; H(y) = \prod_p \left(1 - \frac{1}{p}\right)^y \left(1 - \frac{y}{p}\right)^{-1}. \tag{5.51}$$

We put $y = \frac{1}{4}$, and deduce from (5.48) and (5.50) that

$$L(x) = 4C\left(\frac{1}{4}\right) \int_1^x \left(\log \frac{x}{t}\right)^{\frac{1}{4}} dR(t) + O(1). \tag{5.52}$$

We truncate the sum (5.49) at x^η and $x^{1-\eta}$, $(0 < \eta < \frac{1}{2})$. Since $|r(m)| \le 3^{\Omega(m)}$ we have

$$\begin{aligned}
\sum_{m \le x^\eta} \frac{|r(m)|}{4^{\Omega(m)}} \sum_{h \le x/m} \frac{1}{4^{\Omega(h)}} &\ll \sum_{m \le x^\eta} \left(\frac{3}{4}\right)^{\Omega(m)} \frac{x}{m} \left(\log \frac{x}{m}\right)^{-3/4} \\
&\ll x(\log x)^{-3/4} (\eta \log x)^{\frac{3}{4}} \\
&\ll \eta^{\frac{3}{4}} x.
\end{aligned}$$

Also

$$\begin{aligned}
\sum_{x^{1-\eta} < m \le x} \frac{|r(m)|}{4^{\Omega(m)}} \sum_{h \le x/m} \frac{1}{4^{\Omega(h)}} &\ll \sum_{h \le x^\eta} \left(\frac{1}{4}\right)^{\Omega(h)} \sum_{m \le x/h} \left(\frac{3}{4}\right)^{\Omega(m)} \\
&\ll \sum_{h \le x^\eta} \left(\frac{1}{4}\right)^{\Omega(h)} \frac{x}{h} \left(\log \frac{x}{h}\right)^{-1/4} \\
&\ll \eta^{\frac{1}{4}} x.
\end{aligned}$$

Therefore

$$S(x) = C\left(\frac{1}{4}\right) x \int_{x^\eta}^{x^{1-\eta}} \left(\log \frac{x}{t}\right)^{-3/4} dR(t) + O(\eta^{\frac{1}{4}} x + x(\log x)^{-1/4}) \tag{5.53}$$

where

$$R(t) := \sum_{m \le t} \frac{r(m)}{m.4^{\Omega(m)}}. \tag{5.54}$$

Let $\tilde{L}(s)$ denote the Laplace transform of the function $L(e^v)$, moreover set

$$\tilde{R}(s) := \int_0^\infty e^{-sv} dR(e^v), \ (s > 0). \tag{5.55}$$

From (5.52) we have

$$\begin{aligned} \tilde{L}(s) &= \int_0^\infty e^{-sv} L(e^v) dv \\ &= H\left(\frac{1}{4}\right) \tilde{R}(s) s^{-5/4} + O(s^{-1}). \end{aligned} \tag{5.56}$$

By hypothesis, $L(e^v) = o(v)$ whence $\tilde{L}(s) = o(s^{-2})$ and, from (5.56), $\tilde{R}(s) = o(s^{-3/4})$ as $s \to 0+$. Put $r_1(m) = r(m) + 3^{\Omega(m)}$ so that $r_1(m) \ge 0$ and

$$R_1(t) := \sum_{m \le t} \frac{r_1(m)}{m.4^{\Omega(m)}} \tag{5.57}$$

is increasing, moreover from (5.50) with $y = \frac{3}{4}$, we have

$$R_1(t) = R(t) + \frac{4}{3} C\left(\frac{3}{4}\right) (\log t)^{\frac{3}{4}} + O(1). \tag{5.58}$$

We also have, directly from (5.57),

$$\tilde{R}_1(s) = \tilde{R}(s) + \sum_{m=1}^{\infty} \frac{1}{m^{1+s}} \left(\frac{3}{4}\right)^{\Omega(m)} \tag{5.59}$$

and the second term on the right is

$$\zeta(1+s)^{\frac{3}{4}} \prod_p \left(1 - \frac{1}{p^{1+s}}\right)^{\frac{3}{4}} \left(1 - \frac{3}{4p^{1+s}}\right)^{-1} \sim \frac{H(\frac{3}{4})}{s^{\frac{3}{4}}} \tag{5.60}$$

as $s \to 0+$. Since $\tilde{R}(s) = o(s^{-3/4})$, we deduce from (5.59) and (5.60) that

$$\tilde{R}_1(s) \sim H\left(\frac{3}{4}\right) s^{-3/4}, \ (s \to 0+) \tag{5.61}$$

and the Hardy–Littlewood–Karamata theorem (Lemma 0.4) yields

$$R_1(e^v) \sim \frac{H(\frac{3}{4}) v^{\frac{3}{4}}}{\Gamma(\frac{7}{4})}, \ (v \to \infty). \tag{5.62}$$

Comparison of (5.58) with (5.62) gives

$$R(e^v) = o(v^{\frac{3}{4}}). \tag{5.63}$$

Now consider the integral

$$I(u,\eta) := \int_{\eta u}^{(1-\eta)u} (u-v)^{-3/4} dR(e^v). \tag{5.64}$$

Provided $\eta u \to \infty$, we obtain from (5.63), after integration by parts, that

$$I(u,\eta) = o(1)\eta^{-3/4}. \tag{5.65}$$

We substitute $x = e^u$, $t = e^v$ in (5.53) and provided $\eta \log x \to \infty$, we have

$$\begin{aligned} S(x) &= C\left(\frac{1}{4}\right) xI(u,\eta) + O(\eta^{\frac{1}{4}}x) \\ &= x\{o(1)\eta^{-3/4} + O(\eta^{\frac{1}{4}})\} \end{aligned} \tag{5.66}$$

by (5.65). If $\eta \to 0$ sufficiently slowly as $x \to \infty$, the condition $\eta \log x \to \infty$ is satisfied and the right-hand side of (5.66) is $o(x)$. This is (5.44), and the result follows.

The next result is a necessary and sufficient condition for a given sequence to have divisor density. The sufficiency part is particularly useful and will be the basis of several proofs in this chapter. The result is due to Tenenbaum (1982) and pre-dates Theorem 5.8, which appeared in Hall and Tenenbaum (1986). With hindsight, we may now use Theorem 5.8 to simplify the sufficiency proof given by Tenenbaum – because we need only establish the (apparently) weaker condition (5.38) to obtain $\mathbf{D}\mathscr{A} = z$.

Theorem 5.9 *Let $\mathscr{A}$ be an integer sequence and χ be its characteristic function. Then a necessary and sufficient condition for $\mathbf{D}\mathscr{A} = z$ is that, as x tends to infinity, we have*

$$\sum_{k<x} \left| \sum_{\substack{n<x \\ n\equiv 0 (\mathrm{mod}\ k)}} \frac{\chi(n)-z}{n4^{\Omega(n)}} \right| = o(\sqrt{\log x}). \tag{5.67}$$

We remark that Theorem 5.2 is an immediate corollary of Theorem 5.9. If we restrict k to be a multiple of q, and put $k = hq$, $n = mq$, $x = yq$ in (5.67) we obtain, for fixed q and $y \to \infty$,

$$\sum_{h<y} \left| \sum_{\substack{m<y \\ m\equiv 0 (\mathrm{mod}\ h)}} \frac{\chi(mq)-z}{m4^{\Omega(m)}} \right| = o(\sqrt{\log y}) \tag{5.68}$$

so that $\mathbf{D}\mathscr{A} = z \Rightarrow \mathbf{D}\mathscr{A}(q) = z$ as required.

Proof of Theorem 5.9 We begin by defining, for $\sigma > 0$,

$$f(\sigma) := \sum_{n=1}^{\infty} \frac{(\tau(n,\mathscr{A}) - z\tau(n))^2}{n^{1+\sigma}4^{\Omega(n)}} \tag{5.69}$$

and we claim that $\mathbf{D}\mathscr{A} = z$ if and only if we have

$$f(\sigma) = o(\sigma^{-1}) \text{ as } \sigma \to 0+. \tag{5.70}$$

First, if $\mathbf{D}\mathscr{A} = z$ then the sum $S(x)$ defined in (5.42) satisfies $S(x) = o(x)$, whence (5.70) holds, by integration by parts. Second, if (5.70) holds then, on setting $\sigma = 1/\log x$ we have as $x \to \infty$,

$$L(x) = \sum_{n \le x} \frac{(\tau(n,\mathscr{A}) - z\tau(n))^2}{n4^{\Omega(n)}} = o(\log x) \tag{5.71}$$

from which we deduce (5.38), that is

$$\tau(n,\mathscr{A}) = (z + o(1))\,\tau(n)\, p.p.l., \tag{5.72}$$

and $\mathbf{D}\mathscr{A} = z$ follows from this by Theorem 5.8. (We proceed from (5.71) to (5.72) via (5.39). The step from (5.71) to (5.39) is analogous to that from (5.44) to (5.40) in the proof of Theorem 5.8.)

We require an alternative formula for $f(\sigma)$ and we introduce the functions

$$g(\sigma) := \sum_{n=1}^{\infty} \frac{1}{n^{1+\sigma}4^{\Omega(n)}} \sim H\left(\frac{1}{4}\right)\sigma^{-1/4}, \ (\sigma \to 0+), \tag{5.73}$$

$$f_k(\sigma) := \sum_{t=1}^{\infty} \frac{\chi(kt) - z}{t^{1+\sigma}4^{\Omega(t)}}, \tag{5.74}$$

where H is as in (5.51). We also notice that

$$m^{1+\sigma}4^{\Omega(m)} = \sum_{k|m} k^{1+\sigma}4^{\Omega(k)} \prod_{p|k}\left(1 - \frac{1}{4p^{1+\sigma}}\right) \tag{5.75}$$

and we denote the product on the right by $\lambda(k,\sigma)$. We have

$$\begin{aligned} f(\sigma) &= \sum_{n=1}^{\infty} \frac{1}{n^{1+\sigma}4^{\Omega(n)}} \sum_{d,d'|n} (\chi(d) - z)(\chi(d') - z) \\ &= g(\sigma) \sum_{d=1}^{\infty}\sum_{d'=1}^{\infty} \frac{(\chi(d) - z)(\chi(d') - z)}{[d,d']^{1+\sigma}4^{\Omega([d,d'])}} \\ &= g(\sigma) \sum_{d=1}^{\infty}\sum_{d'=1}^{\infty} \frac{\chi(d) - z)(\chi(d') - z)}{(dd')^{1+\sigma}4^{\Omega(dd')}} \sum_{k|(d,d')} k^{1+\sigma}4^{\Omega(k)}\lambda(k,\sigma) \end{aligned} \tag{5.76}$$

by (5.73) and (5.75). We change the order of summation, writing $d = kt$, $d' = kt'$ to obtain

$$f(\sigma) = g(\sigma) \sum_{k=1}^{\infty} \frac{\lambda(k,\sigma)}{k^{1+\sigma}4^{\Omega(k)}} f_k(\sigma)^2. \tag{5.77}$$

We are now in a position to prove the sufficiency part of our theorem. We write

$$A_k(u) := \sum_{\substack{n<e^u \\ n\equiv 0 (\mathrm{mod}\ k)}} \frac{\chi(n) - z}{n4^{\Omega(n)}} \tag{5.78}$$

so that (5.68) reads

$$\sum_{k=1}^{\infty} |A_k(u)| = o(\sqrt{u}), \ (u \to \infty). \tag{5.79}$$

We have

$$\begin{aligned} \sum_{k=1}^{\infty} \frac{|f_k(\sigma)|}{k^{1+\sigma}4^{\Omega(k)}} &= \sum_{k=1}^{\infty} \left| \int_0^{\infty} e^{-\sigma u} dA_k(u) \right| \\ &= \sigma \sum_{k=1}^{\infty} \left| \int_0^{\infty} e^{-\sigma u} A_k(u) du \right| \\ &\leq \int_0^{\infty} e^{-\sigma u} \sum_{k=1}^{\infty} |A_k(u)|\, du \\ &= o(\sigma^{-1/2}) \end{aligned} \tag{5.80}$$

as $\sigma \to 0+$. From (5.73) and (5.74) we have $|f_k(\sigma)| \leq g(\sigma)$; moreover $0 < \lambda(k,\sigma) \leq 1$ whence (5.77) yields

$$f(\sigma) \leq g(\sigma)^2 \sum_{k=1}^{\infty} \frac{|f_k(\sigma)|}{k^{1+\sigma}4^{\Omega(k)}} = o(\sigma^{-1}) \tag{5.81}$$

by (5.73) and (5.80). This is (5.70), and we have seen that $\mathbf{D}\mathscr{A} = z$ follows.

Next, we remark on the proof of necessity: we may assume that (5.70) holds, whence from (5.73) and (5.77) we have

$$\sum_{k=1}^{\infty} \frac{\lambda(k,\sigma)}{k^{1+\sigma}4^{\Omega(k)}} f_k(\sigma)^2 = o(\sigma^{-3/4}), \ (\sigma \to 0+). \tag{5.82}$$

It is not difficult to check that we have

$$\sum_{k=1}^{\infty} \frac{\lambda(k,\sigma)^{-1}}{k^{1+\sigma}4^{\Omega(k)}} \ll \sigma^{-1/4}, \ (\sigma \to 0+). \tag{5.83}$$

and (5.82), (5.83) and the Cauchy-Schwarz inequality yield

$$\sum_{k=1}^{\infty} \frac{|f_k(\sigma)|}{k^{1+\sigma}4^{\Omega(k)}} = o(\sigma^{-1/2}), \ (\sigma \to 0+), \tag{5.84}$$

that is, see (5.80),

$$\sum_{k=1}^{\infty} \left| \int_0^{\infty} e^{-\sigma u} dA_k(u) \right| = o(\sigma^{-1/2}), \ (\sigma \to 0+). \tag{5.85}$$

We are going to show that this implies (5.79) and our argument follows the main lines of Karamata's proof of the Hardy–Littlewood–Karamata theorem. The details are not quite straightforward and we procced in stages. Firstly, we deduce from (5.85) that for each fixed polynomial P we have

$$\sum_{k=1}^{\infty} \left| \int_0^{\infty} e^{-\sigma u} P(e^{-\sigma u}) dA_k(u) \right| = o(\sigma^{-1/2}). \tag{5.86}$$

We introduce the familiar function

$$Q_o(v) = \begin{cases} 0, & 0 \le v \le e^{-1} \\ v^{-1}, & e^{-1} < v \le 1 \end{cases} \tag{5.87}$$

and we approximate Q_0 in a certain sense by polynomials. More specifically, for each $\varepsilon > 0$ we construct, by Weierstrass' approximation theorem, a polynomial $Q^{(\varepsilon)}(v)$ such that

$$0 \le Q_0(v) - Q^{(\varepsilon)}(v) \le 1, \qquad (0 \le v \le 1) \tag{5.88}$$

$$Q_0(v) - Q^{(\varepsilon)}(v) \le \varepsilon \qquad \text{if } n \notin (e^{-1-\varepsilon}, e^{-1}). \tag{5.89}$$

From (5.86) and (5.87) we have

$$\begin{aligned} \sum_{k=1}^{\infty} \left| A_k\left(\frac{1}{\sigma}\right) \right| &= \sum_{k=1}^{\infty} \left| \int_0^{\infty} e^{-\sigma u} Q_0(e^{-\sigma u}) dA_k(u) \right| \\ &\le J_1(\sigma, \varepsilon) + J_2(\sigma, \varepsilon) \end{aligned} \tag{5.90}$$

where

$$\begin{aligned} J_1(\sigma, \varepsilon) &= \sum_{k=1}^{\infty} \left| \int_0^{\infty} e^{-\sigma u} Q^{(\varepsilon)}(e^{-\sigma u}) dA_k(u) \right| \\ &= o(\sigma^{-1/2}) \end{aligned} \tag{5.91}$$

$$\begin{aligned} J_2(\sigma, \varepsilon) &= \sum_{k=1}^{\infty} \left| \int_0^{\infty} e^{-\sigma u} \{Q_0(e^{-\sigma u}) - Q^{(\varepsilon)}(e^{-\sigma u})\} dA_k(u) \right| \\ &\le \int_0^{\infty} e^{-\sigma u} \{Q_0(e^{-\sigma u}) - Q^{(\varepsilon)}(e^{-\sigma u})\} dA(u), \end{aligned} \tag{5.92}$$

where

$$A(u) := \sum_{n<e^u} \frac{\tau(n)}{n4^{\Omega(n)}}. \tag{5.93}$$

From (5.88), (5.89) and (5.92) we deduce that

$$J_2(\sigma,\varepsilon) \le \varepsilon \sum_{k=1}^{\infty} \frac{\tau(n)}{n^{1+\sigma}4^{\Omega(n)}} + J_3(\sigma,\varepsilon) \tag{5.94}$$

where

$$J_3(\sigma,\varepsilon) := \int_{1/\sigma}^{(1+\varepsilon)/\sigma} dA(u) = A\left(\frac{1+\varepsilon}{\sigma}\right) - A\left(\frac{1}{\sigma}\right). \tag{5.95}$$

The first term on the right of (5.94) is $\ll \varepsilon\sigma^{-1/2}$. Also $A(u) \sim A\sqrt{u}$ for a suitable constant A, whence from (5.95), $J_3(\sigma,\varepsilon) \ll \varepsilon\sigma^{-1/2}$ for $0 < \sigma \le \sigma_0(\varepsilon)$. From (5.90), (5.91), (5.94) and (5.95) we now have

$$\sum_{k=1}^{\infty} \left| A_k\left(\frac{1}{\sigma}\right)\right| \ll \varepsilon\sigma^{-1/2}, \quad 0 < \sigma \le \sigma_1(\varepsilon) \tag{5.96}$$

that is, (5.79) holds – and this is just (5.68) as required. This completes the proof.

One of the striking differences between the definitions of asymptotic and logarithmic densities on the one hand, and divisor density on the other, is that in the former cases the existence of the density $\mathbf{d}\mathscr{A} = z$ or $\boldsymbol{\delta}\mathscr{A} = z$ as it may be, is expressed immediately and simply by an asymptotic relation of the form

$$\sum_{n\le x} \chi(n)f(n) = (z + o(1)) \sum_{n\le x} f(n) \tag{5.97}$$

in which χ is the characteristic function of $\mathscr{A}$ and f is a simple weight, 1 for asymptotic, and n^{-1} for logarithmic density; whereas in the latter case, no such relation appears in the definition of divisor density, and it is not obvious that (5.97), with some suitable f, either implies or is implied by the proposition $\mathbf{D}\mathscr{A} = z$. Returning to Theorem 5.9, certainly condition (5.67) has some resemblance to (5.97) but of course the imposition that we must add the absolute values of the separate sums arising from different k makes it much stronger: we cannot disentangle the z as in (5.97). However, an immediately corollary of Theorem 5.9 and the triangle inequality is that whenever $\mathbf{D}\mathscr{A} = z$ we must have

$$\sum_{n\le x} \frac{\chi(n)\tau(n)}{n4^{\Omega(n)}} = (z + o(1)) \sum_{n\le x} \frac{\tau(n)}{n4^{\Omega(n)}}, \tag{5.98}$$

so that we have a necessary condition with the same shape as (5.97). There is no loss in the triangle inequality when $z = 0$ or 1 and so in these cases (5.98) is also sufficient for $\mathbf{D}\mathscr{A} = z$. For $0 < z < 1$ this condition cannot be sufficient as it would assign a divisor density to every arithmetic progression, and to the sequence of squarefree numbers.

In the original paper (Hall (1978)) on divisor density it was shown in Theorem 3 that $\mathbf{D}\mathscr{A} = z$ implies (5.97) for a certain positive multiplicative function f, different from that which appears in (5.98). At first sight this discrepancy seems surprising, because we cannot readily interchange distinct multiplicative functions in a relation like (5.97); and although the explanation is simple enough we are led to an interesting general question about multiplicative functions which we shall discuss later. For the moment we concentrate on our next result which provides a necessary condition for a sequence to have divisor density: if $\mathbf{D}\mathscr{A} = z$ then (5.97) holds for a large class of functions f; and the two functions mentioned above are just particular examples. Such a class first appeared in Hall (1981), and a wider class was given in Tenenbaum (1982). The following theorem includes the latter result.

Theorem 5.10 *Let $g : \mathbf{Z}^+ \to \mathbf{R}^+ \cup \{0\}$ be multiplicative, and such that*

(i) $\sum \{|g(p) - 1|/p$: *all primes* $p\} < \infty$,
(ii) *there exist fixed θ_1 and θ_2, with $\theta_1 > 0$, $2 > \theta_2 > 0$, such that $0 \le g(p^v) \le \theta_1\theta_2^v$, for all prime powers p^v.*

Then if $\mathscr{A}$ is an integer sequence having divisor density $\mathbf{D}\mathscr{A} = z$, we have

$$\sum_{n\le x} \frac{\chi(n)g(n)}{n\tau(n)} = (z + o(1)) \sum_{n\le x} \frac{g(n)}{n\tau(n)}, \tag{5.99}$$

where χ is the characteristic function of $\mathscr{A}$.

The condition (5.99) is certainly not sufficient to imply $\mathbf{D}\mathscr{A} = z$ when $0 < z < 1$, for any g in the class. In general it is also insufficient when $z = 0$ or 1, an exception being the g implicit in (5.98). To justify our first remark, notice that for every g in the class there exists a constant $c_0(g)$ such that

$$\sum_{n\le x} \frac{g(n)}{n\tau(n)} \sim c_0(g)(\log x)^{\frac{1}{2}}. \tag{5.100}$$

To see this, we observe that

$$\begin{aligned}
&\int_0^\infty e^{-su} d\Big\{\sum_{n\le e^u} \frac{g(n)}{n\tau(n)}\Big\} \\
&= \sum_{n=1}^{\infty} \frac{g(n)}{n^{1+s}\tau(n)} \\
&= \prod_p \left(1 + \frac{g(p)}{2p^{1+s}} + \frac{g(p^2)}{3p^{2+2s}} + \cdots\right) \\
&\sim s^{-1/2} H\left(\frac{1}{2}\right) \prod_p \left(1 - \frac{1}{2p}\right)\left(1 + \frac{g(p)}{2p} + \frac{g(p^2)}{3p^2} + \cdots\right) \qquad (5.101)
\end{aligned}$$

as $s \to 0+$, with H as in (5.51). We apply the Hardy–Littlewood–Karamata theorem to deduce (5.100). Now let $\mathscr{P}$ be a set of primes for which

$$\sum\{p^{-1} : p \in \mathscr{P}\} < \infty \qquad (5.102)$$

and let $\mathscr{A}$ be the sequence of integers with no prime factor in $\mathscr{P}$. Then

$$\sum_{n\le x} \frac{\chi(n)g(n)}{n\tau(n)} \sim wc_0(g)(\log x)^{\frac{1}{2}} \qquad (5.103)$$

where

$$w = \prod_{p\in\mathscr{P}} \left(1 + \frac{g(p)}{2p} + \frac{g(p^2)}{3p^2} + \cdots\right)^{-1} \qquad (5.104)$$

and we can arrange $\mathscr{P}$ so that w takes any prescribed value $z \in (0,1)$. Thus (5.99) holds, but $\mathbf{D}\mathscr{A}$ does not exist.

Our second remark is easier to justify. We put $g(n) = \mu(n)^2$ and $\mathscr{A}$ the sequence of squarefree numbers, so that (5.99) holds with $z = 1$, or $z = 0$ if we take the complement of $\mathscr{A}$.

In Hall (1978) there appears the conjecture that there is no function f, multiplicative or otherwise, such that (5.97) standing alone implies that $\mathscr{A}$ possesses divisor density z. This conjecture has not received much attention – it is after all negative – but it quickly leads into questions about multiplicative functions and their mean values which are of independent interest (indeed would seem to be fairly basic). We return to this question and its ramifications after the following proof.

Proof of Theorem 5.10 In view of (5.100) we have to show that

$$\sum_{n\le x} \frac{(\chi(n) - z)g(n)}{n\tau(n)} = o(\sqrt{\log x}). \qquad (5.105)$$

We are going to derive this from Theorem 5.9, i.e. (5.67), and to this end we write

$$\frac{4^{\Omega(n)}g(n)}{\tau(n)} = \sum_{k|n} h(k) \tag{5.106}$$

which defines the multiplicative function h via the Möbius Inversion Formula. From (5.106) we have $h(p) = 2g(p) - 1$ so that

$$\sum_p \frac{|h(p)-1|}{p} = 2\sum_p \frac{|g(p)-1|}{p} < \infty \tag{5.107}$$

and we also have, for $v \geq 2$ and every prime p,

$$|h(p^v)| \leq \theta_1(4\theta_3)^v, \ \theta_3 := \max(1, \theta_2). \tag{5.108}$$

We fix $r > 1$ such that $(4\theta_3)^r < 8$ and put $s = r/(r-1)$. Also we write

$$A_k := \sum_{\substack{n \leq x \\ n \equiv 0 (\mathrm{mod}\ k)}} \frac{\chi(n) - z}{n^{\Omega(n)}} \tag{5.109}$$

and we note that, uniformly for $k \leq x$,

$$A_k \ll \frac{(\log x)^{\frac{1}{4}}}{k4^{\Omega(k)}}. \tag{5.110}$$

The sum on the left of (5.105) is

$$\sum_{k \leq x} h(k)A_k \tag{5.111}$$

and we apply Hölder's inequality, with exponents r and s (as specified above) to obtain

$$\left|\sum_{k\leq x} h(k)A_k\right| \leq \left(\sum_{k\leq x} |h(k)|^r\,|A_k|\right)^{1/r}\left(\sum_{k\leq x}|A_k|\right)^{1/s}. \tag{5.112}$$

By hypothesis, $\mathbf{D}\mathscr{A} = z$ and so by Theorem 5.9, (5.67) holds. Hence the second factor on the right of (5.112) is $o((\log x)^{1/2s})$. We have, by (5.108) and (5.110),

$$\begin{aligned}\sum_{k\leq x} |h(k)|^r\,|A_k| &\ll (\log x)^{\frac{1}{4}} \sum_{k\leq x} \frac{|h(k)|^r}{k4^{\Omega(k)}} \\ &\ll (\log x)^{\frac{1}{4}} \prod_{p\leq x}\left(1 + \frac{|h(p)|^r}{4p} + \frac{|h(p^2)|^r}{16p^2} + \cdots\right) \\ &\ll (\log x)^{\frac{1}{2}} \prod_{p\leq x}\left(1 + \frac{r\,|h(p)-1|}{4p} + \frac{r(r-1)|h(p)-1|^2}{8p}\right)\end{aligned} \tag{5.113}$$

since $r < 2$ (automatically) and so for $u \geq 0$,

$$(1+u)^r \leq 1 + ru + \frac{1}{2}r(r-1)u^2.$$

The product on the right of (5.113) is bounded by (5.107) and so the first factor on the right of (5.112) is $\ll (\log x)^{1/2r}$, whence the sum on the left-hand side is $o(\sqrt{\log x})$. This establishes (5.105) and completes the proof of the theorem.

We return to the question whether or not the relation

$$\sum_{n \leq x} \chi(n) f(n) = (z + o(1)) \sum_{n \leq x} f(n) \qquad (5.114)$$

in which χ is the characteristic function of the integer sequence $\mathscr{A}$ can imply $\mathbf{D}\mathscr{A} = z$. For simplicity we restrict f to be non-negative, but we do not suppose at this point that it is multiplicative. We must have

$$\sum_{n=1}^{\infty} f(n) = \infty, \qquad (5.115)$$

for if the infinite series were convergent we could set $\mathscr{A} = \mathbf{Z}^+ \setminus \{n_0\}$ for some n_0 such that $f(n_0) > 0$: (5.114) would then hold with $z < 1$, whereas $\mathbf{D}\mathscr{A} = 1$.

Next, we have seen that (5.98) implies $\mathbf{D}\mathscr{A} = z$ when $z = 0$ or 1, and so our question becomes nugatory unless (5.114) can hold for some $z \in (0, 1)$. The necessary condition for this is

$$f(n) = o\left(\sum_{m<n} f(m)\right), \text{ as } n \to \infty, \qquad (5.116)$$

moreover (5.116) implies that for every z, $0 < z < 1$, there exists χ or, what is the same thing, a sequence $\mathscr{A}$ for which (5.114) holds.

Definition 5.11 *We denote by $\mathscr{F}^*$ the class of non-negative functions f satisfying (5.115) and (5.116), and by $\mathscr{M}\mathscr{F}^*$ the sub-class of multiplicative functions belonging to $\mathscr{F}^*$. For each $f \in \mathscr{F}^*$ and $z \in (0,1)$, $K(f,z)$ denotes the class of integer sequences $\mathscr{A}$ such that (5.114) holds.*

Notice that if g satisfies the hypotheses (i), (ii) of Theorem 5.10 and $f_1(n) = g(n)/(n\tau(n))$ then $f_1 \in \mathscr{M}\mathscr{F}^*$.

Suppose now that there exists an $f \in \mathscr{F}^*$ with the property that (5.114) implies $\mathbf{D}\mathscr{A} = z$. By Theorem 5.10, (5.114) holds with f replaced by any of the functions f_1 described above, that is

$$K(f,z) \subseteq \cap\{K(f_1,z), \text{ all } f_1\}. \qquad (5.117)$$

The function f must be very oscillatory, because $K(f,z)$ may not contain any sequence, such as the sequence of squarefree numbers or an arithmetic progression, which does not possess divisor density. It seems likely that this must admit some bizarre elements into $K(f,z)$ which do not belong to the class on the right of (5.117).

Conjecture 5.12 *There is no function $f \in \mathscr{F}^*$ such that (5.114) implies $\mathbf{D}\mathscr{A} = z$.*

Failing this, we make the following conjecture, which is a question purely concerning multiplicative functions.

Conjecture 5.13 *Let $e, f \in \mathscr{M}\mathscr{F}^*$ and $K(f,z) \subseteq K(e,z)$ for all $z \in (0,1)$. Then there exists a constant γ such that*

$$e(n) \equiv n^{\gamma} f(n). \tag{5.118}$$

Clearly if Conjecture 5.13 is true then any gegenbeispiel to Conjecture 5.12 lies outside $\mathscr{M}\mathscr{F}^*$.

5.3 Slowly switching sequences

Initially Erdös believed that the relation

$$\tau(n,\mathscr{A}) = (z + o(1))\,\tau(n)\ p.p., \tag{5.119}$$

could not hold for $0 < z < 1$, but he was soon convinced that (5.119) was possible, by the example (5.7) and, more significantly, by the example

$$\mathscr{A} = \{d : \log d \leq z (\text{mod } 1)\}. \tag{5.120}$$

At this point Erdös made the following conjecture, perhaps the most important in the development of this subject: *let*

(i) $u_{j+1} > cv_j,\ v_j > cu_j$ *for* $j \in \mathbf{Z}^+$ *and some fixed* $c > 1$,
(ii) $\mathscr{A} = \bigcup\{\mathbf{Z}^+ \cap (u_j, v_j] : j = 1, 2, 3, \ldots\}$
(iii) $\delta\mathscr{A} = z$.

Then $\mathscr{A}$ has divisor density $\mathbf{D}\mathscr{A} = z$.

Of course, if true, this includes (5.120) as a special case. At first sight the conjecture flies in the face of Theorem 5.3 which states that divisor density and logarithmic densities are independent: however we recall that Theorem 5.3 depends on Theorem 5.4, and we cannot impose this sort of restriction on $\omega(d)$ for the integers d contained in a relatively long block such as $(u_j, v_j]$; it is essentially a Poisson variable. Blocks of consecutive integers occur elsewhere in this book, e.g. in §1.3, where we are concerned with certain kinds of Behrend sequences: the point is that the multiplicative structure of the integers in a long block is in effect random, or random enough; depending on the precise context the blocks can sometimes have quite moderate length. For example the blocks which appear in Corollary 1.10 are of type $(u, u + u^\alpha]$ where α is arbitrarily small.

A proof of the above conjecture of Erdös may be found in Hall (1978). Since then it has been realized that it is not so much the length of the blocks but their frequency which matters for a result of this sort, in other words we have to limit the number of *switches* in and out of $\mathscr{A}$ which can occur up to x by some function of x. If we extend the characteristic function of $\mathscr{A}$ onto $\mathbf{R}^+$ by setting

$$\chi(w) = \begin{cases} 1 & \text{if } w \in \cup(u_j, v_j] \\ 0 & \text{if } w \leq u_1 \text{ or } w \in \cup(v_j, u_{j+1}] \end{cases} \tag{5.121}$$

then our condition becomes an upper bound on the total variation of χ on $[0, x]$. The following is a result of this type.

Theorem 5.14 (Hall & Tenenbaum (1986)) *Let*

(iii) $\mathscr{A} = \bigcup\{\mathbf{Z}^+ \cap (u_j, v_j] \,:\, j = 1, 2, 3, \ldots\}$

(iii) $\boldsymbol{\delta}\mathscr{A} = z$,

(iii) $\operatorname{card}\{j \,:\, u_j \leq x\} \leq R(x^\eta)$

where $\eta(x) \to 0$ *as* $x \to \infty$, *and* $R(x)$ *is an increasing function such that* $R(x) \leq x$ *and such that there exists a corresponding Stieltjes measure* $\lambda(t)$ *on* $[0, \frac{1}{2}]$ *satisfying both*

(iv) $\displaystyle\sum_{n \leq x} (\frac{1}{4})^{\Omega(n)} = \int_0^{1/2} x^{1-t} d\lambda(t) + O(x/R(x))$,

and

(v) $|d\lambda(t)| \ll t^{-1/4} dt, \; (0 < t \leq \frac{1}{2})$.

Then $\mathscr{A}$ *has divisor density* $\mathbf{D}\mathscr{A} = z$.

Notice that (iii) implies $u_j/j \to \infty$. Since

$$\sum_{u_j < a \leq v_j} a^{-1} = \int_{u_j}^{v_j} \frac{dw}{w} + O(u_j^{-1}),$$

condition (ii) translates immediately into

(ii′) $\int_1^x \frac{\chi(w)-z}{w}\,dw = o(\log x)$.

At first sight the theorem looks rather technical, and it may be helpful for us to see immediately how the truth of Erdös' conjecture follows from it. We set

$$d\lambda(t) = \frac{H(\frac{1}{4})t^{-1/4}dt}{\Gamma(\frac{1}{4})\Gamma(\frac{3}{4})} \tag{5.122}$$

with H as in (5.51). By (5.50), condition (iv) is satisfied with $R(x) = (\log x)^{7/4}$. We choose $\eta(x)$ such that $\eta(x) \to 0$, $\log x \ll (\eta(x)\log x)^{7/4}$, and conditions (i)–(v) are all satisfied. By the theorem, $\mathbf{D}\mathscr{A} = z$.

The formula (5.122) is far from optimal, and we have

Theorem 5.15 (Hall & Tenenbaum (1986)) *There exists a constant $c > 0$ and a Stieltjes measure λ such that conditions (iv) and (v) of Theorem 5.14 are satisfied with*

$$R(x) = \exp\left(\frac{c_1(\log x)^{\frac{3}{5}}}{(\log\log x)^{\frac{1}{5}}}\right). \tag{5.123}$$

Proof of Theorem 5.14 We write

$$A_k := \sum_{\substack{n\le x\\ n\equiv 0 (\mathrm{mod}\ k)}} \frac{\chi(n)-z}{n4^{\Omega(n)}} = \frac{1}{k4^{\Omega(k)}}B_k \tag{5.124}$$

where

$$B_k := \sum_{m\le x/k} \frac{\chi(mk)-z}{m4^{\Omega(m)}} \ll \left(\log\frac{x}{k}\right)^{\frac{1}{4}}. \tag{5.125}$$

We assume that $\eta(x) \to 0$ as $x \to \infty$ so slowly that condition (iii) above holds; and we set

$$K = K(x) = x^{1-\eta(x)}. \tag{5.126}$$

For $k > K$, the trivial bound on the right of (5.125) yields

$$B_k \ll \eta(x)^{\frac{1}{4}}(\log x)^{\frac{1}{4}}. \tag{5.127}$$

We write

$$M(x) := \sum_{m\le x}\left(\frac{1}{4}\right)^{\Omega(m)} =: \int_0^{\frac{1}{2}} x^{1-t}d\lambda(t) + E(x) \tag{5.128}$$

say, and we have for $k \le K$,

$$B_k = O\left(\eta(x)^{\frac{1}{4}}(\log x)^{\frac{1}{4}}\right) + \int_{x^\eta}^{x/k} \frac{\chi(uk) - z}{u} dM(u). \tag{5.129}$$

We split the integral on the right into two parts J_1 and J_2 arising from the two terms on the right of (5.128). The first part is

$$\begin{aligned} J_1 &:= \int_{x^\eta}^{x/k} \frac{\chi(uk) - z}{u} \int_0^{\frac{1}{2}} (1-t)u^{-t} d\lambda(t)\, du \\ &= \int_0^{\frac{1}{2}} (1-t)k^t d\lambda(t) \int_{kx^\eta}^{x} \frac{\chi(v) - z}{v^{1+t}} dv \end{aligned} \tag{5.130}$$

on substituting $u = v/k$. By hypothesis (ii'), there is a function $\varepsilon(x) \downarrow 0$ such that

$$\int_1^x \frac{\chi(w) - z}{w} dw \ll \varepsilon(x) \log x \tag{5.131}$$

and, by (5.131) and the Second Mean Value Theorem, the innermost integral on the extreme right of (5.130) is $\ll (kx^\eta)^{-t}\varepsilon(x)\log x$ if, as we may assume, $\varepsilon(x)\log x$ is an increasing function of x. Hence

$$\begin{aligned} J_1 &\ll \varepsilon(x) \log x \int_0^{\frac{1}{2}} x^{-\eta t}\, |d\lambda(t)| \\ &\ll \varepsilon(x)\eta(x)^{-3/4}(\log x)^{\frac{1}{4}} \end{aligned} \tag{5.132}$$

by condition (v) above. Next,

$$\begin{aligned} J_2 &= \int_{x^\eta}^{x/k} \frac{\chi(uk) - z}{u} dE(u) \\ &= \left[\frac{\chi(uk) - z}{u} E(u)\right]_{x^\eta}^{x/k} + \int_{x^\eta}^{x/k} \frac{\chi(uk) - z}{u^2} E(u) du \\ &\quad + \int_{x^\eta}^{x/k} \frac{1}{u} E(u) d\chi(uk) \end{aligned} \tag{5.133}$$

after an integration by parts. By condition (iv), $E(u) \ll u/R(u)$, where R is increasing, whence

$$J_2 \ll \int_{x^\eta}^{\infty} \frac{du}{uR(u)} + \frac{1}{R(x^\eta)} \int_0^{x/k} d\chi(uk). \tag{5.134}$$

We may assume $R(u) \ge (\log u)^{7/4}$, so that the integral is convergent. Hence by hypothesis (iii), we have $J_2 = O(1)$. We deduce from (5.127), (5.129) and (5.132) that

$$B_k \ll \eta(x)^{1/4}(\log x)^{1/4} + \varepsilon(x^\eta)\eta(x)^{-3/4}(\log x)^{1/4} + 1, \tag{5.135}$$

and we may assume that $\eta(x) \to 0$ as $x \to \infty$ so slowly that the right-hand side is $o((\log x)^{\frac{1}{4}})$. Hence

$$\begin{aligned} \sum_{k \le x} |A_k| &= o\left((\log x)^{\frac{1}{4}}\right) \sum_{k \le x} \frac{1}{k 4^{\Omega(k)}} \\ &= o\left((\log x)^{\frac{1}{2}}\right), \end{aligned} \tag{5.136}$$

and we apply Theorem 5.9 to obtain our result.

This proof is shorter and easier than the one appearing in Hall (1978), which was limited to establishing the truth of Erdös' conjecture. The simplification comes from Tenenbaum's theorem (Theorem 5.9) and arises essentially because the sums A_k are *linear* in χ. If we were to estimate the sum

$$\sum_{n \le x} \left(\frac{\tau(n, \mathscr{A})}{\tau(n)} - z \right)^2 \tag{5.137}$$

or its 'logarithmic' variant (5.39) directly we should not be able to avoid awkward and lengthy calculations involving l.c.m.$[a, a']$ for pairs $a, a' \in \mathscr{A}$. Of course an upper bound for (5.137) is implicit somewhere within the proof – in the presentation here, Tenenbaum's method 'linearizes' the sum (5.39) and the step up to (5.137) is the content of Theorem 5.8.

Proof of Theorem 5.15 We obtain the following more general result: *for complex y, such that $|y| < 1$, there exists a Stieltjes measure λ_y on $[0, \frac{1}{2}]$ such that*

$$|d\lambda_y(t)| \ll_y t^{-\operatorname{Re} y}, \quad 0 < t \le \frac{1}{2} \tag{5.138}$$

and

$$\sum_{n \le x} y^{\Omega(n)} = \int_0^{\frac{1}{2}} x^{1-t} d\lambda_y(t) + O_y\left(\frac{x}{R(x)}\right) \tag{5.139}$$

where R is given by (5.123).

We begin with Perron's formula, (Titchmarsh (1951) Lemma 3.12), with x restricted to have fractional part $\frac{1}{2}$: this restriction may be dropped in the final result. Because $|y| < 1$ this gives

$$\sum_{n \le x} y^{\Omega(n)} = \frac{1}{2\pi i} \int_{c-iT}^{c+iT} \zeta(s, y) \frac{x^s}{s} ds + O\left(\frac{x \log x}{T}\right) \tag{5.140}$$

where $c = 1 + (1/\log x)$, T is at our disposal, and

$$\zeta(s, y) = \sum_{n=1}^{\infty} \frac{y^{\Omega(n)}}{n^s} = \prod_p \left(1 - \frac{y}{p^s}\right)^{-1}. \tag{5.141}$$

Let

$$H(s, y) := \exp\left\{\sum_p \sum_m \frac{y^m - y}{mp^{ms}}\right\}. \tag{5.142}$$

Then H is analytic in the half-plane $\operatorname{Re} s > \frac{1}{2}$, and if $\mathscr{D}$ is a simply connected sub-domain of this half-plane in which ζ has no zeros or pole, then for $s \in \mathscr{D}$ we have

$$\zeta(s, y) = H(s, y)\zeta(s)^y. \tag{5.143}$$

Richert (1967) deduced from Vinogradov's method (Vinogradov (1958)) that

$$\zeta(\sigma + it) \ll |t|^{100(1-\sigma)^{3/2}} (\log|t|)^{\frac{2}{3}}, \tag{5.144}$$

uniformly for $\sigma \geq \frac{1}{2}$, $|t| \geq 2$. Theorem 3.10 of Titchmarsh (1951) then gives the zero-free region $\{\sigma + it : \sigma > 1 - 2\theta(t)\}$ where we may take $\theta(t) = c_0 \left(\log(|t| + 2)\right)^{-2/3} \left(\log\log\left(|t| + 3\right)\right)^{-1/3}$ with a suitable $c_0 > 0$. (We assume $\theta(0) < \frac{1}{4}$.) We obtain $\mathscr{D}$ by inserting a cut in the real axis from $\sigma = 1$ to $1 - 2\theta(0)$. In $\mathscr{D}$ we have

$$|\zeta(s)| \ll \left(\log(|t| + 2)\right)^{\frac{2}{3}}, \quad |s - 1| \gg 1 \tag{5.145}$$

by (5.144), and we deduce that in the smaller region $\{\sigma + it : \sigma \geq 1 - \theta(t)\}$ we have

$$|\log\zeta(s)| \ll \log\log(|t| + 3) \quad |s - 1| \gg 1. \tag{5.146}$$

For this we apply the Borel–Carathéodory inequality (Titchmarsh (1939), p.174) to the function $g(z) := \log\zeta(1 + it + z)$, with concentric circles of radii $r_1 = 1 - \sigma$, $r_2 = c3r_1/2$ centred at $z = 0$. Let us assume $|t| \geq 3$. We have

$$M(r_1, g) \leq \frac{2r_1}{r_2 - r_1} A(r_2, g) + \frac{r_2 + r_1}{r_2 - r_1} |(0)|$$

where $M(r, g) := \sup\{|g(z)| : |z| = r\}$, $A(r, g) := \sup\{\operatorname{Re} g(z) : |z| = r\}$. We have $|g(0)| \ll \log\log|t|$, $A(r_2, g) \ll \log\log|t|$ by (5.145) and (5.146) follows.

Now we move the contour of integration in (5.140) to the contour G comprising the curves $\sigma = 1 - \theta(t)$, $-T \leq t < 0$ and $0 < t \leq T$, horizontal segments $1 - \theta(T) \leq \sigma \leq c$ at $t = \pm T$ and a lacet from

$s_0 := 1 - \theta(0)$ winding positively around $s = 1$. We have assumed (adjusting c_0 if necessary) that $\theta(t_0) < \frac{1}{4}$, and we (approximately) optimize T by requiring that $T \cong x^{\theta(T)}$ or

$$T = \exp\{c(\log x)^{3/5}(\log\log x)^{-1/5}\}, \tag{5.147}$$

say. The contribution to (5.140) from the error term $O(T^{-1}x\log x)$ and all of G except the lacet is absorbed by the error term given in (5.139). We collapse the lacet to two real line segments joining s_0 to a small circle of radius ε, centre 1. Near $s = 1$ we may write

$$\zeta(s, y) =: h(s, y)(s-1)^{-y} \tag{5.148}$$

where $\arg(s-1) := 0$ at $s = 1 + \varepsilon$ and, for $|s-1| < \frac{1}{2}$,

$$h(s, y) = H(s, y)\{1 + \gamma(s-1) + \cdots\}^y. \tag{5.149}$$

(The radius of convergence of the series for $(s-1)\zeta(s)$ is infinite but H is only analytic for $\operatorname{Re} s > \frac{1}{2}$; $(s-1)\zeta(s) \neq 0$ for $|s-1| < 14.13\ldots$). Since $\operatorname{Re} y < 1$ the contribution to the integral from the small circle $\to 0$ as $\varepsilon \to 0$ and so we may finally collapse the lacet to two line segments, on which we substitute $s = 1 - t$, to obtain a main term

$$\int_0^{\frac{1}{2}} x^{1-t}\, d\lambda_y(t) \tag{5.150}$$

in which

$$d\lambda_y(t) = \frac{\sin \pi y}{\pi(1-t)t^y} h(1-t, y) \tag{5.151}$$

for $0 < t \le \theta(t_0)$, $d\lambda_y(t) = 0$ for $\theta(t_0) < t \le \frac{1}{2}$. The factor $\sin \pi y$ arises in the usual way when we evaluate $\arg(s-1)$ on the two edges of the lacet. The reader should compare (5.151) with (5.122), in which we may replace $\frac{1}{4}$ by y, noting of course that $\Gamma(y)\Gamma(1-y) = \pi/\sin \pi y$.

The error term in (5.139) is effectively the same as that in the Prime Number Theorem. However the problems are different in that in the Prime Number Theorem we are committed to a main term *lix* whereas here we may choose, and theoretically optimize, the measure $d\lambda_y(t)$. We are unable to make any tangible advantage of this apparent freedom at the time of writing.

5.4 Slowly switching sequences: a reformulation

We recall that a slowly switching sequence has the form

$$\mathscr{A} = \bigcup\{(u_j, v_j] \cap \mathbf{Z}^+ : j = 1, 2, 3, \ldots\} \tag{5.152}$$

where $\operatorname{card}\{j : u_j \le x\}$ has limited growth as a function of x. Let $z \in (0,1)$ be arbitrary. Then provided $u_j < v_j < u_{j+1}$ for all j, as we assume, there exists an infinite class of continuous, strictly increasing functions f such that

$$\begin{aligned} f(u_j) &= j, \quad j = 1,2,3,\ldots \\ f(v_j) &= j+z, \quad j = 0,1,2,\ldots, \end{aligned} \tag{5.153}$$

where we have set $v_0 := 0$ for convenience. We may then write

$$\mathscr{A} = \{d : f(d) \in (0,z](\text{mod } 1)\}. \tag{5.154}$$

Often we write the condition on d as $f(d) \le z$ (mod 1): notice that we exclude $f(d) \equiv 0$ (mod 1).

The natural choice for z is $\boldsymbol{\delta}\mathscr{A}$ if the logarithmic density exists. If it does not exist, or we choose $z \ne \boldsymbol{\delta}\mathscr{A}$, f will be too oscillatory to serve any useful purpose. Evidently we can employ (5.154) to describe a given sequence or we can choose any strictly increasing f and regard (5.154) as defining $\mathscr{A}$. An example of this latter procedure is the sequence (5.120). If we do this, with the end in view that we shall be able to deduce that $\mathbf{D}\mathscr{A} = z$ from Theorem 5.14, it is clear that we need sufficient conditions on f for the sequence (5.154) to have $\boldsymbol{\delta}\mathscr{A} = z$.

Since the sequence $\{u_j\}$ will have broadly exponential growth it will be convenient to write $f(w) = F(\log w)$, where F is defined on $\mathbf{R}$.

Theorem 5.16 *Let $F : \mathbf{R}^+ \to \mathbf{R}$, $F \to \infty$, and there exist T_0 such that $F'(t) > 0$, $t \ge T_0$, $F \in C^2([T_0,\infty))$. Let*

$$\int_{T_0}^{T} \frac{|F''(t)|}{F'(t)^2}\,dt = o(T), \quad T \to \infty \tag{5.155}$$

and $F(t) = o(e^t)$. Then the sequence

$$\mathscr{A} = \{d : 0 < F(\log d) \le z(\text{mod } 1)\}, \tag{5.156}$$

has logarithmic density z, for each $z \in (0,1]$. If in addition

$$F(t) = \exp\left(o(1).\frac{t^{3/5}}{(\log t)^{1/5}}\right) \tag{5.157}$$

then $\mathscr{A}$ has divisor density z.

The second part of the theorem is a corollary of the first part together with Theorems 5.14 and 5.15. We may set $F(t) = t^\alpha$, $\alpha > 0$ or $F(t) = (\log t)^\beta$, $\beta > 1$ for example, and have $\mathbf{D}\mathscr{A} = z$. Thus the sequence (5.120) has $\mathbf{D}\mathscr{A} = z$.

In the proof, it will be convenient to have $F \in C^2(\mathbf{R})$, $F'(t) > 0$ for all t, with F increasing from z to ∞ on $(-\infty, \infty)$. We may alter F, say for $t < T_1$, to achieve this without upsetting either conclusion.

Proof of Theorem 5.16 We have to prove the first part. Let $f(w) = F(\log w)$, and $g : (z, \infty) \to \mathbf{R}^+$ be the inverse of f. For given x, put $\zeta = f(x)$, $N = [\zeta]$. Then

$$\sum_{\substack{d \le x \\ d \in \mathscr{A}}} d^{-1} = \sum_{n<N} \log\left(\frac{g(n+z)}{g(n)}\right) + \log\left(\frac{\min(x, g(N+z))}{g(N)}\right)$$
$$+ O\left(\sum_{n \le N} g(n)^{-1}\right). \tag{5.158}$$

Since $f(w) = o(w)$ we have $g(n)/n \to \infty$ as $n \to \infty$ and the error term above is $o(\log x)$. We put $h(u) = \log g(u)$, $\Psi(u) = [u] - u + \frac{1}{2}$ and we have

$$\sum_{n=1}^{N} h(n) = \int_1^N h(u)du + \frac{1}{2}h(N) - \int_1^N \Psi(u)h'(u)du. \tag{5.159}$$

There is a similar formula with $h(u)$ replaced by $H(u+z)$ and N by $N-1$. We subtract these, and deduce that the main term on the right of (5.158) is

$$\min\{h(\zeta), h(N+z)\} - \int_{N-1+z}^{N} h(u)du + \frac{1}{2}\{h(N-1+z) - h(N)\}$$
$$- \int_1^{N-1} \Psi(u)\{h'(u+z) - h'(u)\}du + \int_{N-1}^{N} \Psi(u)h'(u)du + O(1).$$
$$\tag{5.160}$$

We approximate $h(u)$, wherever it occurs, by $h(\zeta) = \log x$, and note that

$$\left| \int_1^{N-1} \Psi(u)\{h'(u+z) - h'(u)\}du \right| \le \frac{z}{2} \int_1^N |h''(u)|\, du. \tag{5.161}$$

So (5.160) becomes

$$z \log x + O\left(\int_1^N |h''(u)|\, du + \int_{N-1}^{\zeta} |h'(u)|\, du + 1 \right) \tag{5.162}$$

and since

$$|h'(u)| \leq \int_1^{\zeta} |h''(v)|\,dv + O(1),\ (u \leq \zeta)$$

this gives $(z + o(1)) \log x$ in (5.158) provided

$$\int_1^{\zeta} |h''(u)|\,du = o(\log x). \tag{5.163}$$

We substitute $u = f(w)$ so that the left-hand side becomes

$$\int_{u_1}^{x} \left| d \frac{g'(f(w))}{g(f(w))} \right| = \int_{u_1}^{x} \left| d \left(wf'(w)\right)^{-1} \right| = \int_{u_1}^{x} \left| \frac{1}{w} \frac{F''(\log w)}{F'(\log w)^2} \right| dw \tag{5.164}$$

since $f(w) = F(\log w)$. Finally we substitute $w = e^t$, $x = e^T$ to deduce from (5.155) that the integral in (5.164) is $o(\log x)$. Hence (5.163) holds and $\boldsymbol{\delta}\mathscr{A} = z$ as required. This completes the proof of Theorem 5.16.

The sequence

$$\mathscr{A} = \{d : (\log\log d)^{\beta} \leq z (\mathrm{mod}\ 1)\} \tag{5.165}$$

has logarithmic density if and only if $\beta > 1$, so that in this sense condition (5.155) is best possible; and the theorem shows that this sequence also has divisor density equal to z when $\beta > 1$. It is conceivable that when $\beta \leq 1$ the sequence manages to have divisor density: of course this would not necessarily be equal to z. We may rule this possibility out, for $z \in (0, 1)$, by reference to Theorem 5.10. The proof is by contradiction. Suppose then that $\mathscr{A}$ (as in (5.165)) has $\mathbf{D}\mathscr{A} = w$, and set $g(n) = 1$ identically in the theorem, so that by (5.99),

$$\sum_{\substack{n \leq x \\ n \in \mathscr{A}}} \frac{1}{n\tau(n)} = (w + o(1)) \sum_{n \leq x} \frac{1}{n\tau(n)} = (c_0 w + o(1)) \sqrt{\log x} \tag{5.166}$$

for a suitable (positive) constant c_0. Let $m \to \infty$ through $\mathbf{Z}^+$ and associate with each m, the numbers x, u, v such that

$$\begin{aligned} (\log\log x)^{\beta} &= m + z \\ (\log\log u)^{\beta} &= m \\ (\log\log v)^{\beta} &= m + 1. \end{aligned} \tag{5.167}$$

By hypothesis, $\beta \leq 1$ and $0 < z < 1$ whence

$$\log\left(\frac{x}{u}\right) \gg \log x, \quad \log\left(\frac{v}{x}\right) \gg \log x, \tag{5.168}$$

and we deduce from (5.166), substituting $x = u$, $x = v$, that both

$$\sum_{\substack{u<n\leq x \\ n\in\mathscr{A}}} \frac{1}{n\tau(n)} = (c_0 w + o(1))\left(\sqrt{\log x} - \sqrt{\log u}\right) \tag{5.169}$$

and

$$\sum_{\substack{x<n\leq v \\ n\in\mathscr{A}}} \frac{1}{n\tau(n)} = (c_0 w + o(1))\left(\sqrt{\log v} - \sqrt{\log x}\right). \tag{5.170}$$

For $u < n \leq x$ we have $(\log\log n)^\beta \leq z \pmod 1$ by (5.168), and (5.169) implies that $w = 1$. For $x < n \leq v$ we have $(\log\log n)^\beta > z \pmod 1$, and (5.170) implies that $w = 0$. This is the required contradiction and the proof is complete.

It therefore seems that the first condition, (5.155) of Theorem 5.16, is realistic, and the theorem is a useful generator of sequences with divisor density. The status of condition (5.157) is less clear.

6

Divisor uniform distribution

6.1 Introduction

Let f be a real-valued function defined on the positive integers. We shall be concerned in this chapter with the distribution (mod 1) of $f(d)$ as d runs through the divisors of a large integer n. We are therefore interested in the counting function

$$\operatorname{card}\{d : d|n. \quad 0 < f(d) \le z (\text{mod } 1)\}, \tag{6.1}$$

where $0 < z \le 1$. We confine our attention to uniform distribution, and we ask whether the cardinality in (6.1) is approximated by $z\tau(n)$, and if so then how good is the approximation. We put

$$\mathscr{A}(z) = \mathscr{A}(z; f) = \{a : 0 < f(a) \le z (\text{mod } 1)\} \tag{6.2}$$

so that we may write this cardinality more simply as $\tau(n, \mathscr{A}(z))$ and we define the *discrepancy*

$$\Delta(n; f) = \sup_{0<w<z\le 1} |\tau(n, \mathscr{A}(z)) - \tau(n, \mathscr{A}(w)) - (z - w)\tau(n)|. \tag{6.3}$$

Definition 6.1 (Hall (1976)) *The function f is divisor uniformly distributed if and only if, for every $\varepsilon > 0$, we have*

$$\Delta(n; f) < \varepsilon\tau(n)\, p.p. \tag{6.4}$$

An equivalent definition is that on a suitable sequence of asymptotic density 1 we have

$$\Delta(n; f) = o(\tau(n)). \tag{6.5}$$

Although the formal definition of divisor uniform distribution was given in 1976 it had been shown previously in Hall (1974) that the function $f(d) = \log d$ has this property, moreover that for this particular function we have

$$\Delta(n; f) < \tau(n)^{\frac{1}{2}+\varepsilon} \text{ p.p.} \tag{6.6}$$

There is a warning note for the reader at this point. The function $\log n$ ($n \in \mathbf{Z}^+$) is *not* uniformly distributed (mod 1) in the traditional sense; and we shall soon have other examples of functions f such that $f(n)$ is not uniformly distributed (mod 1) for $n \in \mathbf{Z}^+$ but nevertheless f is divisor uniformly distributed. These functions present themselves naturally. In the opposite direction we shall construct a function which does possess the former property and not the latter. Uniform distribution and divisor uniform distribution are therefore independent, but related, properties.

Let us set $z = 1$ in (6.3) and apply (6.5). We deduce that if f is divisor uniformly distributed then for every $w \in (0, 1)$ we have

$$\tau(n, \mathscr{A}(w)) = (w + o(1))\, \tau(n) \text{ p.p.}, \tag{6.7}$$

equivalently $\mathbf{D}\mathscr{A}(w) = w$. That is, a necessary condition for f to be divisor uniformly distributed is that the sequence $\mathscr{A}(w; f)$ should have divisor density w for every $w \in (0, 1)$. This condition (which also holds, trivially, for $w = 1$) is also sufficient. Let $\mathbf{D}\mathscr{A}(w) = w$, $0 < w \leq 1$ and let $q \in \mathbf{Z}^+$. Then for $1 \leq l \leq q$ we have $\mathbf{D}\mathscr{A}(l/q) = l/q$, whence

$$\left|\tau\left(n, \mathscr{A}\left(\frac{l}{q}\right)\right) - \frac{l}{q}\tau(n)\right| < \frac{1}{q}\tau(n), \quad 1 \leq l \leq q, \text{ p.p.} \tag{6.8}$$

Because $(\tau(n, \mathscr{A}(w))$ is a non-decreasing function of w, we may deduce from (6.8) that

$$|\tau(n, \mathscr{A}(w)) - w\tau(n)| < \frac{2}{q}\tau(n), \quad 0 < w \leq 1, \text{ p.p.}, \tag{6.9}$$

and so (6.4) holds, with $\varepsilon = 4/q$. Since q is arbitrary, we conclude that f is divisor uniformly distributed.

The ideas of divisor uniform distribution and divisor density are closely connected: we note that divisor density was defined later (Hall (1978)), so that examples of sequences possessing divisor density, in particular that in (5.3), were already available when the definition was made. In studying divisor uniform distribution we may distinguish, broadly, two approaches: the application of results about divisor density, and the direct method, employing classical techniques from the theory of uniform distribution. The first approach yields, very quickly, a wide class of divisor uniformly distributed functions but does not readily give explicit upper bounds for the discrepancy $\Delta(n; f)$. For this task the second approach is almost obligatory; the reason for this will be explained at the beginning of §6.4.

Before proceeding further, let us notice a conclusion which may be drawn from (6.6). By the Hardy–Ramanujan theorem, (6.6) implies that for $f(d) = \log d$, we have

$$\Delta(n; f) < (\log n)^{\frac{1}{2}\log 2+\varepsilon}\ p.p. \tag{6.10}$$

whence we may assert that, provided $\mu < \frac{1}{2}\log 2$, almost all integers n possess a divisor d such that

$$0 < \log d \leq (\log n)^{-\mu} (\mathrm{mod}\ 1). \tag{6.11}$$

Hence for the same values of μ the sequence

$$\mathscr{A}_\mu = \{a : 0 < \log a \leq (\log a)^{-\mu} (\mathrm{mod}\ 1)\} \tag{6.12}$$

is Behrend. This result should be compared with Theorem 4.13, which shows that $\mathscr{A}_\mu$ is Behrend if and only if $\mu < \log 2$ (and is even more precise). Now (6.10) is not best possible, and we give a better result later. However it is certainly not known whether or not we have

$$\Delta(n; \log) < (\log n)^{\varepsilon}\ p.p., \tag{6.13}$$

indeed this is one of the longstanding open problems in the area; and we should need (6.13) to hold to be able to infer that $\mathscr{A}_\mu$ is Behrend for every $\mu < \log 2$ by this method. Of course a bound for the discrepancy of any distribution contains more information than the existence of points in short intervals; we may be able to construct Behrend sequences from bounds for discrepancy but such constructions should be viewed as corollaries rather than the main objectives of this theory.

6.2 Applications of divisor density

In this section we show that various functions are divisor uniformly distributed, as a consequence of results in the last chapter.

Theorem 6.2 *Let $F \in C^2(\mathbf{R}^+)$ be increasing from some point onward, and for a suitable T_0 let*

$$\int_{T_0}^{T} \frac{|F''(t)|}{F'(t)^2}\,dt = o(T), \quad T \to \infty \tag{6.14}$$

moreover let

$$F(t) < \exp\left(o(1)\frac{t^{3/5}}{(\log t)^{1/5}}\right). \tag{6.15}$$

Then $f(d) = F(\log d)$ is divisor uniformly distributed.

This is a corollary of Theorem 5.16, which establishes that the sequence $\mathscr{A}(z)$ in (6.2) has divisor density z for each z. It follows that the functions

(i) $(\log d)^\alpha, \quad \alpha > 0$
(ii) $(\log\log d)^\beta, \quad \beta > 1$

are divisor uniformly distributed. Moreover the condition on β is sharp, by the argument following (5.165). We may also prove this directly as follows. Let $1 < u \le 2$. By Dickman's theorem, the integers n for which $P^+(n) > n^{1/u}$ have asymptotic density $\log u$. Half the divisors of such an integer have the form $d = P^+(n)m$, and for these d, and $0 < \beta \le 1$, we have

$$\begin{aligned}(\log\log d)^\beta &= \left(\log\log P^+(n)\right)^\beta + O\left(\frac{\log m}{\log n(\log\log n)^{1-\beta}}\right)\\ &= \left(\log\log P^+(n)\right)^\beta + O\left(1 - \frac{1}{u}\right),\end{aligned} \tag{6.16}$$

because $m \le n/P^+(n) < n^{1-\frac{1}{u}}$. If u is sufficiently close to 1, we may deduce that $(\log\log d)^\beta \in I_n$ for at least $\tau(n)/2$ divisors of n, where I_n is an interval of length $< \frac{1}{3}$. Hence $(\log\log d)^\beta$ is not divisor uniformly distributed.

We shall see later in this chapter that if f is divisor uniformly distributed but has a suitably small derivative, this argument will give a useful lower bound for the discrepancy.

There has been a great deal of work on uniform distribution of sequences $\{F(n) : n \in \mathbf{Z}^+\}$ (see e.g. Kuipers and Niederreiter (1974)), and we take advantage of this in the next theorem. Hall (1978) (Theorem 1, Cor.) proved that if $F \in C^1(\mathbf{R}^+)$ satisfies the conditions

$$\begin{aligned}&\text{(i)} \qquad \{F(n) : n \in \mathbf{Z}^+\} \text{ is uniformly distributed (mod 1)},\\ &\text{(ii)} \qquad F'(x) \to 0 \text{ as } x \to \infty,\end{aligned} \tag{6.17}$$

then the sequence

$$\mathscr{A}(z) = \{a : F(\log a) \le z (\text{mod } 1)\} \tag{6.18}$$

has divisor density $\mathbf{D}\mathscr{A}(z) = z$, $0 < z \le 1$. Hence the function $f(d) = F(\log d)$ is divisor uniformly distributed. Hall and Tenenbaum (1986)

(Théorème 6, Cor. 1) showed that we may take, more generally, $f(d) = F(\theta(d))$ where θ satisfies suitable conditions (not quite stated precisely). We develop this idea a little further here. It is more convenient to work with a function of $\log d$.

Theorem 6.3 *Let F satisfy the two conditions in (6.17) and $f(d) = F(g(\log d))$, where g is such that*

(i) $g'(u)/g(u) \gg 1/u$,

(this condition need hold for $u \geq u_0$ only),

(ii) $\displaystyle\int_{u_0}^{U} \frac{|g(u)g''(u)|}{g'(u)^2}\,du \ll U, \quad U \to \infty,$

and

(iii) $g(u) < \exp\left(o(1).\dfrac{u^{3/5}}{(\log u)^{1/5}}\right), \quad u \to \infty.$

Then f is divisor uniformly distributed.

Thus we may take $g(u) = u^\alpha$, $\alpha > 0$ or $g(u) = \exp(u^\beta)$, $0 < \beta < 3/5$. Condition (iii), like condition (6.15) in Theorem 6.2, goes back to Theorem 5.15, and ultimately depends on the zero-free region for the Riemann ζ-function. Condition (ii) is a quasi-convexity condition: if g'' is ultimately of fixed sign it is weaker than condition (i).

Proof of Theorem 6.3 For $u \geq u_0$, g has an inverse which we denote by h. We put $v_0 = g(u_0)$ and we note that conditions (i) and (ii) translate into

(i′) $\dfrac{h'(v)}{h(v)} \ll \dfrac{1}{v}, \quad v \geq v_0$

(ii′) $\displaystyle\int_{v_0}^{V} |vh''(v)|\,dv \ll h(V), \quad V \to \infty.$

We shall have to consider the second forward difference of the function h, and we note that

$$h(m+2) - 2h(m+1) + h(m) = \int_m^{m+2} (1 - |v - m - 1|)h''(v)dv. \qquad (6.19)$$

It follows that we have, for integers M,

$$\begin{aligned}\sum_{v_0 \leq m < M} m\,|h(m+2) - 2h(m+1) - h(m)| \;&\leq\; \int_{v_0}^{M+1} v\,|h''(v)|\,dv \\ &\ll\; h(M+1) \\ &\ll\; h(M). \qquad (6.20)\end{aligned}$$

The first inequality follows from (6.19), the second from condition (ii′), and the third from condition (i′). After these preliminaries, we are able to follow the proof in Hall (1978) fairly closely. Let $z \in (0,1)$ and $\xi < \min(z, 1-z)$. We consider the sequences

$$\begin{aligned} \mathscr{A}^{+} &= \{a : F([g(\log a)] \leq z + \xi (\mathrm{mod}\ 1)\} \\ \mathscr{A}^{-} &= \{a : F([g(\log a)] \leq z - \xi (\mathrm{mod}\ 1)\}. \end{aligned} \tag{6.21}$$

Our first objective is to prove that $\boldsymbol{\delta}\mathscr{A}^{\pm} = z \pm \xi$, and we restrict our attention to $\mathscr{A}^{+}$, writing $z + \xi = w$ for convenience. We denote by $\chi(n)$ the characteristic function of the sequence $\{n : F(n) \leq w (\mathrm{mod}\ 1)\}$ and we put

$$\lambda(m) = \sum \left\{ \frac{1}{a} : m \leq g(\log a) < m+1 \right\}, \tag{6.22}$$

$$M = [g(\log x)]. \tag{6.23}$$

We have from (6.21)–(6.23) that

$$\sum_{a \leq x\ a \in \mathscr{A}^{+}} \frac{1}{a} = \sum_{m<M} \chi(m)\lambda(m) + O\left(\lambda(M)\right) \tag{6.24}$$

and, by hypothesis that F is uniformly distributed, we also have

$$\sum_{n \leq m} \chi(n) = wm + \varepsilon_m m, \tag{6.25}$$

where $\varepsilon_m \to 0$ as $m \to \infty$. By partial summation, we deduce from (6.24)–(6.25) that

$$\begin{aligned} \sum_{a \leq x\ a \in \mathscr{A}^{+}} \frac{1}{a} &= w \log x + O\left(\lambda(M) + \varepsilon_M M \lambda(M)\right) \\ &\quad + O\left(\sum_{m<M} \varepsilon_m m \left|\lambda(m) - \lambda(m+1)\right| \right). \end{aligned} \tag{6.26}$$

From (6.22), for $m \geq v_0$,

$$\begin{aligned} \lambda(m) &= \sum \left\{ \frac{1}{a} : e^{h(m)} \leq a < e^{h(m+1)} \right\} \\ &= h(m+1) - h(m) + O(e^{-h(m)}). \end{aligned} \tag{6.27}$$

Condition (i′) implies that

$$\begin{aligned} h(m+1) - h(m) &= \int_m^{m+1} h'(v)dv \ll h(m+1) \int_m^{m+1} \frac{dv}{v} \\ &\ll \frac{h(m+1)}{m} \ll \frac{h(m)}{m}. \end{aligned} \tag{6.28}$$

Condition (iii) implies that $h(m) \gg (\log m)^{5/3}$ and together with (6.27) and (6.28) this yields

$$m\lambda(m) \ll h(m), \quad m \geq v_0. \tag{6.29}$$

Hence the first error term on the right of (6.26) is

$$\ll (M^{-1} + \varepsilon_M)h(M) = o(\log x) \tag{6.30}$$

because $h(M) \leq \log x$ by (6.23). As for the second error term, we have from (6.27),

$$\lambda(m+1) - \lambda(m) = h(m+2) - 2h(m+1) + h(m) + O\left(\frac{1}{m^3}\right), \tag{6.31}$$

using the lower bound $h(m) \gg (\log m)^{\frac{5}{3}}$, whence by (6.20),

$$\sum_{v_0 \leq m < M} m\,|\lambda(m) - \lambda(m+1)| \ll h(M) + 1. \tag{6.32}$$

It follows that the second error term in (6.26) is $o\,(h(M)) = o(\log x)$, and $\boldsymbol{\delta}\mathscr{A}^+ = w = z + \xi$ as required. Clearly $\boldsymbol{\delta}\mathscr{A}^- = z - \xi$ as well.

Next, we notice that $\mathscr{A}^+$ and $\mathscr{A}^-$ are slowly switching sequences, in the sense of §5.3: the switches can only take place when $g(\log a)$ increases through an integer value and by condition (iii), Theorem 5.14 applies with the R given in Theorem 5.15. Therefore $\mathbf{D}\mathscr{A}^{\pm} = z \pm \xi$, that is both

$$\tau(n, \mathscr{A}^{\pm}) = (z \pm \xi + o(1))\,\tau(n)\,p.p. \tag{6.33}$$

By hypothesis, $F' \to 0$. Hence there exists $a_0 = a_0(\xi, F, g)$ such that for all $a > a_0$ we have

$$|F\,(g(\log a)) - F([g(\log a)])| < \xi. \tag{6.34}$$

Therefore for all n, we have

$$\tau(n, \mathscr{A}^-) - a_0 \leq \tau\,(n, \mathscr{A}(z)) \leq \tau(n, \mathscr{A}^+) + a_0 \tag{6.35}$$

where

$$\mathscr{A}(z) = \{a : 0 < F\,(g(\log a)) \leq z (\mathrm{mod}\ 1)\}. \tag{6.36}$$

Now consider the integers $n \leq x$. We can let $\xi = \xi(x) \to 0$ as $x \to \infty$ sufficiently slowly so that a_0, as above, does not exceed $\sqrt{\log x}$. For all but $o(x)$ exceptional integers $n \leq x$ we have $\tau(n) > (\log x)^{3/5}$ by the Hardy–Ramanujan theorem, whence $a_0 < \tau(n)^{5/6}$. From (6.35), we have, with $o(x)$ exceptions,

$$|\tau\,(n, \mathscr{A}(z)) - z\tau(n)| < (\xi + o(1))\,\tau(n) + \tau(n)^{5/6}, \tag{6.37}$$

whence $\mathbf{D}\mathscr{A}(z) = z$. Since this holds for all $z \in (0,1)$ (and trivially for $z = 1$), we conclude that $f(d) = F(g(\log d))$ is divisor uniformly distributed. This completes the proof of Theorem 6.3.

Next, we construct an example to show that the sequence $\{f(n) : n \in \mathbf{Z}^+\}$ may be uniformly distributed (mod 1) and yet f is not divisor uniformly distributed. This is modelled on the proof of Theorem 5.3 (that divisor and logarithmic density are independent). Put

$$f(n) = \begin{cases} \sqrt{n}, & \text{if } \omega(n) > \frac{2}{3}\log\log n \\ 1 & \text{else.} \end{cases} \tag{6.38}$$

Since $\{\sqrt{n} : n \in \mathbf{Z}^+\}$ is uniformly distributed, so is $\{f(n) : n \in \mathbf{Z}^+\}$ because $f(n) = \sqrt{n}\, p.p.$ On the other hand, by Theorem 5.4 we have $\mathbf{D}\mathscr{A}(z;f) = 0$, with $\mathscr{A}(z;f)$ as in (6.2), for every $z \in (0,1)$. Hence f is not divisor uniformly distributed.

The theorem which follows is a rather brisk corollary of Theorem 5.8. It is potentially very useful because logarithmic density is often technically easier to cope with than asymptotic density.

Theorem 6.4 *In order that f should be divisor uniformly distributed it is sufficient that we have $\Delta(n;f) = o(\tau(n))$ on a sequence of logarithmic density* 1.

Proof On this hypothesis, (5.38) holds for $\mathscr{A} = \mathscr{A}(z;f)$, $(0 < z < 1)$, whence by Theorem 5.8, $\mathbf{D}\mathscr{A}(z;f) = z$. Hence f is divisor uniformly distributed.

The last theorem in this section is a necessary and sufficient condition for a function f to be divisor uniformly distributed. It is due to Tenenbaum, and appeared in the same paper, Tenenbaum (1982), as Theorem 5.9. It is motivated by Weyl's criterion for the uniform distribution (mod 1) of an integer sequence, in that it involves the functions $e(vf(n))$, for each fixed $v \in \mathbf{Z}^+$. We employ the familiar notation $e(x) = \exp(2\pi i x)$.

Theorem 6.5 *A necessary and sufficient condition that f should be divisor uniformly distributed (mod 1) is that for each fixed $v \in \mathbf{Z}^+$ we should have*

$$\sum_{k<x} \left| \sum_{\substack{n<x \\ n\equiv 0 (\mathrm{mod}\, k)}} \frac{e(vf(n))}{n 4^{\Omega(n)}} \right| = o(\sqrt{\log x}). \tag{6.39}$$

Proof We begin with necessity. Let f be divisor uniformly distributed, so that for each $z \in (0,1]$ we have

$$\mathbf{D}\{a : 0 < f(a) \leq z(\text{mod } 1)\} = z. \tag{6.40}$$

For $k \in \mathbf{Z}^+$ and each z, put

$$\Phi_{x,k}(z) = \frac{1}{\sqrt{\log x}} \sum_{\substack{n<x \\ n\equiv 0(\text{mod} k)}} \left\{ \frac{1}{n4^{\Omega(n)}} : 0 < f(n) \leq z(\text{mod } 1) \right\}. \tag{6.41}$$

Then by Theorem 5.9 and (6.40) we have, for each z

$$\sum_{k<x} |\Phi_{x,k}(z) - z\Phi_{x,k}(1)| = o(1). \tag{6.42}$$

For each $v \in \mathbf{Z}^+$ we have

$$\int_0^1 e(vz)d\Phi_{x,k}(z) = 2\pi i v \int_0^1 e(vz)\{\Phi_{x,k}(z) - z\Phi_{x,k}(1)\}dz. \tag{6.43}$$

The left-hand side of (6.42) is $\ll 1$, whence we may apply Lebesgue's dominated convergence theorem to (6.42) and (6.43) to obtain, for each v,

$$\sum_{k<x} \left| \int_0^1 e(vz)d\Phi_{x,k}(z) \right| = o(1), \quad x \to \infty \tag{6.44}$$

which is (6.39).

Next, we prove sufficiency. Let (6.39) hold for each fixed $v \in \mathbf{Z}^+$. We are going to show that for each z_1 and z_2 such that $0 \leq z_1 < z_2 \leq 1$ we have, for almost all n, that

$$\text{card}\{d : d|n, z_1 < f(d) \leq z_2(\text{mod } 1)\} \leq (z_2 - z_1 + o(1))\,\tau(n). \tag{6.45}$$

The opposite inequality will then follow by considering the complementary set. We employ Fejér means, defined as follows. Let $\varphi(t)$ be real, integrable and periodic on $\mathbf{R}$ with period 1, and its Fourier series be

$$\varphi(t) \sim \sum_{v=-\infty}^{\infty} c_v e(vt). \tag{6.46}$$

Then the Nth Fejér mean $\sigma_N(t,\varphi)$ of φ is the (real, $c_{-v} = \bar{c}_v$) trigonometric polynomial

$$\begin{aligned} \sigma_N(t,\varphi) &= \sum_{v=-N}^{N} \left(1 - \frac{|v|}{N}\right) c_v e(vt) \\ &= \int_0^{\frac{1}{2}} \mathscr{F}_N(u)\{\varphi(t+u) + \varphi(t-u)\}du, \end{aligned} \tag{6.47}$$

in which $\mathscr{F}_N(u)$ is the familiar Fejér kernel

$$\mathscr{F}_N(u) = \frac{\sin^2 N\pi u}{N \sin^2 \pi u}. \tag{6.48}$$

Let φ be the characteristic function of an interval I (mod 1). From (6.47)–(6.48), $\sigma_N(t, \varphi) \geq 0$ for all t, moreover if $t \in I$ and the distance from t to ∂I is at least ξ, then we have

$$\sigma_N(t, \varphi) \geq 1 - 2 \int_{\xi}^{\frac{1}{2}} \mathscr{F}_N(u) du \geq 1 - \frac{1}{2N\xi}. \tag{6.49}$$

We may assume that $z_2 - z_1 < 1$ in (6.45), else there is nothing to prove, and we suppose $2\xi < 1 - (z_2 - z_1)$ and let I be the interval $(z_1 - \xi, z_2 + \xi)$ on the torus $\mathbf{T} = \mathbf{R}/\mathbf{Z}$. Then the cardinality in (6.45) does not exceed

$$\Lambda \sum_{d|n} \sigma_N(f(d), \varphi), \quad \Lambda = \left(1 - \frac{1}{2N\xi}\right)^{-1}. \tag{6.50}$$

We require a (straightforward) generalization of the sufficiency part of Theorem 5.9, in which χ, the characteristic function of $\mathscr{A}$, is replaced by a weight function $w : \mathbf{Z}^+ \to [0, 1]$, in (5.68), and the conclusion is that we have

$$\sum_{d|n} w(d) = (z + o(1))\, \tau(n)\, p.p. \tag{6.51}$$

We shall apply this with $w(d) = \sigma_N(f(d), \varphi)$ and $z = z_2 - z_1 + 2\xi = c_0$ in (6.47). We have to show that the hypothesis of the modified Theorem 5.9 holds, viz.

$$\sum_{k<x} \left| \sum_{\substack{n<x \\ n\equiv 0 (\mathrm{mod}\ k)}} \frac{\sigma_N(f(n), \varphi) - c_0}{n 4^{\Omega(n)}} \right| = o(\sqrt{\log x}). \tag{6.52}$$

By (6.47), the left-hand side of (6.52) is

$$\sum_{k<x} \left| \sum_{\substack{n<x \\ n\equiv 0 (\mathrm{mod}\ k)}} \frac{1}{n 4^{\Omega(n)}} \sum_{\substack{\nu=-N \\ \nu\neq 0}}^{N} \left(1 - \frac{|\nu|}{N}\right) c_\nu e(\nu f(n)) \right|$$

$$\leq 2 \sum_{\nu=1}^{N} \left(1 - \frac{\nu}{N}\right) |c_\nu| \sum_{k<x} \left| \sum_{\substack{n<x \\ n\equiv 0 (\mathrm{mod}\ k)}} \frac{e(\nu f(n))}{n . 4^{\Omega(n)}} \right| = o(\sqrt{\log x})$$

by the hypothesis (6.39). Hence (6.52) holds, and implies (6.51), that is

$$\sum_{d|n} \sigma_N(f(d), \varphi) = (z_2 - z_1 + 2\xi + o(1))\, \tau(n)\, p.p., \tag{6.53}$$

from which we deduce that

$$\begin{aligned} &\operatorname{card}\{d : d|n,\ z_1 < f(d) \le z_2 (\mathrm{mod}\ 1)\} \\ &\le \left(1 - \frac{1}{2N\xi}\right)^{-1} (z_2 - z_1 + 2\xi + o(1))\, \tau(n)\, p.p. \end{aligned} \tag{6.54}$$

The Fourier coefficients c_ν are uniformly bounded, indeed $|c_\nu| \le 1$ from (6.46) since $0 \le \varphi \le 1$. We may therefore let $N = N(x) \to \infty$ sufficiently slowly so that (6.52) holds, and put $\xi = 1/\sqrt{N}$. Hence (6.45) holds, and this is all we need.

By means of Theorem 6.5, it was shown in Dupain, Hall and Tenenbaum (1982) and Hall and Tenenbaum (1986) that the functions $f(d) = \theta d$, θ irrational, and $f(d) = d^\alpha$, $\alpha > 0$, $\alpha \notin \mathbf{Z}$ are divisor uniformly distributed. These results do not follow from Theorems 6.2, 6.3 or 6.9 (below), because f increases too fast: Theorem 5.15 is insufficient for us to cope with the weights $y^{\Omega(n)}$ by partial summation. No estimates for the discrepancy were given in the above proofs, but recently Tenenbaum has obtained rather strong results of this type, which are to be published elsewhere. For this reason we do not pursue these quick f here, and refer the reader to the literature and Tenenbaum's forthcoming paper (199y).

6.3 Weyl sums and additive functions

A natural approach to questions about uniform distribution (mod 1) is to consider the associated Weyl sums. Accordingly for real-valued arithmetic functions f and $\nu \in \mathbf{Z} \setminus \{0\}$ we define

$$S_\nu(n; f) = \sum_{d|n} e(\nu f(d)). \tag{6.55}$$

Since $S_{-\nu}(n; f) = \overline{S_\nu(n; f)}$ we shall for the most part be concerned with $\nu \in \mathbf{Z}^+$.

In this and the next section it is to be understood that when we write $\operatorname{card}\{d : d|n,\ w < f(d) \le z(\mathrm{mod} 1)\}$, we count divisors for which $f(d) \equiv w$ or $z(\mathrm{mod} 1)$ with weight $\frac{1}{2}$ each in the cardinality.

There is a well-known quantitive upper bound for the discrepancy in terms of Weyl sums due to Erdös and Turán (1948). In the present context this reads as follows.

Lemma 6.6 *Let $\Delta(n;f)$ be as in (6.3) but with the end-point rule above. Then we have, uniformly for $N \in \mathbf{Z}^+$,*

$$\Delta(n;f) \ll \frac{\tau(n)}{N} + \sum_{\nu=1}^{N} \frac{1}{\nu} |S_\nu(n;f)|. \tag{6.56}$$

This provides us with a standard method to show that f is divisor uniformly distributed, and moreover to estimate the discrepancy, provided we can find *p.p.* bounds for the Weyl sums.

The easiest case, which we study in this section, is when the arithmetical function f is additive, for then $S_\nu(n;f)$ is multiplicative. We begin with a theorem of Kátai (1976).

Theorem 6.7 *A necessary and sufficient condition that the additive function f should be divisor uniformly distributed is that we should have*

$$\sum_p \frac{1}{p} \|\nu f(p)\|^2 = \infty \tag{6.57}$$

for every $\nu \in \mathbf{Z}^+$, where $\|x\| = \min(|x-m|, m \in \mathbf{Z}$.

We note that there is a mistake in Kátai's statement of his theorem, in that he has 2ν instead of ν in (6.57). (The redundant 2 makes its appearance in the formula for $g_m(p)$ on p.210 of his paper.)

Proof We begin with sufficiency, deducing from (6.56) that

$$\sum_{n\le x} \frac{\Delta(n;f)}{\tau(n)} \ll \frac{x}{N} + \sum_{\nu=1}^{N} \frac{1}{\nu} \sum_{n\le x} \frac{|S_\nu(n;f)|}{\tau(n)}. \tag{6.58}$$

We apply the Halberstam–Richert inequality (*Divisors*, Theorem 01) to the inner sum, which is therefore

$$\begin{aligned} &\ll x \prod_{p\le x}\left(1-\frac{1}{p}\right)\left(1+\frac{|S_\nu(p;f)|}{2p}+\frac{|S_\nu(p^2;f)|}{3p^2}+\cdots\right) \\ &\ll x \prod_{p\le x}\left(1+\frac{|\cos \pi\nu f(p)|-1}{p}\right), \end{aligned}$$

estimating $|S_\nu(p^r;f)| \le r+1$ trivially, for $r \ge 2$,

$$\begin{aligned} &\ll x\exp\left\{-\sum_{P\le x}\frac{1-|\cos \pi\nu f(p)|}{p}\right\} \\ &\ll x\exp\left\{-4\sum_{p\le x}\frac{1}{p}\|\nu f(p)\|^2\right\}. \end{aligned}$$

By hypothesis (6.57) holds, whence each inner sum on the right of (6.58) is $o(x)$, and we may let $N = N(x) \to \infty$ in such a way that the sum over v is also $o(x)$. Hence $\Delta(n;f) = o(\tau(n))$ *p.p.* by Markoff's inequality.

We give a slightly longer and more complicated proof of neccesity than Kátai's; this gives a little extra information. We begin with the identity

$$\int_0^1 \Bigg| \sum_{\substack{d|n \\ u-\frac{t}{2}<f(d)\le u+\frac{t}{2} \pmod 1}} 1 - t\tau(n) \Bigg|^2 du = \sum_{v\neq 0} \left(\frac{\sin \pi v t}{\pi v}\right)^2 |S_v(n;f)|^2, \tag{6.59}$$

of which the proof may be left to the reader. We denote the integral on the left by $\Delta_2(n;f,t)$, and we observe that for all t,

$$\Delta_2(n;f,t) \le \Delta(n;f)^2. \tag{6.60}$$

For $\sigma > 1$, we have

$$\begin{aligned}
\sum_{n=1}^{\infty} \frac{|S_v(n;f)|^2}{n^\sigma \tau(n)^2} &\ge \prod_p \left(1 + \frac{|S_v(p;f)|^2}{4p^\sigma}\right) \\
&\ge \frac{\zeta(\sigma)}{\zeta(2\sigma)} \prod_p \left(1 + \frac{|S_v(p;f)|^2 - 4}{4p^\sigma}\left(1 - \frac{1}{2p^\sigma}\right)\right) \\
&\ge \frac{\zeta(\sigma)}{\zeta(2\sigma)} \prod_p \left(1 + \frac{|S_v(p;f)| - 2}{p^\sigma}\left(1 - \frac{1}{2p^\sigma}\right)\right) \\
&\ge \frac{\zeta(\sigma)}{\zeta(2\sigma)} \prod_p \left(1 - \frac{\pi^2}{p^\sigma}\left(1 - \frac{1}{2p^\sigma}\right) \|vf(p)\|^2\right).
\end{aligned} \tag{6.61}$$

Since $p^{-\sigma} < \frac{1}{2}$, $\|vf(p)\| \le \frac{1}{2}$, the factors in this product exceed $1 - 3\pi^2/32 > 0$. Now suppose that for some v, say $v = \mu$, the series in (6.57) is convergent. Then from (6.61) we have, uniformly for $\sigma > 1$,

$$\sum_{n=1}^{\infty} \frac{|S_\mu(n;f)|^2}{n^\sigma \tau(n)^2} \gg_\mu \zeta(\sigma), \tag{6.62}$$

whence from (6.59), we have

$$\sum_{n=1}^{\infty} \frac{\Delta_2(n;f,t)}{n^\sigma \tau(n)^2} \gg_{\mu,t} \zeta(\sigma), \quad \sigma > 1 \tag{6.63}$$

for any t such that μt is not an integer. By hypothesis f is divisor uniformly distributed and $\Delta(n;f)/\tau(n) \to 0$ on a sequence of asymptotic, and hence logarithmic density 1. (Of course there is no distinction, by Theorem 6.4.) Therefore

$$\sum_{n=1}^{\infty} \frac{\Delta(n;f)^2}{n^{\sigma}\tau(n)^2} = o\left(\zeta(\sigma)\right), \quad \sigma \to 1+ \tag{6.64}$$

and (6.63)–(6.64) are in contradiction by (6.60). Hence the series (6.57) diverges for every v as required.

This proof raises a question about the status of the proposition

$$\Delta_2(n;f,t) = o\left(\tau(n)\right)^2 \ p.p., \tag{6.65}$$

whether or not f be additive. If t is rational then (6.65) does not imply that f is divisor uniformly distributed. For example if $t = h/k$ then the distribution of $f(d)$ (mod 1) might be, approximately

$$\text{card}\{d : d|n, \ f(d) \le z(\text{mod } 1)\} = (z + \lambda \sin 2\pi kz)\tau(n), \tag{6.66}$$

where $2\pi k\,|\lambda| \le 1$. If t is irrational, then (6.59) and (6.65) imply that for every $v \in \mathbf{Z}^+$,

$$\sum_{n=1}^{\infty} \frac{|S_v(n;f)|^2}{n^{\sigma}\tau(n)^2} = o\left(\zeta(\sigma)\right), \ (\sigma \to 1^+). \tag{6.67}$$

We put $\sigma = 1 + (1/\log x)$, and apply Cauchy's inequality which yields, for each fixed v,

$$\sum_{n \le x} \frac{|S_v(n;f)|}{n\tau(n)} = o(\log x), \quad x \to \infty. \tag{6.68}$$

Hence f is divisor uniformly distributed, by the Erdös–Turán inequality (Lemma 6.6) and Theorem 6.4.

Perhaps the simplest additive function is $f(d) = \log d$. Either of Theorems 6.2,6.3 show that it is divisor uniformly distributed (the latter slightly less obviously). To apply Kátai's theorem we need that

$$\sum_{p} \frac{1}{p} \|v \log p\|^2 = \infty, \quad v \in \mathbf{Z}^+ \tag{6.69}$$

which is an easy deduction from the Prime Number Theorem: the more precise result

$$\sum_{p \le x} \frac{1}{p} \|v \log p\|^2 = \frac{1}{12} \log\log x + O\left(\log\log(3 + v)\right) \tag{6.70}$$

follows from Lemma 6.15, see (6.153). If we inserted (6.70) into the proof of Theorem 6.7 we should obtain a *p.p.* upper bound for the discrepancy from (6.58), which in this case reads

$$\sum_{n\le x} \frac{\Delta(n;\log)}{\tau(n)} \ll x(\log x)^{-1/3} \log\log x. \tag{6.71}$$

This leads to a weaker result than (6.6) because $\frac{1}{3} < \frac{1}{2}\log 2$, but if we bypass the inequality

$$1 - |\cos \pi v f(p)| \ge 4\,\|v f(p)\|^2 \tag{6.72}$$

which is wasteful when $\|vf(p)\| < \frac{1}{2}$, we may replace the $\frac{1}{3}$ in (6.71) by $1 - \frac{2}{\pi} > \frac{1}{2}\log 2$. The next theorem, proved independently by Hall (1975) and Kátai (1976), gives a stronger result, obtained by using a weight different to $1/\tau(n)$ in (6.58) or (6.71).

Theorem 6.8 *Let $\xi(n) \to \infty$ as $n \to \infty$. Then for almost all n,*

$$\Delta(n;\log) < (\log n)^{\log(4/\pi)} \exp\{\xi(n)\sqrt{\log\log n}\}. \tag{6.73}$$

The right-hand factor arises from the Hardy–Ramanujan theorem. We shall actually prove that

$$\Delta(n;\log) < \left(\frac{4}{\pi}\right)^{\omega(n)} (\log\log n)^A \ p.p. \tag{6.74}$$

for a suitable A.

Proof We write

$$\tau(n,\theta) = \sum_{d|n} d^{i\theta}, \quad \theta \in \mathbf{R} \tag{6.75}$$

as in Hall (1975); *Divisors,* Chapter 3. Then

$$S_v(n;\log) = \tau(n, 2\pi v). \tag{6.76}$$

By Lemma 30.3 of *Divisors* we have, uniformly for $0 < y \le y_0$ and $v \in \mathbf{Z}^+$,

$$\sum_{n\le x} |\tau(n, 2\pi v)|\, y^{\omega(n)} \ll x(\log x)^{4(y/\pi)-1} \log^B(2+v) \tag{6.77}$$

where $B = B(y_0)$. We put $y = y_0 = \pi/4$ and deduce from (6.77) and Lemma 6.6 that, uniformly for $N \ge 2$, we have

$$\sum_{n\le x} \Delta(n;\log) \left(\frac{\pi}{4}\right)^{\omega(n)} \ll N^{-1} x \log x + x \log^{B+1} N. \tag{6.78}$$

(The sum involving the factor N^{-1} is estimated trivially as if we had $y = 1$. If this term were critical it would be better to insert a *p.p.* upper bound for $\tau(n)$ at the outset.) We put $N = [\log x]$, and the result, that is (6.74), follows from Markoff's inequality, for any $A > B + 1$.

6.4 Weyl sums and discrepancy

We made no attempt in the previous chapter to obtain an explicit, quantitative error term in the formula

$$\tau(n, \mathscr{A}) = (z + o(1))\,\tau(n)\,p.p. \tag{6.79}$$

in any of the cases except (5.5) in which we established that $\mathbf{D}\mathscr{A} = z$. Much of the theory of divisor density depends on the Hardy–Littlewood–Karamata theorem, and although a best possible quantitative form of this has been obtained by Freud (1952–4) the remainder term is, even in the best circumstances, very weak.

For this among other reasons it is usually better, if we want explicit upper bounds for the discrepancy $\Delta(n; f)$, to tackle the problem directly. The Erdös–Turán inequality, Lemma 6.6, employed in §6.3 is a natural approach. However, unless f is additive, so that the Weyl sums $S_\nu(n; f)$ are multiplicative, it is difficult to make any headway with (weighted) average orders of $|S_\nu(n; f)|$. This means that an almost obligatory initial step in any problem is to apply Cauchy's inequality to that of Erdös–Turán, to obtain

$$\Delta(n; f)^2 \ll \frac{\tau(n)^2}{N^2} + \log(N+1) \sum_{\nu=1}^{N} \frac{1}{\nu} |S_\nu(n; f)|^2 . \tag{6.80}$$

This undoubtedly introduces a weakness into the method, and the reader may convince himself of this by deriving a *p.p.* upper bound for $\Delta(n; \log)$ from (6.80) as starting point. (He will arrive at (6.6), or (6.72) with $\sqrt{2}$ instead of $4/\pi$.) But, often, (6.80) is forced; and we now give a general result based upon it. This is a slight refinement of Théorème 5 of Hall and Tenenbaum (1986).

Theorem 6.9 *Let $f \in C^1(\mathbf{R}^+)$ be such that $uf'(u)$ is ultimately monotonic. Let $F, R_1 : \mathbf{R}^+ \to \mathbf{R}^+$ be such that both $F(t)$ and $F_1(t) = t/F(t)$ tend monotonically to ∞, moreover $F(t) \gg \log^2 t$, and let*

$$\frac{1}{F(\log u)} \ll |uf'(u)| \ll R_1\,(F(\log u)) \tag{6.81}$$

where R_1 is such that the function $R(t) = R_1(\log t)$ satisfies (5.138) *and*

(5.139) *and in addition*

$$R_1\left(\frac{1}{2}t\right) \le R_1(t)^{1-\kappa} \tag{6.82}$$

for some fixed $\kappa > 0$. *In particular we may take*

$$R_1(t) = \exp\left\{\frac{ct^{3/5}}{(\log(2+t))^{1/5}}\right\} \tag{6.83}$$

with a suitably small positive c. Then f is divisor uniformly distributed, and for each fixed y, $0 < y < 1$, *we have*

$$\sum_{n\le x} \frac{y^{\Omega(n)}}{n}\Delta(n;f)^2 \ll (\log x)^{4y}\frac{(\log F_1(\log x))^2}{F_1(\log x)^{2y}}. \tag{6.84}$$

In fact the conclusion that f is divisor uniformly distributed follows from (6.84) if we put $y = \frac{1}{4}$. For we may deduce that

$$\Delta(n;f)^2 < 4^{\Omega(n)}F_1(\log n)^{-1/3}$$

on a sequence of logarithmic density 1, and the upper bound on the right is $o(\tau(n)^2)$ *p.p.* The conclusion is then implied by Theorem 6.4. The optimal y may not be $\frac{1}{4}$, and depends on the circumstances. We discuss this point and give corollaries after the proof.

We remark that for any reasonable choice of R_1, such as (6.83), condition (6.82) is easily satisfied.

Proof of Theorem 6.9 We consider the sum

$$Z(\sigma,v) = \sum_{n=1}^{\infty}\frac{y^{\Omega(n)}}{n^\sigma}|S_v(n;f)|^2, \quad \sigma > 1 \tag{6.85}$$

where $S_v(n;f)$ is the Weyl sum introduced in (6.55). We put

$$\zeta(s,y) = \prod_p\left(1-\frac{y}{p^s}\right)^{-1}, \quad \varphi(n;\sigma,y) = n^\sigma\prod_{p|n}\left(1-\frac{y}{p^\sigma}\right) \tag{6.86}$$

and we have

$$\begin{aligned} Z(\sigma,v) &= \zeta(\sigma,y)\sum_{d=1}^{\infty}\sum_{t=1}^{\infty} e(vf(d)-vf(t))\frac{y^{\Omega([d,t])}}{[d,t]^\sigma} \\ &= \zeta(\sigma,y)\sum_{d=1}^{\infty}\sum_{t=1}^{\infty}\frac{y^{\Omega(dt)}}{(dt)^\sigma}e(vf(d)-vf(t))\sum_{r|(d,t)}\frac{\varphi(r;\sigma,y)}{y^{\Omega(r)}} \end{aligned}$$

$$= \zeta(\sigma, y) \sum_{r=1}^{\infty} \frac{y^{\Omega(r)}\varphi(r;\sigma,y)}{r^{2\sigma}} \left| \sum_{m=1}^{\infty} \frac{y^{\Omega(m)} e(vf(mr))}{m^{\sigma}} \right|^2$$

$$< \zeta(\sigma, y) \sum_{r=1}^{\infty} \frac{y^{\Omega(r)}}{r^{\sigma}} \left| \sum_{m=1}^{\infty} \frac{y^{\Omega(m)} e(vf(mr))}{m^{\sigma}} \right|^2 \tag{6.87}$$

by (6.86). We want an upper bound for the inner sum, which we denote by T_v. We put $g(u) = e(vf(ur))$. By partial summation and (5.139) we have, for any $M \geq 2$,

$$T_v = O((\log M)^y) + \int_M^{\infty} -\frac{d}{du}\left(\frac{g(u)}{u^{\sigma}}\right) \left\{ \int_0^{\frac{1}{2}} u^{1-v} d\lambda_y(v) + O\left(\frac{u}{R_1(\log u)}\right) \right\} du. \tag{6.88}$$

At this point we fix $M = M(r)$ in such a way that

$$F(\log Mr) \leq \frac{1}{2} \log M. \tag{6.89}$$

This is in order because we have $F(t) = (t/F_1(t)) = o(t)$ by hypothesis. Notice that for $u \geq M$ we have

$$\begin{aligned} F(\log ur) &= \frac{\log ur}{F_1(\log ur)} \leq \frac{\log ur}{F_1(\log Mr)} = F(\log Mr) \frac{\log ur}{\log Mr} \\ &\leq \frac{1}{2} \log M . \frac{\log u + \log r}{\log M + \log r} \leq \frac{1}{2} \log u. \end{aligned} \tag{6.90}$$

The error term arising from the integral in (6.88) is

$$\begin{aligned} &\ll \int_M^{\infty} \frac{|ug'(u)| + |g(u)|}{u^{\sigma}} . \frac{du}{R_1(\log u)} \\ &\ll v \int_M^{\infty} \frac{R_1(F(\log ur))\, du}{u^{\sigma} R_1(\log u)} \\ &\ll v \int_M^{\infty} \frac{R_1(\frac{1}{2}\log u) du}{u^{\sigma} R_1(\log u)} \end{aligned} \tag{6.91}$$

by (6.81) and (6.90). By condition (6.82) on R_1, this is

$$\ll v \int_{\log M}^{\infty} R_1(t)^{-\kappa} dt. \tag{6.92}$$

It is a straightforward consequence of (6.82) that for every $\varepsilon > 0$, the integral

$$\int_0^{\infty} R_1(t)^{-\varepsilon} dt$$

is convergent. Hence the integral (6.92) is

$$\ll \quad vR_1(\log M)^{-\kappa/2} \int_{\log M}^{\infty} R_1(t)^{-\kappa/2} dt$$
$$\ll \quad vR_1(\log M)^{-\kappa/2},$$

and we have, after an integration by parts, that

$$T_v = O\left((\log M)^y + vR_1(\log M)^{-\kappa/2}\right) + \int_0^{\frac{1}{2}} (1-v)d\lambda_y(v) \int_M^{\infty} \frac{g(u)}{u^{\sigma+v}} du. \tag{6.93}$$

We substitute $u = Me^t$ in the inner integral here, which becomes

$$M^{-(\sigma+v-1)} \int_0^{\infty} e^{-(\sigma+v-1)t} e\left(vf(rMe^t)\right) dt$$
$$= M^{-(\sigma+v-1)} \int_0^{\infty} e^{-(\sigma+v-1)} G'(t) dt \tag{6.94}$$

where

$$G(t) = \int_0^t e\left(vf(rMe^w)\right) dw. \tag{6.95}$$

By hypothesis, the function

$$\frac{d}{dw}\{vf(rMe^w)\} = vrMe^w f'(rMe^w)$$

is monotonic, moreover (6.81) implies that

$$\left|\frac{d}{dw}\{vf(rMe^w)\}\right| \gg \frac{v}{F\left(w + \log(rM)\right)} \tag{6.96}$$

whence Lemma 4.2 of Titchmarsh (1951) yields

$$G(t) \ll \frac{1}{v} F\left(t + \log(rM)\right), \tag{6.97}$$

and the right-hand member of (6.94) is therefore

$$\ll \frac{1}{v} \int_0^{\infty} \frac{(\sigma+v-1)}{M^{\sigma+v-1}} e^{-(\sigma+v-1)t} F\left(t + \log(rM)\right) dt. \tag{6.98}$$

Now F is subadditive, because for $\xi_1, \xi_2 > 0$,

$$F(\xi_1+\xi_2) = \frac{\xi_1+\xi_2}{F_1(\xi_1+\xi_2)} \le \frac{\xi_1}{F_1(\xi_1)} + \frac{\xi_2}{F(\xi_2)} = F(\xi_1) + F(\xi_2) \tag{6.99}$$

whence the integral in (6.98) is

$$\ll \frac{F\left(\log(rM)\right)}{vM^v} + \frac{1}{vM^v} \int_0^{\infty} (\sigma+v-1)e^{-(\sigma+v-1)t} F(t) dt. \tag{6.100}$$

Substitute $t = s/(\sigma + v - 1)$. We have

$$\begin{aligned}
&\int_0^\infty e^{-s} F\left(\frac{s}{\sigma+v-1}\right) ds \\
&\quad \le F\left(\frac{1}{\sigma+v-1}\right) + \int_1^\infty e^{-s} F\left(\frac{s}{\sigma+v-1}\right) ds \\
&\quad \le F\left(\frac{1}{\sigma+v-1}\right)\left\{1 + \int_1^\infty e^{-s} s ds\right\} \\
&\quad \ll F\left(\frac{1}{\sigma+v-1}\right). \qquad (6.101)
\end{aligned}$$

By (5.138) we have $|d\lambda_y(v)| \ll v^{-y} dv$, and we insert this into (6.93) and employ (6.98), (6.100–6.101). We obtain

$$\begin{aligned}
T_v \ll\ & (\log M)^y + v R_1 (\log M)^{-\kappa/2} \\
& + \frac{1}{v}\int_0^{\frac{1}{2}} v^{-y} M^{-v}\left\{F(\log(rM)) + F\left(\frac{1}{\sigma+v-1}\right)\right\} dv \\
\ll\ & (\log M)^y + v R_1 (\log M)^{-\kappa/2} + \frac{1}{v}(\log M)^{y-1} F\left(\frac{1}{\sigma-1}\right),
\end{aligned} \qquad (6.102)$$

because $F(\log(rM)) \le \frac{1}{2}\log M$ by (6.89). This is a lower bound constraint on M, and we now determine $M = M(r)$ by setting

$$\log M = 3F(\log r) + F\left(\frac{1}{\sigma-1}\right). \qquad (6.103)$$

Notice that if σ is sufficiently close to 1, say $\sigma \le \sigma_0$, M is so large that $F(\log M) \le \frac{1}{6}\log M$. Then $F(\log(rM)) \le F(\log r) + F(\log M) \le (\frac{1}{3} + \frac{1}{6})\log M$, that is (6.89) holds. We deduce from (6.102)–(6.103) that for $\sigma \le \sigma_0$,

$$T_v \ll F(\log r)^y + F\left(\frac{1}{\sigma-1}\right)^y + v R_1 \left(F\left(\frac{1}{\sigma-1}\right)\right)^{-\kappa/2} \qquad (6.104)$$

and we insert this into (6.87), which yields

$$\begin{aligned}
Z(\sigma, v) \ll\ & \zeta(\sigma, y)^2 \left\{F\left(\frac{1}{\sigma-1}\right)^{2y} + v^2 R_1 \left(F\left(\frac{1}{\sigma-1}\right)\right)^{-\kappa}\right\} \\
& + \zeta(\sigma, y)\sum_{r=1}^{\infty} \frac{y^{\Omega(r)}}{r^\sigma} F(\log r)^{2y}.
\end{aligned} \qquad (6.105)$$

This last term on the right is absorbed by the first one. This is clearly true of that part in which $\log r \leq 1/(\sigma - 1)$. The remainder does not exceed

$$\frac{\zeta(\sigma, y)}{F_1(\frac{1}{\sigma-1})^{2y}} \sum_{r=1}^{\infty} \frac{y^{\Omega(r)}}{r^{\sigma}} (\log r)^{2y} \ll \frac{\zeta(\sigma, y)}{F_1(\frac{1}{\sigma-1})^{2y}} \cdot \frac{1}{(\sigma-1)^{3y}}$$
$$\ll \zeta(\sigma, y)^2 F\left(\frac{1}{\sigma-1}\right)^{2y}$$

because $\zeta(\sigma, y) \asymp (\sigma - 1)^{-y}$ for $1 < \sigma \leq 2$.

Now consider the sum on the left of (6.84). We put $\sigma = 1 + (1/\log x)$, and employ (6.80). The sum is therefore

$$\leq \sum_{n=1}^{\infty} \frac{y^{\Omega(n)}}{n^{\sigma}} \Delta(n; f)^2$$
$$\leq \frac{1}{N^2} \sum_{n=1}^{\infty} \frac{y^{\Omega(n)}}{n^{\sigma}} \tau(n)^2 + \log(N+1) \sum_{v=1}^{N} \frac{1}{v} Z(\sigma, v)$$
$$\leq \frac{1}{N^2}(\sigma-1)^{-4y} + \log^2(N+1)(\sigma-1)^{-2y} F\left(\frac{1}{\sigma-1}\right)^{2y}$$
$$+ (\sigma-1)^{-2y} N^2 R_1\left(F\left(\frac{1}{\sigma-1}\right)\right)^{-\kappa}. \quad (6.106)$$

We put $N = [F_1(\log x)^y]$. Since $F(\log x) \gg (\log\log x)^2$ and we may take $R_1(u) = \exp(b\sqrt{u})$ for any b, the term involving R_1 in (6.106) is negligible. This gives (6.84) and completes the proof.

Corollary 6.10 *Let f be as in the statement of the theorem, and $\xi(n) \to \infty$. Then*

$$\Delta(n; f) < \xi(n)\tau(n) \left(1 - \frac{\log F_1(\log n)}{2 \log\log n}\right)^{\Omega(n)/2} \log F_1(\log n) \quad (6.107)$$

on a sequence of logarithmic density 1.

We note that, by the Hardy–Ramanujan theorem, the right-hand side does not exceed

$$\tau(n) F_1(\log n)^{(-1/4)+\varepsilon} \text{ p.p.} \quad (6.108)$$

so that (6.107) is always non-trivial.

Corollary 6.11 *For $\alpha > 0$, we have*

$$\Delta(n; \log^{\alpha}) < \tau(n)(\log n)^{-\eta(\alpha)+\varepsilon} \quad (6.109)$$

on a sequence of logarithmic density 1, *where*

$$\eta(\alpha) = -\frac{1}{2}\log\left(1 - \frac{\min(\alpha, 1)}{2}\right). \tag{6.110}$$

For $\beta > 1$ we have

$$\Delta(n; (\log\log)^{\beta}) < \tau(n)(\log\log n)^{((1-\beta)/4)+\varepsilon} \tag{6.111}$$

on a sequence of logarithmic density 1.

Regarding $\log^{\alpha}$, see also Hall and Tenenbaum (1986), p. 135. We prove Corollary 6.10 and leave 6.11 as an exercise.

Proof of Corollary 6.10 We consider the integers $n \leq x$, and we show that (6.107) is satisfied unless n belongs to an exceptional set $E = E(x)$ such that

$$\sum_{n \leq x,\ n \in E} \frac{1}{n} = o(\log x). \tag{6.112}$$

We put

$$y = \frac{1}{2}\left(1 + \frac{\log F(\log x)}{\log\log x}\right)^{-1} = \frac{1}{4}\left(1 - \frac{\log F_1(\log x)}{2\log\log x}\right)^{-1} \tag{6.113}$$

and, with this value of y, (6.84) becomes

$$\sum_{n \leq x} \frac{y^{\Omega(n)}}{n} \Delta(n; f)^2 \ll (\log F_1(\log x))^2 \log x. \tag{6.114}$$

Hence if $\xi_1(x) \to \infty$, we have

$$\Delta(n; f) < \xi_1(x) 2^{\Omega(n)} \left(1 - \frac{\log F_1(\log x)}{2\log\log x}\right)^{\Omega(n)/2} \log F_1(\log x), \tag{6.115}$$

for all integers $n \leq x$ not belonging to E. We now require that n has all the following properties:

(i) $n > x^{\varepsilon(x)}$, $\varepsilon(x) \to 0$.
(ii) $\frac{1}{2}\log\log x < \Omega(n) < 2\log\log x$.
(iii) $8 < \log F_1(\log n) \leq \log\log n$.
(iv) $2^{\Omega(n)} < \xi_1(x)\tau(n)$.

We include the exceptional n in E. This is clearly permissable in case (i), and in case (ii) by the Hardy–Ramanujan theorem; also (iii) holds for $n \geq n_0$ because $F_1 \to \infty$, $F_1(\log n) = \log n / F(\log n)$, and $F \to \infty$. Finally we have, for fixed $\lambda < 1$,

$$\sum_{n \le x} \frac{1}{n} \left(\frac{2^{\Omega(n)}}{\tau(n)} \right)^{\lambda} \ll_{\lambda} \log x \tag{6.116}$$

whence (iv) is in order as well.

Since $\Omega(n) > \frac{1}{2} \log\log x$ by (ii), the right-hand side of (6.115) is a decreasing function of $\log F_1$ for $\log F_1 > 8$. By (iii), we may therefore replace $F_1(\log x)$ by $F_1(\log n)$. We apply (iv) to obtain

$$\Delta(n;f) < \xi_1(x)^2 \tau(n) \left(1 - \frac{\log F_1(\log n)}{2\log\log x}\right)^{\Omega(n)/2} \log F_1(\log n). \tag{6.117}$$

We have, by (i) and (iii),

$$1 - \frac{\log F_1(\log n)}{2\log\log x} \le \left(1 - \frac{\log F_1(\log n)}{2\log\log n}\right)\left(1 - \frac{\log \varepsilon(x)}{\log\log x}\right) \tag{6.118}$$

and we insert this into (6.117) and apply the right-hand inequality in (ii), which yields

$$\Delta(n;f) < \left(\frac{\xi_1(x)}{\varepsilon(x)}\right)^2 \tau(n) \left(1 - \frac{\log F_1(\log n)}{2\log\log n}\right)^{\Omega(n)/2} \log F_1(\log n). \tag{6.119}$$

We set

$$\xi_1(x) = \varepsilon(x) \min\{\sqrt{\xi(n)} : n > x^{\varepsilon(x)}\} \tag{6.120}$$

and let $\varepsilon(x) \to 0$ sufficiently slowly that $\xi_1(x) \to \infty$. Hence (6.107) holds for the n not counted in (6.112). This completes the proof.

As we indicated in the introduction to this chapter, we can employ upper bounds for $\Delta(n;f)$ to construct Behrend sequences. Let $\{z_n : n \in \mathbf{Z}^+\}$ be a positive decreasing sequence such that $z_n \to 0$ and, on a subsequence of $\mathbf{Z}^+$ of logarithmic density 1, $z_n\tau(n) > \Delta(n;f) + 1$. For such n there exists a divisor $d > 1$ such that

$$0 < f(d) \le z_n \le z_d \pmod 1, \tag{6.121}$$

whence the sequence

$$\mathscr{A} = \{a > 1 : 0 < f(a) \le z_a \pmod 1\} \tag{6.122}$$

is Behrend. As an example, we apply this idea to the functions $f(d) = (\log d)^{\alpha}$, $(\log\log d)^{\beta}$ using the results contained in Corollary 6.11.

We begin with $f(d) = (\log d)^\alpha$, and we may take $z_n = (\log n)^{-\eta}$ for any $\eta < \eta(\alpha)$. We put $\lambda = 1/\alpha$ and we conclude that

$$\begin{aligned}\mathscr{A} &= \bigcup_{k=1}^{\infty}\{a : k \le (\log a)^\alpha \le k + (\log a)^{-\eta}\} \\ &\subseteq \bigcup_{k=1}^{\infty}\{a : k^\lambda \le \log a \le k^\lambda + \lambda(k+1)^{\lambda-1}k^{-\eta\lambda}\} \qquad (6.123)\end{aligned}$$

is Behrend. This is always non-trivial, and it includes Erdös' conjecture concerning the sequence $\mathscr{A}_\lambda$ defined in (1.110) if

$$\lambda(k+1)^{\lambda-1}k^{-\eta\lambda} \le \log 2, \quad k \ge k_1; \qquad (6.124)$$

we need only consider the tail of $\mathscr{A}$ by Corollary 0.14. For Erdös' conjecture we are interested in the case $\lambda > 1$ only, and (6.124) reduces to

$$1 + \frac{1}{2}\log\left(1 - \frac{\alpha}{2}\right) < \alpha \qquad (6.125)$$

or $\alpha > .76070\ldots$, or $\lambda < 1.31457\ldots$; this is the result of Hall and Tenenbaum (1986) mentioned in Chapter 1. As we pointed out there, it has been superseded by the sharp condition $\lambda \le 1/(1 - \log 2)$.

We examine the case $f(d) = (\log\log d)^\beta$, $\beta > 1$. We may take $z_n = (\log\log n)^{((1-\beta)/4)+\varepsilon}$, and we conclude that

$$\mathscr{A} = \bigcup_k \{\mathbf{Z}^+ \cap (T_k, H_k T_k]\} \qquad (6.126)$$

is Behrend, where

$$\begin{aligned}\log T_k &= \exp(k^{1/\beta}) \qquad (6.127)\\ \log H_k &= k^{(5(1-\beta)/4\beta)+\varepsilon}\log T_k. \qquad (6.128)\end{aligned}$$

This comes within the ambit of Theorem 1.15. If $\beta \le 5$, condition (1.159) is satisfied and the theorem applies to show that $\mathscr{A}$ is Behrend. As β increases, the exponent of k in (6.128) decreases towards $-5/4$, so that condition (1.160) is easily satisfied. However (1.161) fails: the T_k do not increase sufficiently rapidly (cf. (1.175)). Thus $\mathscr{A}$ is a new Behrend sequence whenever $\beta > 5$.

6.5 Lower bounds for discrepancy

Ω-theorems are useful in that they put a limit on what we can expect to achieve with upper bounds. Perhaps of more interest is that they usually involve completely different methods; we may be able to see further. We begin with a simple extension of the proof given at the start of this chapter that for $\beta \le 1$, $(\log\log d)^\beta$ is not divisor uniformly distributed.

Theorem 6.12 *Let* $f(d) = F(\log d)$ *where* F *is differentiable and* $|F'(u)| \ll u^{-\kappa}$ *for fixed* $\kappa > \log 2$. *Then*

$$\Delta(n; f) > \tau(n)^{\mu-\varepsilon}\ p.p., \tag{6.129}$$

where $\mu = (\kappa - \log 2)/(1 - \log 2)$.

Proof Let $\lambda \in (0, \kappa)$ be at our disposal. Let $m = m(\lambda)$ be the largest divisor of n such that

$$P^+(m(\lambda)) < \exp\left((\log n)^\lambda\right) \tag{6.130}$$

and $\xi(n) \to \infty$ as $n \to \infty$, Then we have

$$\log P^+(n) > \xi(n)^{-1} \log n\ p.p.; \tag{6.131}$$

moreover by Theorem 07 of *Divisors* also

$$\log m(\lambda) < \xi(n)(\log n)^\lambda\ p.p. \tag{6.132}$$

Now consider the divisors $d = P^+(n)t$ where $t|m$. We have

$$\begin{aligned} F(\log d) &= F\left(\log P^+(n)\right) + O\left(\frac{\log m}{(\log P^+(n))^\kappa}\right) \\ &= F\left(\log P^+(n)\right) + O\left(\xi(n)^{1+\kappa}(\log n)^{\lambda-\kappa}\right) \end{aligned} \tag{6.133}$$

that is $f(d) \in E$ where

$$|E| \ll \xi(n)^{1+\kappa}(\log n)^{\lambda-\kappa}. \tag{6.134}$$

Let $\xi(n) = (\log n)^{o(1)}$. Then we deduce from (6.134) and the Hardy–Ramanujan theorem that

$$\tau(n)\,|E| < (\log n)^{\log 2+\lambda-\kappa+o(1)}. \tag{6.135}$$

By Turán's method, $\omega\,(m(\lambda))$ has normal order $\lambda \log\log n$, whence the number of divisors d of n such that $f(d) \in E$ is at least

$$\tau(m) \ge 2^{\omega(m)} > (\log n)^{\lambda \log 2+o(1)}. \tag{6.136}$$

We obtain a lower bound for the discrepancy, for example $\Delta(n; f) > \frac{1}{2}\tau(m)$, if $\tau(m) > 2\tau(n)\,|E|$ and so we choose λ so that $\lambda \log 2 > \log 2 + \lambda - \kappa$, that is $\lambda < (\kappa - \log 2)/(1 - \log 2)$. Our result follows from (6.136).

We could if we wished replace $(\log n)^{o(1)}$ throughout by $\exp(\xi_1(n)\sqrt{\log\log n})$, $\xi_1(n) \to \infty$. Of greater interest would be a precise lower bound for Δ in the case $|F'(u)| \ll u^{-1}\Psi(u)$ when Ψ is slowly increasing.

Theorem 6.12 shows that $\Delta(n;\log^\alpha) \to \infty$ *p.p.* for every $\alpha \in (0, 1-\log 2)$. For larger values of α we do not have a lower bound for the discrepancy, and this includes the case $\alpha = 1$. The function $f(d) = \log d$ is of particular interest, as it is both additive and the restriction to $\mathbf{Z}^+$ of an analytic function, and it has been an open problem since 1974 to prove that $\Delta(n;\log) \to \infty$ *p.p.*; in the opposite direction the problem of improving Theorem 6.8 is almost as old. The last result in this chapter comes quite close to, but does not succeed in showing that $\Delta(n;\log)$ is large. We modify the problem slightly by introducing a weight on the divisors, $r^{\Omega(d)}$, where $r < 1$. That is we consider the distribution function

$$\sum\{r^{\Omega(d)} : d|n,\ w < \log d \le z (\text{mod } 1)\} \tag{6.137}$$

with the usual rule about the endpoints: any divisors d such that $\log d \equiv w$ or z (mod 1) count $\frac{1}{2}r^{\Omega(d)}$. We can if we wish replace $\Omega(d)$ by $\omega(d)$ with no material difference. We write

$$\tau^{(r)}(n) = \sum_{d|n} r^{\Omega(d)} \tag{6.138}$$

and we have the following result.

Theorem 6.13 *Let $\varepsilon > 0$, and $\varepsilon(x) \to 0$ as $x \to \infty$. Let r and t be such that*

$$\frac{1}{2} \le r < 1 - \exp\left(\frac{-\varepsilon(x)\sqrt{\log\log x}}{\log\log\log x}\right) \tag{6.139}$$

and

$$0 < t < (\log\log x)^{-2-\varepsilon}. \tag{6.140}$$

Then for all but at most $o(x)$ exceptional integers $n \le x$ we have

$$\int_0^1 \left| \sum_{\substack{d|n \\ u-\frac{t}{2} < \log d \le u+\frac{t}{2} (\text{mod } 1)}} r^{\Omega(d)} - t\tau^{(r)}(n) \right|^2 du \ge t^2 \exp\{\varepsilon(x)\sqrt{\log\log x}\}. \tag{6.141}$$

There is an alternative 'for almost all n' formulation in which we change x to n in each of (6.139–6.141). Although r is very close to 1 it does not seem easy to remove it altogether. Perhaps a fairly simple extra idea has been overlooked.

Proof of Theorem 6.13 The Weyl sums, which we write as divisor functions, take the modified form

$$\tau(n, 2\pi v; r) = \sum_{d|n} r^{\Omega(d)} e(v \log d) \tag{6.142}$$

and we seek a lower bound for the sum

$$\sum_{v=1}^{N} |\tau(n, 2\pi v; r)|^2 \tag{6.143}$$

to hold for $x + o(x)$ integers $n \leq x$. We put

$$N = N(x) = \left[\frac{1}{2}(\log\log x)^{2+\varepsilon}\right]. \tag{6.144}$$

Our argument is based on that used in Hall (1994) (proof of Theorem 2) concerning the function $\max\{|\tau(n,\theta)| : a \leq \theta \leq b\}$ and we set out to show that, normally

$$\sum_{v=1}^{N} \log^+ |\tau(n, 2\pi v; r)| \tag{6.145}$$

is large (in which $\log^+ u := \max(\log u, 0)$). Initially we have to exclude small prime factors and so we introduce $w = w(x) \to \infty$ as $x \to \infty$ and write

$$\tau(n, \theta; r, w) = \prod_{p|n,\ p>w} (1 + rp^{i\theta}). \tag{6.146}$$

We may assume that n has no repeated prime factor $> w$ since the number of exceptions is $\ll x/w \log w$. We put

$$L(n) = \sum_{v=1}^{N} \log |\tau(n, 2\pi v; r, w)| \tag{6.147}$$

and

$$M_k(n) = \sum_{v=1}^{N} |\log |\tau(n, 2\pi v; r, w)||^k, \quad k \in \mathbf{Z}^+, \tag{6.148}$$

and we show that, normally, $L(n)$ is much smaller than $M_1(n)$. We begin with an upper bound for $L(n)$.

Lemma 6.14 *Let*

$$f_N(t) = \sum_{\nu=1}^{N} \log|1 + re(\nu t)| \tag{6.149}$$

and $E(T) = \{t \in [0,1) : |f_N(t)| > T\}$. *Then* $E(T)$ *comprises at most* $\frac{1}{2}N(N+1)$ *disjoint intervals and, uniformly for* $r < 1$, *we have*

$$|E(T)| = \text{meas}E(T) \ll \frac{N}{T^2}. \tag{6.150}$$

This holds for all $N \in \mathbf{Z}^+$, independently of (6.144).

Proof We may write $f_N(t) = \log\left|P(e^{2\pi i t})\right|$ where P is a polynomial of degree $\frac{1}{2}N(N+1)$, whence the equation $|f_N(t)| = T$ has at most $N(N+1)$ roots in $[0,1)$. The first part follows from this. Next we have (we leave the calculation to the reader),

$$\int_0^1 \log|1 + re(\mu t)| \log|1 + re(\nu t)|\, dt \le \frac{\pi^2}{12} \cdot \frac{(\mu,\nu)^2}{\mu\nu} \tag{6.151}$$

whence

$$\int_0^1 |f_N(t)|^2\, dt \ll \sum_{\mu \le N} \sum_{\nu \le N} \frac{(\mu,\nu)^2}{\mu\nu} \ll N,$$

and (6.150) follows.

Lemma 6.15 *Let* g *be a periodic function of period* 1, *having bounded variation over the period. Let*

$$\overline{g} = \int_0^1 g(u)du,\ I(g) = \int_0^1 |g(u) - \overline{g}|\, du,\ V(g) = \int_0^1 |dg(u)|. \tag{6.152}$$

Then for real $\phi \ne 0$, $2 \le w < z$ *and every* c, *we have*

$$\sum_{w<p\le z} p^{-1} g(\phi \log p) = \overline{g} \log\left(\frac{\log z}{\log w}\right) + O\left(\frac{I(g)}{|\phi| \log w}\right)$$
$$+ O_c\left\{\left(M(g) + \left(1 + |\phi|\right)V(g)\right) e^{-c\sqrt{\log w}}\right\}, \tag{6.153}$$

where $M(g) = \sup|g|$.

This is Lemma 30.1 of *Divisors*, except that the period is 1 and we have replaced $V(g)$ by $I(g)$ in the first error term (cf. *Divisors* (3.6)).

Lemma 6.16 *Let* $\xi(x) \to \infty$ *as* $x \to \infty$ *and*

$$w = \exp\left((\log\log x)^{20}\right). \tag{6.154}$$

Then for all but $o(x)$ integers $n \le x$ we have

$$|L(n)| \ll N^{1/2}(\log\log x)^{3/2} + \xi(x)N\log\left(\frac{1}{1-r}\right). \tag{6.155}$$

Proof Set $T = (N\log\log x)^{1/2}$ in Lemma 6.14 so that we have $|E(T)| \ll 1/\log\log x$ by (6.150). By Lemma 6.15, with g the characteristic function of E, we have

$$\sum_{w<p\le x}\left\{\frac{1}{p} : \log p \in E(\text{mod } 1)\right\}$$
$$= |E|\left\{\log\left(\frac{\log x}{\log w}\right) + O\left(\frac{1}{\log w}\right)\right\} + O(N^2 e^{-\sqrt{\log w}}) \ll 1$$

by (6.144), (6.154) and the bound for $|E|$. Let $\omega^*(n, E)$ denote the number of prime factors of n such that $p > w$, $\log p \in E$ (mod 1). Then we have

$$\sum_{n\le x}\omega^*(n,E) \le x\sum_{w<p\le x}\left\{\frac{1}{p} : \log p \in E(\text{mod } 1)\right\} \ll x \tag{6.156}$$

whence $\omega^*(n, E) \le \xi(x)$ with $o(x)$ exceptions. By (6.147) and (6.149) we have

$$L(n) = \sum_{p|n,\ p>w} f_N(\log p) \tag{6.157}$$

and we infer that for the unexceptional n there holds

$$|L(n)| \le \omega(n)T + \omega^*(n,E)N\log\left(\frac{1}{1-r}\right) \tag{6.158}$$

inserting the trivial upper bound for $|f_N|$ on the right. We may assume $\omega(n) < 2\log\log x$ by Hardy–Ramanujan, and (6.155) follows.

Lemma 6.17 *Let r, N and w be as in* (6.139), (6.144) *and* (6.154). *Then for $x + o(x)$ integers $n \le x$ we have*

$$|L(n)| < \xi(x)^{-1}N\sqrt{\log\log x} \tag{6.159}$$

provided $\xi(x) \to \infty$ sufficiently slowly.

This is an immediate deduction from the previous lemma.

Next, we want a lower bound for $M_1(n)$. For this we employ the moment inequality used in Hall (1994) which is of course a familiar application of the Cauchy–Hölder inequality, viz.

$$M_2(n) \le M_1(n)^{2/3}M_4(n)^{1/3}. \tag{6.160}$$

Thus we require a lower bound for $M_2(n)$ and an upper bound for $M_4(n)$. We select the easier task first.

Lemma 6.18 *Let r, N, w and ξ be as above. Then for each fixed k we have, with $o(x)$ exceptions,*

$$M_{2k}(n) < \xi(x)N(\log\log x)^k. \tag{6.161}$$

Proof We consider the sum

$$H(x,\lambda) = \sum_{n\le x} |\tau(n,\theta;r,w)|^\lambda \tag{6.162}$$

where $\theta = 2\pi\nu$ for $1 \le \nu \le N$ and $\lambda = \pm(\log\log x)^{-1/2}$. Thus $|\lambda\log(1-r)| \ll 1$ and the Halberstam–Richert inequality (Theorem 01, *Divisors*) yields

$$H(x,\lambda) \ll \frac{x}{\log x}\exp\left\{\sum_{w<p\le x}\frac{1}{p}\left|1+rp^{i\theta}\right|^\lambda\right\}. \tag{6.163}$$

We apply Lemma 6.15, with $\phi = \nu$, $c = 1$ and

$$g(u) = |1+re(u)|^\lambda, \quad M(g), I(g), V(g) \ll 1.$$

Hence the sum in the exponential in (6.163) equals

$$F(\lambda,r)\log\left(\frac{\log x}{\log w}\right) + O\left(\frac{1}{\nu\log w} + \nu e^{-\sqrt{\log w}}\right) \tag{6.164}$$

where

$$F(\lambda,r) = \int_0^1 |1+re(u)|^\lambda\,du = \frac{1}{2\pi}\int_0^{2\pi}\left|1+re^{it}\right|dt. \tag{6.165}$$

The error term in (6.164) is negligible by (6.154). A classical theorem of Hardy (1915) on integral means implies that $F(\lambda,r)$ increases with r, whence

$$F(\lambda,r) \le F(\lambda,1) = \frac{2^\lambda}{\sqrt{\pi}}\frac{\Gamma(\frac{1+\lambda}{2})}{\Gamma(1+\frac{\lambda}{2})} = 1 + O(\lambda^2) \tag{6.166}$$

for $|\lambda| \le \frac{1}{2}$. (For F is meromorphic, with poles at $-1, -3, -5, \ldots$, moreover

$$\left.\frac{\partial}{\partial\lambda}F(\lambda,1)\right|_{\lambda=0} = \frac{1}{2\pi}\int_0^{2\pi}\log\left|1+e^{it}\right|dt = 0, \tag{6.167}$$

whence the right-hand inequality in (6.166) follows from Taylor's theorem.)

We deduce from (6.163–6.164) and (6.166) that for $|\lambda| \le 1/\sqrt{\log\log x}$ and x large enough, we have $|H(x,\lambda)| \ll x$. We put $\lambda = \pm 1/\sqrt{\log\log x}$, to obtain

$$\sum_{n\le x} \cosh\left(\frac{\log|\tau(n,\theta;r,w)|}{\sqrt{\log\log x}}\right) \ll x \tag{6.168}$$

whence

$$\sum_{n\le x}\sum_{v=1}^{N} \log^{2k}|\tau(n,\theta;r,w)| \ll_k Nx(\log\log x)^k \tag{6.169}$$

and (6.161) follows.

It is quite likely that the factor $\xi(x)$ in (6.161) is actually redundant, provided we replace $<$ by $\ll_k$. Indeed it is possible that an asymptotic formula, holding *p.p.*, could be given for $M_{2k}(n)$. We shall be content with a lower bound for $M_2(n)$ but it will be clear that we could obtain an asymptotic formula with more work; there is no reason to suppose that the higher moments behave differently. As it is, we obtain a slightly inferior lower bound for $M_1(n)$ when we apply (6.160). Another method to avoid this loss is indicated, albeit for an easier problem, in Hall (1994), involving third moments.

It remains to deal with $M_2(n)$.

Lemma 6.19 (Double Variance) *Let Q be a finite index set with cardinality $|Q|$, and for $\mu \in Q$ and $n \le x$ let*

$$f(n,\mu) = \sum_{p|n} f(p,\mu), \tag{6.170}$$

where $|f(p,\mu)| \le R$ always, and $f(p,\mu) = 0$ when $p > x^{1/4}$. Let

$$F(n) = \sum_{\mu\in Q} f(n,\mu)^2 \tag{6.171}$$

and put

$$\begin{aligned} V(\mu) &= \sum_p \frac{f(p,\mu)}{p} \\ X_s(\mu,v) &= \sum_p \frac{f(p,\mu)f(p,v)}{p^s} \\ Y_s(\mu,v) &= \sum_p \frac{f(p,\mu)^2 f(p,v)}{p^s} \end{aligned} \tag{6.172}$$

$$W_s(\mu, \nu) = \sum_p \frac{f(p, \mu)^2 f(p, \nu)^2}{p^s}$$
$$Z_s(\mu) = X_s(\mu, \mu)$$

and

$$\overline{F} = \sum_{\mu \in Q} \{Z_1(\mu) + V(\mu)^2 - Z_2(\mu)\}. \tag{6.173}$$

Then we have, for $R \geq 1$,

$$\sum_{n \leq x} \left(F(n) - \overline{F}\right)^2 = xJ + O\left(\frac{|Q|^2 R^4 x}{\log^4 x}\right) \tag{6.174}$$

where

$$\begin{aligned} J = \sum_{\mu \in Q} \sum_{\nu \in Q} \{ & W_1(\mu, \nu) - 7W_2(\mu, \nu) + 12W_3(\mu, \nu) - 6W_4(\mu, \nu) \\ & +2\,(X_1(\mu, \nu) - X_2(\mu, \nu))^2 \\ & +2V(\mu)\{Y_1(\nu, \mu) - 3Y_2(\nu, \mu) + 2Y_3(\nu, \mu)\} \\ & +2V(\nu)\{Y_1(\mu, \nu) - 3Y_2(\mu, \nu) + 2Y_3(\mu, \nu)\} \\ & +4V(\mu)V(\nu)\,(X_1(\mu, \nu) - X_2(\mu, \nu))\,\}. \end{aligned} \tag{6.175}$$

Proof We expand the square in (6.174) so that the sum becomes

$$\sum_{n \leq x} F(n)^2 - 2\overline{F} \sum_{n \leq x} F(n) + [x]\overline{F}^2. \tag{6.176}$$

We have

$$\begin{aligned} \sum_{n \leq x} F(n) &= \sum_{\mu \in Q} \sum_{n \leq x} \sum_{p|n} \sum_{q|n} f(p, \mu) f(q, \mu) \\ &= \sum_{\mu \in Q} \sum_{p} \sum_{q} f(p, \mu) f(q, \mu) \left\{ \frac{x}{[p, q]} + O(1) \right\} \end{aligned} \tag{6.177}$$

and we note that the error term here is, trivially, $\ll |Q|\, R^2 \pi(x^{1/4})^2$. The main term is

$$\begin{aligned} & x \sum_{\mu \in Q} \left\{ \sum_{p} \frac{f(p, \mu)^2}{p} + \sum_{p \neq q} \sum \frac{f(p, \mu) f(q, \mu)}{pq} \right\} \\ & \qquad = x \sum_{\mu \in Q} \{Z_1(\mu) + V(\mu)^2 - Z_2(\mu)\} = x\overline{F} \end{aligned} \tag{6.178}$$

by (6.172). Trivially, $|\overline{F}| \ll |Q|\, R^2(\log\log x)^2$ so that the error term arising from this sum is absorbed by that given in (6.174). Next, we have

$$\begin{aligned}\sum_{n\le x} F(n)^2 &= \sum_{n\le x}\sum_{\mu\in Q}\sum_{\nu\in Q} f(n,\mu)^2 f(n,\nu)^2 \\ &= \sum_{\mu\in Q}\sum_{\nu\in Q}\sum_{p}\sum_{q}\sum_{r}\sum_{s} f(p,\mu)f(q,\mu)f(r,\nu)f(s,\nu)\times \\ &\qquad \left\{\frac{x}{[p,q,r,s]} + O(1)\right\} \qquad (6.179)\end{aligned}$$

and again we make a trivial estimate of the error term, which does not exceed that given in (6.174). We write the main term in the form

$$x\overline{F}^2 + x\sum_{\mu\in Q}\sum_{\nu\in Q} K(\mu,\nu) \qquad (6.180)$$

so that

$$K(\mu,\nu) = \sum_{p}\sum_{q}\sum_{\substack{r \\ (pq,rs)>1}}\sum_{s} f(p,\mu)\cdots f(s,\nu)\left(\frac{1}{[p,q,r,s]} - \frac{1}{[pq][rs]}\right). \qquad (6.181)$$

The proof will be complete if we show that the inner sum in (6.175) is $K(\mu,\nu)$, and we leave this to the reader.

We note that the condition $f(p,\mu) = 0$ for $p > x^{\frac{1}{4}}$ is a luxury which we could forego if the application demanded it.

To apply the lemma we put

$$f(p,\mu) = \begin{cases} \log|1 + re(\mu\log p)| & \text{if } w < p \le x^{1/4}, \\ 0 & \text{else}, \end{cases} \qquad (6.182)$$

$$Q = \{1,2,3,\ldots,N\} \qquad (6.183)$$

$$R = \log\left(\frac{1}{1-r}\right). \qquad (6.184)$$

We observe that for $n \le x$, n may have at most three prime factors $> x^{1/4}$ whence

$$\log|\tau(n, 2\pi\mu; r, w)| = f(n,\mu) + O(R), \qquad (6.185)$$

if as we have assumed, n has no repeated prime factor $> w$. The inequality $(a+b)^2 \le 2a^2 + 2b^2$ now yields

$$F(n) = \sum_{\mu\in Q} f(n,\mu)^2 \le 2M_2(n) + O(NR^2). \qquad (6.186)$$

We need an approximate formula for $\overline{F}$, defined in (6.173), and an upper bound for J, defined in (6.175). We begin with the observation that all the sums in (6.172) in which $s \geq 2$ are negligible: they do not exceed

$$\sum_{p>w} \frac{R^4}{p^2} \ll \frac{R^4}{w} \ll \frac{1}{\log x}. \tag{6.187}$$

Next, we consider the sums $V(\mu)$ to which we apply Lemma 6.15. We have

$$\int_0^1 \log|1 + re(u)|\, du = 0, \quad r \leq 1, \tag{6.188}$$

so that we obtain, uniformly for $\mu \leq N$, that

$$|V(\mu)| \ll \frac{1}{\mu \log w} + \mu R e^{-\sqrt{\log w}} \ll (\log\log x)^{-20}, \tag{6.189}$$

estimating $I(g)$ by Cauchy's inequality and (6.151). We see from this that all the terms in (6.175) which involve a factor $V(\mu)$ are negligible: for example $Y_1(\mu, \nu) \ll R^3 \log\log x \ll (\log\log x)^{5/2}$ etc. Hence we are only concerned with the terms $Z_1(\mu)$ in (6.173), $W_1(\mu, \nu)$ and $X_1(\mu, \nu)^2$ in (6.175).

Let us define

$$\Lambda_k(r) = \int_0^1 \log^k |1 + re(u)|\, du. \tag{6.190}$$

We have, with F as in (6.166),

$$\sum_{h=0}^{\infty} \frac{1}{4^h (2h)!} \Lambda_{2h}(r) = \frac{1}{2}\left\{F\left(\frac{1}{2}, r\right) + F\left(-\frac{1}{2}, r\right)\right\}$$

whence $\Lambda_{2h}(r) \ll_h 1$, $h \in \mathbf{Z}^+$. Lemma 6.15 yields

$$Z_1(\mu) = \Lambda_2(r) \log\left(\frac{\log x}{\log w}\right) + O\left((\log\log x)^{-20}\right) \tag{6.191}$$

and

$$W_1(\mu, \mu) = \Lambda_4(r) \log\left(\frac{\log x}{\log w}\right) + O\left((\log\log x)^{-20}\right) \tag{6.192}$$

uniformly for $\mu \leq N$, and then Cauchy's inequality and (6.192) imply that

$$W_1(\mu, \nu) \ll \log\log x, \quad 1 \leq \mu, \nu \leq N. \tag{6.193}$$

To cope with $X_1(\mu, \nu)$ we have to consider the function

$$g(u) = \log|1 + re(\mu u)| \log|1 + re(\nu u)|, \tag{6.194}$$

and we treat this as having period 1. (There would be a slight saving in Lemma 6.15 if we replaced $g(u)$ by $g(u/(\mu, v))$ and took $\phi = (\mu, v)$, but this is unimportant in this application.) The mean value $\overline{g}$ has been estimated already in (6.151): we have

$$|\overline{g}| \ll \frac{(\mu, v)^2}{\mu v} \tag{6.195}$$

uniformly for $r \leq 1$. The total variation may be estimated by Leibniz' rule and is

$$\int_0^1 |dg| \ll R(\mu R) + R(v R) \ll NR^2 \tag{6.196}$$

whence we have

$$|X_1(\mu, v)| \ll \frac{(\mu, v)^2}{\mu v} \log\log x. \tag{6.197}$$

We note that

$$\sum_{\mu \leq N} \sum_{v \leq N} \frac{(\mu, v)^4}{\mu^2 v^2} \ll N, \tag{6.198}$$

and so we obtain from (6.193), (6.198) that

$$J \ll N^2 \log\log x + N(\log\log x)^2. \tag{6.199}$$

Next, from (6.190), we have

$$\Lambda_2(r) = \frac{1}{2} \sum_{m=1}^{\infty} \frac{r^{2m}}{m^2} > \frac{1}{8}, \quad \frac{1}{2} \leq r \leq 1. \tag{6.200}$$

Hence from (6.191) we have

$$\overline{F} \geq cN \log\log x, \quad c > 0 \tag{6.201}$$

subject to our constraints on r, N and w. We have $J = o(N^2\, (\log\log x)^2)$ from (6.199), whence we deduce from Lemma 6.19, (6.174), that

$$F(n) > \frac{1}{2} cN \log\log x \tag{6.202}$$

for all but $o(x)$ integers $n \leq x$. Hence (6.186) implies

$$M_2(n) > \frac{1}{5} cN \log\log x \tag{6.203}$$

for these same integers. Together with Lemma 6.18, this gives, via (6.160),

$$M_1(n) > \xi(x)^{-1/2} N \sqrt{\log\log x} \tag{6.204}$$

for $x + o(x)$ integers $n \le x$. Together with (6.159) this implies for large x,

$$\sum_{\nu=1}^{N} \log^{+} |\tau(n, 2\pi\nu; r, w)| > \frac{1}{3}\xi(x)^{-1/2} N \sqrt{\log\log x} \tag{6.205}$$

with at most $o(x)$ exceptions, and we may strike out the w at the cost of an error term

$$\begin{aligned} &\le 2N \log\log w . \log\left(\frac{1}{1-r}\right) \\ &\le 40\varepsilon(x) N \sqrt{\log\log x} \end{aligned} \tag{6.206}$$

by the definitions (6.139), (6.154) of r and w. Thus if $\xi(x) \to \infty$ sufficiently slowly, we have from (6.205–6.206) that, with $o(x)$ exceptions,

$$\sum_{\nu=1}^{N} \log^{+} |\tau(n, 2\pi\nu; r)| \ge \frac{1}{2}\varepsilon(x) N \sqrt{\log\log x} \tag{6.207}$$

whence by Jensen's inequality,

$$\sum_{\nu=1}^{N} |\tau(n, 2\pi\nu; r)|^2 \ge N \exp\left\{\varepsilon(x)\sqrt{\log\log x}\right\} \tag{6.208}$$

Finally, we have an inequality, in the weighted case, analogous to (6.59). We restrict the sum on the right to the range $1 \le \nu \le N$: for these ν and the values of t specified in (6.140) we have

$$\frac{\sin \pi\nu t}{\pi\nu} \ge \frac{2}{\pi} t \tag{6.209}$$

and since we may assume that $N \ge \pi^2/4$, we obtain (6.141) as stated.

A result of Hall (1975b) is that for each fixed real $\theta \ne 0$ we have

$$|\tau(n, \theta)| < \exp\left\{\xi(n)\sqrt{\log\log n}\right\} \ p.p. \tag{6.210}$$

provided $\xi(n) \to \infty$. No direct method of combining this with the Erdös–Turán inequality to bound $\Delta(n; \log)$ has been found, nevertheless, in view of (6.210) and Theorem 6.13 it seems reasonable to suppose that the order of magnitude of $\Delta(n; \log)$ is something like $\exp\{\sqrt{\log\log n}\}$, *p.p.*

7

$H(x, y, z)$

7.1 Introduction

We consider the important special case

$$\mathscr{A} = \mathscr{A}(y, z) = (y, z] \cap \mathbf{Z}. \tag{7.1}$$

We denote the counting function of $\mathscr{M}(\mathscr{A}(y, z))$ by $H(x, y, z)$, following Tenenbaum (1984). Thus

$$H(x, y, z) = \text{card}\{n : n \leq x, \exists d : d|n, y < d \leq z\}. \tag{7.2}$$

This function may be studied independently of sets of multiples. In a general setting, y and z may be functions of x, perhaps the most interesting case being when they are powers of x. Tenenbaum (1980) proved the following theorem: let

$$y = x^{(1-u)/t}, \ z = x^{1/t}, \quad 0 \leq u \leq 1, \ t \geq 1. \tag{7.3}$$

Then

$$\lim_{x \to \infty} x^{-1} H(x, y, z) =: h(u, t) \tag{7.4}$$

exists. Properties of the function $h(u, t)$ were given. (We have used the notation employed in *Divisors* §2.4 which contains a proof; extra information is contained in *Divisors*, and in the original paper.)

Within the framework of sets of multiples y and z are necessarily independent of x. $\mathscr{A}$ is finite and is therefore a Besicovitch sequence, and we are interested in the density $\mathbf{d}\mathscr{M}(\mathscr{A}(y, z))$. We may write this down explicitly, using either the inclusion–exclusion principle or total decomposition sets; however what we usually really want is an asymptotic formula for the function $\mathbf{d}\mathscr{M}(\mathscr{A}(y, z))$ when $y \to \infty$ and z is a suitable function of y. Failing this we seek good upper and lower bounds.

$H(x,y,z)$ is the subject of *Divisors,* Chapter 2 and the work presented here is a sequel, but with a different point of view. Our main purpose in this short chapter is to settle the conjecture which appears in *Divisors* §2.2. This requires some introduction in order that our presentation should be self-contained and comprehensible.

7.2 Short intervals

We refer to the interval $(y,z]$ as short, in the present context, if $z \le 2y$. We define $\beta = \beta(y,z)$ implicitly by the equation

$$z = y + y\log^{-\beta} y, \quad 0 \le \beta < \infty. \tag{7.5}$$

We think of z as a function of y and β as a parameter, and we seek an asymptotic formula for $\mathbf{d}\mathscr{M}(\mathscr{A}(y,z))$. Of course we are interested as well in the question of how large x must be for the formula

$$H(x,y,z) \sim \mathbf{d}\mathscr{M}(\mathscr{A}(y,z))\,x \tag{7.6}$$

to hold as $x, y, z \to \infty$.

Tenenbaum (1984) discovered a *threshold* in the behaviour of $\mathbf{d}\mathscr{M}(\mathscr{A}(y,z))$ at the point $\beta = \log 4 - 1$. The term threshold was introduced in Hall and Tenenbaum (1989) and applies to the circumstance in which we have an asymptotic formula for an arithmetical function containing a parameter (such as β), where the main term discontinuously changes shape as the parameter passes through a critical value (the threshold). Quite often it happens that the formula on one side of the threshold has a simple shape, for which a natural or heuristic explanation is available, while on the other side the formula is more complicated and mysterious or may even be unknown. It may be possible to remove the discontinuity by introducing a further parameter, and this applies in the case under discussion. Let $\xi = \xi(y)$ and

$$\beta = \log 4 - 1 + \frac{\xi}{\sqrt{\log\log y}}. \tag{7.7}$$

Then we deduce from *Divisors,* Theorem 21(i) that if $\xi \to \infty$ as $y \to \infty$, and z is given by (7.5) and (7.7), we have

$$\mathbf{d}\mathscr{M}(\mathscr{A}(y,z)) \sim \log^{-\beta} y; \tag{7.8}$$

moreover provided $x > z^2$, we also have

$$H(x,y,z) \sim x\log^{-\beta} y \sim \sum_{n\le x} \tau(n,\mathscr{A}(y,z)), \tag{7.9}$$

because the sum on the right is, for all x, y, z, equal to $x(\log(z/y) + O(1/y) + O(z-y)$. We can write $\tau(n, \mathscr{A}(y,z))$ in the alternative form $\tau(n; y, z)$ used in *Divisors*. Notice that (7.9) implies that the sum on the right is dominated by the contribution of integers for which $\tau(n; y, z) = 1$. Let

$$Q(\lambda) = \lambda \log \lambda - \lambda + 1, \quad \lambda > 0 \tag{7.10}$$

and define

$$G(\beta) = \begin{cases} Q\left(\frac{1+\beta}{\log 2}\right), & 0 \le \beta \le \log 4 - 1 \\ \beta, & \beta > \log 4 - 1. \end{cases} \tag{7.11}$$

Theorem 21(iii) of *Divisors* applies when $\beta \ge 0$ but $\xi(y)$ is bounded above, and yields

$$L^{-1}(\log y)^{-G(\beta)} \ll \mathbf{d}\mathscr{M}(\mathscr{A}(y,z)) \ll \frac{(\log y)^{-G(\beta)}}{1 + (-\xi)^+} \tag{7.12}$$

where $(-\xi)^+ = \max(-\xi, 0)$ and, for a suitable positive c,

$$L = \exp\{c\sqrt{\log\log y . \log\log\log\log y}\}. \tag{7.13}$$

The corresponding formula for $H(x, y, z)$ is again valid for $x > z^2$. It will emerge later that (7.9) is now false – that is, a necessary and sufficient condition for (7.9) is $\xi(y) \to \infty$.

Notice that for $\beta \le \log 4 - 1$ we have

$$G(\beta) = \beta + 2Q\left(\frac{1+\beta}{\log 4}\right). \tag{7.14}$$

Thus $G(\beta) > \beta$, indeed on the interval $[0, \log 4 - 1]$, $G(\beta) - \beta$ decreases from δ to 0, where $\delta = .086071\ldots$ is the number defined in Lemma 0.3. We have $Q(1) = Q'(1) = 0$, so that $G(\beta)$ is continuously differentiable and convex for $\beta \ge 0$. For $\xi < 0$, Taylor's theorem implies that

$$G(\beta) = \beta + \left(\frac{\xi}{\log 4}\right)^2 (\log\log y)^{-1} + O(\xi^3(\log\log y)^{-3/2}) \tag{7.15}$$

in which the last term on the right is *positive*. If ξ is negative and $\xi = o\left((\log\log y)^{1/6}\right)$ we deduce that

$$(\log y)^{-G(\beta)} \sim \exp\left(-\frac{\xi^2}{\log^2 4}\right) \log^{-\beta} y, \tag{7.16}$$

and we may insert this into (7.12), bearing in mind that

$$\log^{-\beta} y \sim \sum \{a^{-1} : a \in \mathscr{A}(y,z)\}. \tag{7.17}$$

We shall explain later how the factor L^{-1} on the left-hand side of (7.12) arises. If we assume that the upper bound in (7.12) is near the truth then we are led by (7.16) and (7.17) to conjecture that there exists a probability distribution function $F(\xi)$ such that if $\xi \in \mathbf{R}$ is fixed, and $y, z \to \infty$ together, with $z = z(y)$ determined by (7.5) and (7.7), then

$$\mathbf{d}\mathcal{M}(\mathscr{A}(y,z)) \sim F(\xi) \sum \{a^{-1} : a \in \mathscr{A}(y,z)\}. \tag{7.18}$$

If in addition $x \to \infty$ sufficiently fast, we will have

$$H(x, y, z) \sim F(\xi) \sum_{n \le x} \tau(n; y, z). \tag{7.19}$$

The conjecture contained in *Divisors* §2.2 is equivalent to (7.19) except that we specified merely that $x \to \infty$ together with y and z which is plainly wrong (since we could put $x = z$, when $H(x, y, z) = [z] - [y]$, or $x \le y$, when $H(x, y, z) = 0$).

We shall prove that (7.18) and (7.19) hold, with

$$F(\xi) = \frac{1}{\sqrt{\pi}} \int_{-\infty}^{\xi/\log 4} e^{-w^2}\, dw. \tag{7.20}$$

This is the content of Theorem 7.9 below. Rather than set out a formal proof we lead up to our result in stages. We hope that in this way the reader will be able to see more easily which parts of the argument would apply to the case $\beta \in [0, \log 4 - 1)$, β fixed, where it is an open problem to improve on (7.12).

Hall and Tenenbaum (1989) proved that provided $\xi \ge -c_0(\log\log y)^{1/6}$, for any fixed c_0, we have, for $x \ge z^2$,

$$H(x, y, z) \asymp \frac{x(\log y)^{-G(\beta)}}{1 + (-\xi)^+}, \tag{7.21}$$

that is the upper bound in (7.12) is the correct order of magnitude. Our approach now is similar and is based on the principle that for suitable ξ, the dominant contribution to $H(x, y, z)$ comes from integers n for which $\tau(n; y, z) = 1$.

As usual, let $\Omega(n, z)$ denote the number of prime factors of n which do not exceed z, counted according to multiplicity. A very crude first estimate of the probable order of magnitude of $\tau(n; y, z)$ is

$$\tau(n; y, z) \approx 2^{\Omega(n,z)} \frac{\log(z/y)}{\log z}. \tag{7.22}$$

This is of course not supposed to be more than a guide, but it suggests the following strategy: let

$$k_0 = \left[\frac{1+\beta}{\log 2}\log\log z\right]. \tag{7.23}$$

Then if $\Omega(n,z) \geq k_0$, $\tau(n;y,z)$ may well be large and n be counted by $H(x,y,z)$ but these 'safe bets' will be rare. More frequently $\Omega(n,z) < k_0$ and then we expect $\tau(n;y,z) = 0$ or 1 with probabilities which can be estimated.

Notice that if β is approximately $\log 4 - 1$ then k_0 is approximately $2\log\log z$ which is a threshold for the counting function

$$N_k(x,z) = \operatorname{card}\{n : n \leq x,\ \Omega(n,z) = k\}. \tag{7.24}$$

This (irrelevant) difficulty is caused by the prime 2 and we avoid it by working with $\overline{\Omega}(n,z)$ which we define to be the number of odd prime factors of n which do not exceed z. We define $\overline{N}_k(x,z)$ analogously, and we have

$$\sum_{k\geq k_0}\overline{N}_k(x,z) \ll x(\log z)^{-Q((1+\beta)/\log 2)}(\log\log z)^{-1/2}; \tag{7.25}$$

see *Divisors* §2.6 for a detailed calculation. We replace z by y on the right and we notice that for all $\beta \geq 0$ the exponent of $\log y$ is $\leq -G(\beta)$ with equality if and only if $\beta \leq \log 4 - 1$ by (7.11). Provided $(-\xi)^+ = o(\sqrt{\log\log y})$ the sum in (7.25) is negligible compared with the right-hand side of (7.21).

For $k < k_0$ we consider the sums

$$R_k = \sum_{\substack{n\leq x\\ \overline{\Omega}(n,z)=k}} \tau(n;y,z) \tag{7.26}$$

$$S_k = \sum_{\substack{n\leq x\\ \overline{\Omega}(n,z)=k}} \binom{\tau(n;y,z)}{2}. \tag{7.27}$$

Let $H_1(x,y,z)$ denote the contribution to $H(x,y,z)$ from integers n such that $\overline{\Omega}(n,z) < k_0$. We have

$$H_1(x,y,z) \leq \sum_{k<k_0} R_k \tag{7.28}$$

and, for any $h \leq k_0$,

$$H_1(x,y,z) \geq \sum_{k<h} R_k - \sum_{k<h} S_k. \tag{7.29}$$

For technical reasons to do with our upper bound for S_k we have to take h somewhat less than k_0. If the sum over S_k is small relative to that over R_k then we obtain an asymptotic formula for $H_1(x, y, z)$: this clearly requires that usually $\tau(n; y, z) = 0$ or 1.

The method works for all positive ξ and for negative ξ suitably bounded below. As we try to reduce ξ further, we find that we cannot cope satisfactorily with the sum over S_k in (7.29). The problem can be traced back to the crude model (7.22) for determining whether or not we expect $\tau(n; y, z) \leq 1$. For this to hold, we require not only that $\overline{\Omega}(n, z)$ should be suitably small but that the prime factors $\leq z$ of n should have the right kind of distribution. This is rather vague, because this question (which is likely to be the key to the problem) has not yet been resolved. A *sufficient* requirement would be uniformity of distribution, for example

$$\overline{\Omega}(n, t) \leq \frac{1+\beta}{\log 2} \log\log t, \ (3 \leq t \leq z) \tag{7.30}$$

but because of Erdös' law of the iterated logarithm for prime factors (see *Divisors,* Chapter 1; the generalization of this directly germane to (7.30) has been carried out by L. Bastick (1992)), the numbers n satisfying (7.30) are too rare. A practical condition of this sort is employed in *Divisors* §2.7: we add a small function of y on the right of (7.30); $\log L$, with L as in (7.13) is suitable. This leads to the lower bound (7.12).

Our method requires asymptotic formulae for the sums of R_k and upper bounds for those of S_k. We tackle the former problem in the next section.

7.3 The asymptotic formula for $\sum R_k$

Theorem 7.1 *Let R_k be as in (7.26), $\beta = \log 4 - 1 + \xi/\sqrt{\log\log y}$, and k_0 be as in* (7.23). *Let $x > \exp(\log z . \log\log z)$. Then for each fixed c, and $|\xi| \leq c(\log\log y)^{1/6}$, we have*

$$\sum_{k<k_0} R_k = \left(F(\xi) + O_c\left(\frac{E(\xi)}{\sqrt{\log\log y}}\right)\right) x \log^{-\beta} y \tag{7.31}$$

where $F(\xi)$ is the probability distribution function defined by (7.20) and $E(\xi) = 1 + \xi$ if $\xi > 0$, $E(\xi) = (1+\xi^2)\exp(-\xi^2/\log^2 4)$ if $\xi \leq 0$.

The formula is reminiscent of Theorems 4.10 and 4.13 which is to be expected since we are concerned essentially with Poisson variables.

All the formulae for $H(x, y, z)$ in *Divisors* are valid for $x > z^2$ and so it would certainly be desirable to be able to cope with smaller values of x than those given above, although this is not our primary concern.

We require several lemmas. The restriction on x occurs in the first lemma, and we make some further remarks about it.

Lemma 7.2 *Let $A > 0$ and $\eta > 0$ be fixed, and $|w| \le 3 - \eta$, $z \ge z_0(A, \eta)$. Let $x \ge \exp(\frac{1}{2}\log z . \log\log z)$. Then*

$$\sum_{n\le x} w^{\overline{\Omega}(n,z)} = M(w)x + O_{A,\eta}\left(x \exp\left(-A\frac{\log x}{\log z}\right)\right) \tag{7.32}$$

where

$$M(w) = \prod_{3\le p\le z}\left(1-\frac{1}{p}\right)\left(1-\frac{w}{p}\right)^{-1}. \tag{7.33}$$

This is an application of Theorem 02 of *Divisors,* which is an elementary result depending on Rankin's method. Notice that when $w = 0$ the sum on the left of (7.32) is closely related to Buchstab's function $\Phi(x, z)$ and is not asymptotic to $M(0)x$ when $\log x \ll \log z$. These small values of x would require a reappraisal of this section and it is not clear that (7.31) has the correct main term. An examination of our method shows that in fact only values of w close to 1 in (7.32) are critical, when we might expect a wider range of validity. For $|w| < 3$, we define

$$J(w) = \prod_{3\le p\le z}\left(1-\frac{w}{p}\right)^{-1}\left(1-\frac{1}{p}\right)^{w}. \tag{7.34}$$

We put

$$Z = -\sum_{3\le p\le z}\log\left(1-\frac{1}{p}\right). \tag{7.35}$$

Lemma 7.3 *Let $\eta > 0$ be fixed. Then for $z \ge z_1(\eta)$, $j \le (3-\eta)\log\log z$ and $x \ge \exp(\frac{1}{2}\log z . \log\log z)$ we have*

$$\overline{N}_j(x,z) = x\left(J\left(\frac{j}{Z}\right) + O_\eta\left(\frac{1}{Z}\right)\right)\frac{Z^j}{j!}\prod_{3\le p\le z}\left(1-\frac{1}{p}\right). \tag{7.36}$$

Proof of lemma We shall apply Lemma 7.2 with η replaced by $\eta/2$. Let $M(x, w)$ denote the sum on the left of (7.32) so that

$$M(x,w) = \sum_{j=0}^{\infty}\overline{N}_j(x,z)w^j. \tag{7.37}$$

When $j = 0$ we put $w = 0$ in (7.37) and fix $A = 2$. Let $j \ge 1$ and C be the circle $\{w : |w| = r\}$ where we assume $r \le 3 - \frac{1}{2}\eta$. Cauchy's formula gives

$$\overline{N}_j(x, z) = \frac{1}{2\pi i} \int_C M(x, w) \frac{dw}{w^{j+1}}. \tag{7.38}$$

We employ the formula (7.32) for $M(x, w)$. The error term contributes

$$\ll_{A,\eta} r^{-j} x \exp\left(-A \frac{\log x}{\log z}\right). \tag{7.39}$$

We write

$$M(w) = \prod_{3 \le p \le z} \left(1 - \frac{1}{p}\right) J(w) e^{wZ} \tag{7.40}$$

so that the main term in (7.38) becomes

$$x \prod_{3 \le p \le z} \left(1 - \frac{1}{p}\right) \frac{1}{2\pi i} \int_C J(w) e^{wZ} \frac{dw}{w^{j+1}}. \tag{7.41}$$

Let $\mathscr{E}(w, r)$ be defined implicitly by the equation

$$J(w) = J(r) + (w - r) J'(r) + \mathscr{E}(w, r), \tag{7.42}$$

i.e. as the remainder in Taylor's series. We have

$$\begin{aligned} \frac{1}{2\pi i} \int_C J(w) e^{wZ} \frac{dw}{w^{j+1}} &= J(r) \frac{Z^j}{j!} + J'(r) \left(\frac{Z^{j-1}}{(j-1)!} - \frac{r Z^j}{j!}\right) \\ &\quad + \frac{1}{2\pi i} \int_C \mathscr{E}(w, r) e^{wZ} \frac{dw}{w^{j+1}}, \end{aligned} \tag{7.43}$$

and we put $r = j/Z$ so that the term involving $J'(r)$ in (7.43) vanishes. By (7.35), $Z = \log\log z + O(1)$ and we assume that $z_1(A, \eta)$ is both $\ge z_0(A, \frac{\eta}{2})$ and sufficiently large to imply $r \le 3 - \frac{1}{2}\eta$. $J(w)$ and its derivatives are bounded on compact subsets of the disc $\{w : |w| < 3\}$ and so for such r we have

$$\mathscr{E}(w, r) \ll_\eta |w - r|^2. \tag{7.44}$$

We put $w = re^{i\theta}$ and we see that the final term in (7.43) is

$$\begin{aligned} &\ll_\eta \ r^{2-j} \int_0^\pi \left|e^{i\theta} - 1\right|^2 e^{rZ\cos\theta} d\theta \\ &\ll_\eta \left(\frac{j}{Z}\right)^2 \left(\frac{eZ}{j}\right)^j \int_0^\pi \theta^2 e^{-2j\theta^2/\pi^2} d\theta, \end{aligned} \tag{7.45}$$

using the inequality $\cos\theta \le 1 - 2\theta^2/\pi^2$, $0 \le \theta \le \pi$. The integral is $\ll j^{-3/2}$ whence by Stirling's formula this is

$$\ll_\eta \frac{j}{Z^2}\cdot\frac{Z^j}{j!} \ll_\eta \frac{1}{Z}\cdot\frac{Z^j}{j!}. \tag{7.46}$$

Hence the expression in (7.41) is equal to

$$x\left(J\left(\frac{j}{Z}\right)+O_\eta\left(\frac{1}{Z}\right)\right)\frac{Z^j}{j!}\prod_{3\le p\le z}\left(1-\frac{1}{p}\right). \tag{7.47}$$

We substitute $r = j/Z$ in (7.39) and employ Stirling's formula to compare this term with the error term in (7.47) bearing in mind that the product on the right of (7.47) is $\gg 1/\log z$. If we fix $A = 2$, the error term arising from (7.39) is absorbed for the values of x stated. This completes the proof.

Lemma 7.4 *Let $\overline{\Omega}(d)$ denote the number of odd prime factors of d counted according to multiplicity. Let $\eta > 0$ be fixed. Then uniformly for $|w| \le 3-\eta$ we have*

$$\sum_{y<d\le z} w^{\overline{\Omega}(d)} = K(w)(z-y)(\log z)^{w-1} + O_\eta\left(z(\log z)^{\operatorname{Re} w-2}\right) \tag{7.48}$$

where

$$K(w) = \frac{2^{1-w}}{\Gamma(w)}\prod_{p\ge 3}\left(1-\frac{w}{p}\right)^{-1}\left(1-\frac{1}{p}\right)^{w}. \tag{7.49}$$

This result is of a standard type and may be derived from Perron's formula: the factor $1/\Gamma(w)$ arises from Hankel's integral. We omit the proof.

Lemma 7.5 *Let $\eta > 0$ be fixed. Then uniformly for $i \le (3-\eta)\log\log z$ we have*

$$\sum_{\substack{y<d\le z\\ \overline{\Omega}(d)=i}} 1 = \left(K\left(\frac{i}{l}\right)+O_\eta\left(\frac{i}{l^2}\right)\right)\frac{z-y}{\log z}\cdot\frac{(\log\log z)^i}{i!} +$$

$$+O_\eta\left(\frac{z(\log\log z)^i}{i!\log^2 z}\right) \tag{7.50}$$

where $l = \log\log z$ and K is given by (7.49).

This follows from the previous lemma in the same way as Lemma 7.3 follows from Lemma 7.2. Because $K(i/l) \sim 2i/l$ for small i we leave the first error term on the right of (7.50) in this form, analogous to the left-hand term in (7.46).

We proceed to an asymptotic formula for R_k. We assume henceforth that $x > \exp(\log z . \log\log z)$. We have

$$R_k = \sum_{i+j=k} \sum_{\substack{y<d\leq z \\ \overline{\Omega}(d)=i}} \overline{N}_j\left(\frac{x}{d}, z\right) \tag{7.51}$$

and we employ Lemma 7.3. We have

$$\frac{x}{d} = \frac{x}{y}\left(1 + O(\log^{-\beta} y)\right) \tag{7.52}$$

and we notice that the error term here is negligible compared with the term $O_\eta(1/Z)$ in (7.34), since $J(j/Z) \asymp 1$. We put

$$W = \prod_{3\leq p\leq z}\left(1 - \frac{1}{p}\right) = e^{-Z}, \tag{7.53}$$

and we have

$$R_k = xy^{-1}W \sum_{i+j=k}\left(J\left(\frac{j}{Z}\right) + O_\eta\left(\frac{1}{Z}\right)\right)\frac{Z^j}{j!} \sum_{\substack{y<d\leq z \\ \overline{\Omega}(d)=i}} 1. \tag{7.54}$$

We apply Lemma 7.5 to the inner sum: since $K(\frac{i}{l}) \asymp \frac{i}{l}$ (for the i in question) we may write

$$\left(K(\left(\frac{i}{l}\right) + O_\eta\left(\frac{i}{l^2}\right)\right)\frac{z-y}{\log z} = K\left(\frac{i}{l}\right)\left(1 + O_\eta\left(\frac{1}{l}\right)\right)y(\log y)^{-1-\beta} \tag{7.55}$$

and we obtain

$$\begin{aligned} R_k &= xW(\log y)^{-1-\beta}\sum_{i+j=k} J\left(\frac{j}{Z}\right)K\left(\frac{i}{l}\right)\frac{l^iZ^j}{i!j!}\left(1 + O_\eta\left(\frac{1}{l}\right)\right) \\ &\quad + O_\eta\left(\frac{x(2\log\log z)^k}{k!\log^3 z}\right). \end{aligned} \tag{7.56}$$

The functions $J(w)$ and $K(w)$ are analytic for $|w| < 3$ and so their derivatives are bounded in the disc $|w| \leq 3 - \eta$. Hence, when $i + j = k$,

$$J\left(\frac{j}{Z}\right)K\left(\frac{i}{l}\right) - J\left(\frac{k}{2Z}\right)K\left(\frac{k}{2l}\right) \ll_\eta \frac{1}{l}\left|i - \frac{k}{2}\right|. \tag{7.57}$$

We have

$$\sum_{i=0}^{k}\binom{k}{i}\left|i - \frac{k}{2}\right| \ll 2^k\sqrt{k} \tag{7.58}$$

and we employ (7.56)–(7.58) to obtain

$$R_k = xW(\log y)^{-1-\beta}\left(J\left(\frac{k}{2Z}\right)K\left(\frac{k}{2l}\right)+O_\eta\left(\frac{\sqrt{k+1}}{l}\right)\right)\frac{(l+Z)^k}{k!} \quad (7.59)$$

absorbing the final error term from (7.56). We require the following result which is a deduction from Theorem 1.8 of Norton (1978).

Lemma 7.6 *Let* $v > 0$ *and*

$$S_v(\alpha) := \sum_{k \le v+\alpha\sqrt{v}} e^{-v}\frac{v^k}{k!}. \quad (7.60)$$

Then for $|\alpha| \le cv^{\frac{1}{6}}$, *for any fixed* c,

$$S_v(\alpha) = \frac{1}{\sqrt{2\pi}}\int_{-\infty}^{\alpha} e^{-u^2/2}du + O_c\left(\frac{1+\alpha^2}{\sqrt{v}}e^{-\alpha^2/2}\right). \quad (7.61)$$

We shall also need

Lemma 7.7 *Let*

$$S'_v(\alpha) = \sum_{k \le v+\alpha\sqrt{v}} (v+\alpha\sqrt{v}-k)e^{-v}\frac{v^k}{k!}. \quad (7.62)$$

Then for α *as in the previous lemma, we have*

$$S'_v(\alpha) \ll_c v^{\frac{1}{2}}(1+\alpha^+)\exp\left(-\frac{1}{2}\left((-\alpha)^+\right)^2\right) \quad (7.63)$$

Proof of Lemma 7.7 Put $k_1 = v + \alpha\sqrt{v}$. Then

$$S'_v(\alpha) = k_1 S_v(\alpha) - v\sum_{k<k_1} e^{-v}\frac{v^k}{k!} = (k_1 - v)S_v(\alpha) + ve^{-v}\frac{v^{k_1}}{k_1!}. \quad (7.64)$$

We consider two cases. If $\alpha \ge 0$ we apply the bounds

$$S_v(\alpha) \le 1, \quad \frac{v^{k_1}}{k_1!} \ll \frac{e^v}{\sqrt{v}}, \quad (7.65)$$

which lead to (7.63) in this case. If $\alpha < 0$ we deduce from (7.61) that

$$S_v(\alpha) \ll_c \frac{e^{-\alpha^2/2}}{1+|\alpha|} + \frac{1+\alpha^2}{\sqrt{v}}e^{-\alpha^2/2} \ll_c \frac{e^{-\alpha^2/2}}{1+|\alpha|} \quad (7.66)$$

and from Stirling's formula that

$$ve^{-v}\frac{v^{k_1}}{k_1!} \ll_c e^{-\alpha^2/2}\sqrt{v} \quad (7.67)$$

from which (7.63) again follows. This completes the proof.

Let

$$k_1 = l + Z + \alpha\sqrt{l+Z}. \tag{7.68}$$

For $k \le k_1$ we have, uniformly for $k_1 \le 3\log\log z$,

$$J\left(\frac{k}{2Z}\right)K\left(\frac{k}{2l}\right) = J\left(\frac{k_1}{2Z}\right)K\left(\frac{k_1}{2l}\right) + O\left(\frac{k_1-k}{l}\right) \tag{7.69}$$

and we insert this approximation into (7.59) and apply Lemmas 7.6, 7.7 with $v = l + Z$. We may fix $\eta = \frac{1}{2}$ at this point because we shall have $\alpha \ll (\log\log z)^{\frac{1}{6}}$ and $k_1 \le (2+o(1))\log\log z$. We note that, by (7.53),

$$W(\log z)^{-1} = e^{-(l+Z)}. \tag{7.70}$$

We obtain that, for any fixed c and $|\alpha| \le c(\log\log z)^{1/6}$, we have

$$\sum_{k\le k_1} R_k = x\log^{-\beta} y\left\{J\left(\frac{k_1}{2Z}\right)k\left(\frac{k_1}{2l}\right)\cdot\frac{1}{\sqrt{2\pi}}\int_{-\infty}^{\alpha} e^{-u^2/2}du + O_c\left(l^{-1/2}E_1(\alpha)\right)\right\} \tag{7.71}$$

where

$$E_1(\alpha) = \begin{cases} 1+\alpha, & \alpha > 0, \\ (1+\alpha^2)\exp(-\alpha^2/2) & \alpha \le 0 \end{cases}.$$

We approximate $J(k_1/2Z)K(k_1/2l)$ by $J(1)K(1) = 1$ with error $\ll 1 + |\alpha|/\sqrt{l}$: this is absorbed by the error term in (7.71). Finally, we put $k_1 = k_0 - 1$ where k_0 is given by (7.23). We compute α from (7.68), noting that $Z = \log\log z + O(1)$. Thus

$$\alpha = \frac{\xi}{2^{1/2}\log 2} + O\left(\frac{1}{\sqrt{l}}\right). \tag{7.72}$$

We clear our formula of $\sqrt{2}$'s by writing $F(\xi)$ in the form (7.20). The error introduced by the approximation in (7.72) is absorbed. This proves Theorem 7.1.

7.4 The asymptotic formula for $H(x,y,z)$

We now have an asymptotic formula for the sum on the right of (7.28), and (by implication, provided h is suitably close to k_0) for the first sum on the right of (7.29). Provided h is sufficiently close to k_0, these sums will be asymptotically equal. We need an upper bound for S_k.

Lemma 7.8 *For* $0 < v \le 1$ *let*

$$T(v) = \sum_{n \le x} \binom{\tau(n; y, z)}{2} v^{\overline{\Omega}(n,z)}. \tag{7.73}$$

Then provided $x \ge z^2$,

$$T(v) \ll x(\log y)^{-2\beta+2v-2+(2v-1)^+} (\log\log y)^2, \tag{7.74}$$

(where as usual λ^+ *denotes* $\max(\lambda, 0)$*).*

Let us compare the right-hand side of (7.74) with what we should expect if the heuristic approximation (7.22) were exact. We should strike out the unimportant factor $(\log\log y)^2$, and the exponent of $\log y$ would become

$$-2\beta + 4v - 3, \tag{7.75}$$

that is as in (7.74) without the $^+$. It will be clear from the proof that (7.74) is realistic, that is for small k the sum S_k is larger than the over-simplistic model would suggest. We shall see below that this effect begins to bite as soon as k is substantially less than $2\log\log z$, and so the method presented here fails if β is fixed $< \log 4 - 1$, (cf. the remarks following (7.29)).

Proof of lemma We put $\log^{-\beta} y = \theta$ for convenience. Let $d|n$, $d'|n$ where $y < d < d' \le z$. We put $m = (d, d')$ so that $m \le \theta y$ and write $d = mt$, $d' = mt'$. The Halberstam–Richert inequality (*Divisors*, Theorem 00) gives

$$\sum_{\substack{n \le x \\ n \equiv 0 (\mathrm{mod}\ [d,d'])}} v^{\overline{\Omega}(n,z)} \ll v^{\overline{\Omega}(mtt')} \frac{x}{mtt'} (\log X)^{v-1} \tag{7.76}$$

where $X = \min(z, x/[d, d'])$. We have $x/[d, d'] = xm/dd' \ge m$ since $dd' \le z^2 \le x$, whence $X \ge m$. Hence

$$T(v) \ll x \sum_{m \le \theta y} \frac{v^{\overline{\Omega}(m)}}{m} (\log 2m)^{v-1} \Bigg(\sum_{y/m < t \le z/m} \frac{v^{\overline{\Omega}(t)}}{t} \Bigg)^2. \tag{7.77}$$

Since $v \le 1$ we have, trivially

$$\sum_{y/m < t \le z/m} \frac{v^{\overline{\Omega}(t)}}{t} < \frac{m}{y} \left(\left[\frac{z}{m}\right] - \left[\frac{y}{m}\right] \right) < \theta + \frac{m}{y} \le 2\theta. \tag{7.78}$$

This is insufficient for the application and we apply a result of Shiu (1980), (*Divisors,* Theorem 03) (see Lemma 1.9) which yields

$$\sum_{y/m<t\le z/m} v^{\overline{\Omega}(t)} \ll_\kappa \frac{\theta y}{m}\left(\log\frac{y}{m}\right)^{v-1} \tag{7.79}$$

provided $\theta y/m \ge (z/m)^\kappa$ for some fixed $\kappa > 0$. We put $\kappa = \frac{1}{2}$, and the constraint reduces to $y/m \ge \theta^{-2}(1+\theta)$. We roll (7.78) and (7.79) into a universal upper bound, viz.

$$\sum_{y/m<t\le z/m} \frac{v^{\overline{\Omega}(t)}}{t} \ll \theta\left(\log\frac{1}{\theta}\right)^{1-v}\left(\log\frac{y}{m}\right)^{v-1} \tag{7.80}$$

whence

$$T(v) \ll x\log^{-2\beta} y(\log\log y)^{2-2v}\sum_{m\le\theta y}\frac{v^{\overline{\Omega}(m)}}{m}(\log 2m)^{v-1}\left(\log\frac{y}{m}\right)^{2v-2}. \tag{7.81}$$

The inner sum on the right is

$$\begin{aligned}
&\ll \frac{1}{y}\sum_{m\le\theta y} v^{\overline{\Omega}(m)}(\log 2m)^{v-1}\sum_{h\le\frac{y}{m}}(\log 2h)^{2v-2}\\
&\ll \frac{1}{y}\sum_{h\le y}(\log 2h)^{2v-2}\sum_{m\le y/h} v^{\overline{\Omega}(m)}(\log 2m)^{v-1}\\
&\ll \sum_{h\le y}\frac{1}{h}(\log 2h)^{2v-2}\left(\log\frac{2y}{h}\right)^{2v-2}\\
&\ll (\log y)^{v-1+(2v-1)^+}(\log\log y)^B
\end{aligned} \tag{7.82}$$

where $B = 0$ unless $v = \frac{1}{2}$ when $B = 1$. We insert this into (7.81) to obtain the result stated.

Theorem 7.9 *Let*

$$z = y + y\log^{-\beta} y, \quad \beta \ge 0 \tag{7.83}$$

and

$$\beta = \log 4 - 1 + \frac{\xi}{\sqrt{\log\log y}}. \tag{7.84}$$

Then for each fixed c *we have, uniformly for* $\xi \ge -c(\log\log y)^{1/6}$,

$$\mathbf{d}\mathcal{M}(\mathcal{A}(y,z)) = (F(\xi) + E(\xi, y))\log^{-\beta} y \tag{7.85}$$

where

$$F(\xi) = \frac{1}{\sqrt{\pi}} \int_{-\infty}^{\xi/\log 4} e^{-u^2}\, du \tag{7.86}$$

and

$$E(\xi, y) \ll_c \frac{\xi^2 + \log\log\log y}{\sqrt{\log\log y}} e^{-\xi^2/\log^2 4} \tag{7.87}$$

if $\xi \le 0$; *otherwise*

$$E(\xi, y) \ll \frac{\xi + \log\log\log y}{\sqrt{\log\log y}}. \tag{7.88}$$

Moreover, provided $x > \exp(\log z . \log\log z)$, (7.85) *remains valid if we substitute* $x^{-1}H(x,y,z)$ *on the left-hand side.*

Thus (7.18) and (7.19) hold as conjectured. The central problem which remains is to extend the range of negative ξ for which we have an asymptotic formula for $\mathbf{d}\mathcal{M}(\mathcal{A}(y,z))$. There are interesting questions still to be answered for positive ξ. For $\xi \ge 0$, we have

$$F(\xi) = 1 + O\left((1+\xi)^{-1} e^{-\xi^2/\log^2 4}\right) \tag{7.89}$$

and so when ξ exceeds about $\sqrt{\log\log\log y}$ there is no point in writing $F(\xi)$ instead of 1 on the right-hand side of (7.85). As ξ increases further the error term (7.88) becomes unsatisfactory, requiring $\xi = o(\sqrt{\log\log y})$ for (7.85) to be non-trivial. Of course we know from *Divisors*, Theorem 21 that provided $\xi \to \infty$ as $y \to \infty$, we have

$$\mathbf{d}\mathcal{M}(\mathcal{A}(y,z)) \sim \log^{-\beta} y \tag{7.90}$$

and it is reasonable to ask for a uniform error term in this formula, valid for large positive ξ. An easy argument (*Divisors*, p.38) gives

$$\mathbf{d}\mathcal{M}(\mathcal{A}(y,z)) = \left(1 + O(\log^{1-\beta} y)\right) \log^{-\beta} y \tag{7.91}$$

and we bridge the gap between this result and Theorem 7.9 in Theorem 7.10 below.

Proof of Theorem 7.9 We prove the formula for $H(x,y,z)$, from which (7.85) follows. We assume $x > \exp(\log z . \log\log z)$.

We begin by considering the second sum on the right of (7.29). We have, for $v \le 1$,

$$v^h \sum_{k<h} S_k \le T(v) \tag{7.92}$$

with $T(v)$ as in (7.73). We put $v = \frac{1}{2}$ and $h = k_0 - g$. By (7.23), $2^{k_0} \ll (\log y)^{1+\beta}$ and so (7.92) and Lemma 7.8 give

$$\sum_{k<h} S_k \ll 2^{-g} x (\log y)^{-\beta} (\log\log y)^2. \tag{7.93}$$

We put $g = 4\log\log\log y$ if $\xi \geq 0$, $g = 4\log\log\log y + \xi^2$ if $\xi < 0$: in each case the sum (7.93) is absorbed by the error term implicit in (7.85). (There is no point at this stage in ducking under the error term in (7.31) by making g bigger.) Next we consider the first sum on the right of (7.29), and we write

$$\sum_{k<h} R_k = \sum_{k<k_0} R_k - \sum_{h\leq k<k_0} R_k. \tag{7.94}$$

We require an upper bound for the second sum here and we employ (7.59). It is familiar that for all t, $k > 0$ we have, uniformly, $t^k/k! \ll e^t/\sqrt{t}$ whence (7.59) yields, (we may assume $k_0 \leq \frac{5}{2}\log\log z$),

$$R_k \ll \frac{x \log^{-\beta} y}{\sqrt{\log\log y}}. \tag{7.95}$$

We apply this when $\xi \geq 0$ to obtain

$$\sum_{h\leq k<k_0} R_k \ll \frac{\log\log\log y}{\sqrt{\log\log y}} x \log^{-\beta} y \tag{7.96}$$

which is absorbed by the error term in the theorem. When $\xi < 0$ we observe that $k_0 \leq 2\log\log z$, so that (7.59) gives

$$\begin{aligned} \sum_{h\leq k<k_0} R_k &\ll gx(\log y)^{-2-\beta} \frac{(2\log\log z)^{k_0}}{k_0!} \\ &\ll \frac{gx}{\sqrt{\log\log y}} (\log y)^{-Q((1+\beta)/\log 2)}, \end{aligned} \tag{7.97}$$

by Stirling's formula and the definition of Q, that is $Q(t) = t\log t - t + 1$, $(t > 0)$. The exponent of $\log y$ above is $-G(\beta)$, as in (7.15), and we see that the sum (7.97) is again absorbed by the error term in the theorem. We deduce from (7.28), (7.29), (7.96), (7.97) and Theorem 7.1 that

$$H_1(x,y,z) = (F(\xi) + E(\xi,y))\, x \log^{-\beta} y, \tag{7.98}$$

with $E(\xi,y)$ as in the statement of the theorem. We notice that Theorem 7.1 requires $|\xi| \leq c(\log\log y)^{\frac{1}{6}}$ whereas our present hypothesis is $\xi \geq -c(\log\log y)^{1/6}$. The reason for this is that Theorem 7.10 (below) already gives a stronger result than Theorem 7.9 when $\xi \geq 3\sqrt{\log\log\log y}$.

Thus when dealing with positive ξ in this proof we may assume that $\xi \le (\log\log y)^{1/6}$ throughout; hence the error term in (7.88) does not depend on c.

Let $H_2(x, y, z)$ be the contribution to $H(x, y, z)$ from the integers n for which $\overline{\Omega}(n, z) \ge k_0$. By (7.25) we have

$$H_2(x, y, z) \le \sum_{k \ge k_0} \overline{N}_k(x, z) \ll \frac{x}{\sqrt{\log\log y}} (\log y)^{-Q((1+\beta)/\log 2)}. \qquad (7.99)$$

When $\xi < 0$ this is negligible in comparison with the right-hand side of (7.97) and is therefore absorbed. When $\xi \ge 0$ we have only to observe that the exponent of $\log y$ in (7.99) is $\le -\beta$. We assemble (7.98) and (7.99) to obtain our result.

The reader will observe that we could obtain a better result for positive ξ, (we should need g larger) if the error term in Theorem 7.1 were improved.

We conclude with a formula for $H(x, y, z)$ which is more precise than Theorem 7.9 for large positive ξ.

Theorem 7.10 *Let $x \ge z^2$ and $\xi \ge 0$. Let b be fixed and t be determined from*

$$t \ge 0, \quad 2Q(1 + t) + t \log 4 = \frac{\xi}{\sqrt{\log\log y}} \qquad (7.100)$$

and put $A = 2Q(1 + t)$. Then for $\xi \le b\sqrt{\log\log y}$,

$$H(x, y, z) = \left(1 + O_b\left((\log y)^{-A}(\log\log y)^2\right)\right) x \log^{-\beta} y. \qquad (7.101)$$

Put $X = \xi/\sqrt{\log\log y}$ and $A = A(X)$. Then $A(X)$ is an increasing function and $A(X) \sim X$ as $X \to \infty$. We show later that

$$\frac{X^2}{\log^2 4} - \frac{2}{3}X^3 \le A(X) \le \frac{X^2}{\log^2 4}, \quad X \ge 0 \qquad (7.102)$$

so that (7.101) is more precise than the result given by Theorem 7.9 when $\xi \ge 3\sqrt{\log\log\log y}$. (Of course for fixed ξ it is worse than trivial because of the factor $(\log\log y)^2$.)

Proof We have

$$H(x, y, z) \le \sum_{n \le x} \tau(n; y, z) \qquad (7.103)$$

and this is sufficient as an upper bound. For k_1 at our disposal,

$$H(x, y, z) \ge \sum_{k \le k_1} (R_k - S_k) \qquad (7.104)$$

where R_k and S_k are as in (7.26) and (7.27). Hence

$$H(x,y,z) \geq \sum_{n \leq x} \tau(n;y,z) - \sum_{k>k_1} R_k - \sum_{k \leq k_1} S_k. \tag{7.105}$$

We have, uniformly for $w \leq w_0$ for each fixed w_0,

$$\sum_{n \leq x} \tau(n;y,z) w^{\overline{\Omega}(n,z)} \ll x(\log y)^{2w-2-\beta} \tag{7.106}$$

(we leave this as an exercise) whence, provided $w \geq 1$,

$$\sum_{k>k_1} R_k \ll w^{-k_1} x(\log y)^{2w-2-\beta}. \tag{7.107}$$

We put $k_1 = \kappa \log\log y$ and minimize the exponent $2w - 2 - \beta - \kappa \log w$ of $\log y$ on the right by choosing $w = \kappa/2$: we require $\kappa \geq 2$. This gives

$$\sum_{k>k_1} R_k \ll x(\log y)^{-\beta-2Q(\kappa/2)}. \tag{7.108}$$

We apply Lemma 7.8 with $v = \frac{1}{2}$ to obtain

$$\sum_{k \leq k_1} S_k \ll x(\log y)^{-2\beta-1+\kappa\log 2}(\log\log y)^2 \tag{7.109}$$

and we put $\kappa = 2 + 2t$, where $t \geq 0$. The exponents of $\log y$ in (7.108) and (7.109) are equal when (7.100) holds, to $-\beta - A$, and this gives the result stated. The constant implied by Vinogradov's notation is uniform provided $w \leq w_0$ in (7.106) and this is equivalent to $\xi \leq b\sqrt{\log\log y}$, that is $\beta \leq \log 4 - 1 + b$, where $w_0 = w_0(b)$.

Notice that if $\xi \leq c(\log\log y)^{1/6}$ then (7.102) implies

$$(\log y)^{-A} \ll_c e^{-\xi^2/\log^2 4}. \tag{7.110}$$

Moreover in any case

$$(\log y)^{-A} \geq e^{-\xi^2/\log^2 4} \gg 1 - F(\xi) \tag{7.111}$$

by (7.89). We could therefore write $F(\xi)$ instead of 1 in (7.101) if we wished. It would be desirable, and this is a more delicate question, to obtain a result in which $F(\xi)$ appeared as the main term with error $o(1 - F(\xi))$.

It remains to prove (7.102). Let $A(X)$ be defined as above, with X instead of $\xi/\sqrt{\log\log y}$ on the right of (7.100). Then X and t increase from 0 to ∞ together, and

$$\dot{X} = \frac{dX}{dt} = 2\log(1+t) + \log 4 \geq \log 4. \tag{7.112}$$

The function $A(X) \in C^\infty(\mathbf{R}^+)$, and we have

$$A(X) + t \log 4 = X. \tag{7.113}$$

From (7.100), $t \sim X(2 \log X)$ as $X \to \infty$ whence $A(X) \sim X$ as $X \to \infty$ from (7.113). Notice that for X sufficiently large, the error term in (7.101) exceeds that in (7.91). This occurs when $A(X) < X + \log 4 - 2$, that is when $t > (\log 2)^{-1} - 1$ or $X > .78584\ldots$, which is therefore a suitable value of b in the statement of the theorem.

By Taylor's theorem, we have

$$A(X) = A(0) + XA'(0) + \frac{X^2}{2!}A''(0) + \frac{X^3}{3!}A'''(\theta X) \tag{7.114}$$

for some $\theta \in (0, 1)$. We compute the derivatives from (7.113) using the relations

$$\begin{aligned} t' &= \frac{dt}{dX} = \dot{X}^{-1}, \\ t'' &= -\ddot{X}\dot{X}^{-3}, \\ t''' &= -\dddot{X}\dot{X}^{-4} + 3\ddot{X}^2\dot{X}^{-5} = \frac{2}{(1+t)^2}\{\dot{X}^{-4} + 6\dot{X}^{-5}\}. \end{aligned} \tag{7.115}$$

Since $\dot{X}$ increases with t and X, t''' plainly decreases whence $0 < t''' \leq t'''(0)$, and $-t'''(0) \log 4 \leq A'''(\theta X) \leq 0$. We have $t'''(0) \log 4 = 3.9997\ldots < 4$, and (7.102) follows.

Bibliography

L. Bastick (1992) D.Phil Thesis, York University, U.K.

F.A. Behrend (1932–3) Über numeri abundantes I, II. *Sitzungsberichte der Preuß. Akad. phys-math. Klasse.* 322–328, 280–293.

F.A. Behrend (1935) On sequences of integers not divisible one by another. *J. London Math. Soc.* 10, 42–4.

F.A. Behrend (1948) Generalization of an inequality of Heilbronn and Rohrbach. *B. Amer. Math. Soc.* 54, 681–4.

A.S. Besicovitch (1934) On the density of certain sequences. *Math. Annalen.* 110, 336–41.

K. Bognár (1970) On a problem of statistical group theory. *Studia Sci. Math. Hung.* 5, 29–36.

J.D. Bovey (1977) On the size of prime factors of integers. *Acta Arith.* 33, 65–80.

S. Chowla (1934) On abundant numbers. *J. Indian Math. Soc.* (2) 1, 41–4.

H. Davenport (1933) Über numeri abundantes. *Sitzungsberichte Akad. Wiss. Berlin.* 27, 830–7.

H. Davenport & P. Erdös (1937) On sequences of positive integers. *Acta Arith.* 2, 147–51.

H. Davenport & P. Erdös (1951) On sequences of positive integers. *J. Indian Math. Soc.* (2) 15, 19–24.

L.E. Dickson (1931) Even abundant numbers. *Amer. J. Math.* 35, 413–426.

Y. Dupain, R.R. Hall and G. Tenenbaum (1982) Sur l'équirepartition modulo 1 de certaines fonctions de diviseurs. *J. London Math. Soc.* (2) 26, 397–411.

P.D.T.A. Elliott (1979–80) *Probabilistic number theory.* I,II. Springer-Verlag.

P. Erdös (1934) On the density of the abundant numbers. *J. London Math. Soc.* 9, 278–82.

P. Erdös (1935a) On primitive abundant numbers. *J. London Math. Soc.* 10, 49–58.

P. Erdös (1935b) Note on sequences of integers no one of which is divisible by any other. *J. London Math. Soc.* 10, 126–8.

P. Erdös (1936) A generalization of a theorem of Besicovitch. *J. London Math. Soc.* 11, 92–8.

P. Erdös (1946) On the coefficients of the cyclotomic polynomial. *B. Amer. Math. Soc.* 52, 179–84.

P. Erdös (1948a) On the density of some sequences of integers. *B. Amer. Math. Soc.* 54, 685–92.

P. Erdös (1948b) Integers with exactly *k* prime factors. *Ann. Math.* II 49, 53–66.

P. Erdös (1959) Some remarks on prime factors of integers. *Can. J. Math.* 11, 161–7.

P. Erdös (1965) On the distribution of divisors of integers in the residue classes (mod *d*). *B. Math. Soc. Grèce.* 6, 27–36

P. Erdös (1969) On the distribution of prime divisors. *Aequationes Math.* 2, 177–83.

P. Erdös (1979) Some unconventional problems in number theory. *Astérisque.* 61, 73–82.

P. Erdös & R.R. Hall (1974) Some distribution problems concerning the divisors of integers. *Acta Arith.* 26, 175–188.

P. Erdös & R.R. Hall (1976a) Proof of a conjecture about the distribution of divisors of integers in residue classes. *Math. Proc. Camb. Phil. Soc.* 79, 281–9.

P. Erdös & R.R. Hall (1976b) Probabilistic methods in group theory II. *Houston J. Math.* 2, 173–180.

P. Erdös & R.R. Hall (1978) Some new results in probabilistic group theory. *Comment. Math. Helvetici.* 53, 448–457.

P. Erdös & R.R. Hall (1980) On the Möbius function. *J. Reine Angew. Math.* 315, 121–6.

P. Erdös, R.R. Hall. & G. Tenenbaum (1994) On the densities of sets of multiples. *J. Reine Angew. Math.* 454, 119–141.

P. Erdös & M. Kac (1939) On the Gaussian law of errors in the theory of additive functions. *Proc. Nat. Acad. Sci. U.S.A.* 25, 206–7.

P. Erdös & M. Kac (1940) The Gaussian law of errors in the theory of additive number-theoretic functions. *Amer. J. Math.* 62, 738–42.

P. Erdös & A. Rényi (1965) Probabilistic methods in group theory. *J. d'Analyse Math.* 14, 127–138.

P. Erdös, A. Sárközy & E. Szemerédi (1967a) On a theorem of Behrend. *J. Austr. Math. Soc.* 7, 9–16.

P. Erdös, A. Sárközy & E. Szemerédi (1967b) On an extremal problem concerning primitive sequences. *J. London Math. Soc.* 42, 484–8.

P. Erdös & P. Turán (1948) On a problem in the theory of uniform distribution I, II. *Indag. Math.* 10, 370–8, 406–13.

J. Franel (1924) Les suites de Farey et le probléme des nombres premiers. *Göttinger Nachr.* 198–201.

G. Freud (1952–4) Restglied eines Tauberschen Satzes I, II, III. *Acta Math. Acad. Sci. Hungar.* 2, 299–308, 299–307, 275–89.

J. Friedlander & A. Granville (1989) Limitations to the equi-distribution of primes I. *Ann. Math.* 129, 363–382.

J. Friedlander, A. Granville, A. Hildebrand & H. Maier (1991) Oscillation theorems for primes in arithmetic progressions and for sifting functions. *J. Amer. Math. Soc.* 4, 25–86.

G.R.H. Greaves, R.R. Hall, M.N. Huxley & J.C. Wilson (1993) Multiple Franel Integrals. *Mathematika.* 40, 50–69.

H. Halberstam & H-E. Richert (1974) *Sieve methods.* Academic Press, New York.

H. Halberstam & H-E. Richert (1979) On a result of R.R. Hall. *J. Number Theory.* 11, 76–89.

H. Halberstam & K.F. Roth (1966) *Sequences.* Oxford University Press.

R.R. Hall (1972) On a theorem of Erdös and Rényi concerning Abelian groups. *J. London Math. Soc.* (2) 5, 143–53.

R.R. Hall (1974a) Halving an estimate obtained from Selberg's upper bound method. *Acta Arith.* 25, 247–351.

R.R. Hall (1974b) The divisors of integers I. *Acta Arith.* 26, 41–46.

R.R. Hall (1975a) The divisors of integers II. *Acta Arith.* 28, 129–35.

R.R. Hall (1975b) Sums of imaginary powers of the divisors of integers. *J. London Math. Soc.* 9, 571–80.

R.R. Hall (1976) The distribution of $f(d)$ (mod 1). *Acta Arith.* 31, 91–97.

R.R. Hall (1977) Extensions of a theorem of Erdös-=Rényi in probabilistic group theory. *Houston J. Math.* 3, 225–234.

R.R. Hall (1978) A new definition of the density of integer sequences. *J. Austr. Math. Soc.* A. 26, 487–500.

R.R. Hall (1981) The divisor density of integer sequences. *J. London Math. Soc.* (2) 24, 41–53.

R.R. Hall (1989) The distribution of square-free numbers. *J. Reine Angew. Math.* 394, 107–17.

R.R. Hall (1990a) Large irregularities in sets of multiples and sieves. *Mathematika.* 37, 119–135.

R.R. Hall (1990b) Sets of multiples and Behrend sequences, in *A tribute to Paul Erdös.* (ed. A. Baker, B. Bollobás & A. Hajnal.) Cambridge University Press.

R.R. Hall (1992) On some conjectures of Erdös in *Astérisque* I. *J. Number Theory.* 42, 313–319.

R.R. Hall (1994) Ω theorems for the complex divisor function. *Math. Proc. Camb. Phil. Soc.* 115, 145–57.

R.R. Hall & A. Sudbery (1972) On a conjecture of Erdös and Rényi concerning Abelian groups. *J. London Math. Soc.* (2) 6, 177–89.

R.R. Hall & G. Tenenbaum (1986) Les ensembles de multiples et la densité divisorielle. *J. Number Theory.* 22, 308–33.

R.R. Hall & G. Tenenbaum (1988) *Divisors.* Cambridge University Press.

R.R. Hall & G. Tenenbaum (1989) The set of multiples of a short interval, in *Number Theory New York Seminar.* (ed. D.V. Chudnovsky, G.V. Chudnovsky, H. Cohn & M.B. Nathanson.) Springer-Verlag.

R.R. Hall & G. Tenenbaum (1992) On Behrend sequences. *Math. Proc. Camb. Phil. Soc.* 112, 467–82.

G.H. Hardy (1915) The mean value of the modulus of an analytic function. *P. London Math. Soc.* (2) 14, 269–77.

G.H. Hardy (1949) *Divergent series.* Oxford University Press.

G.H. Hardy, J.E. Littlewood & G. Pólya (1934) *Inequalities.* Cambridge University Press.

G.H. Hardy & S. Ramanujan (1917) The normal number of prime factors of a number *n*. *Quart. J. Math.* 48, 76–92.

H. Heilbronn (1937) On an inequality in the elementary theory of numbers. *Proc. Camb. Phil. Soc.* 33, 207–9.

A. Hildebrand & H. Maier (1989) Irregularities in the distribution of primes in short intervals. *J. Reine Angew. Math.* 397, 162–93.

J. Karamata (1931) Neuer Beweis und Verallgemeinerung der Tauberschen Sätze, welche die Laplaceshe und Stieltjesche Transformation betreffen. *J. Reine Angew. Math.* 164, 27–40.

I. Kátai (1976a) The distribution of divisors (mod 1). *Acta Math. Acad. Sci. Hung.* 27, 149–52.

I. Kátai (1976b) Distribution mod 1 of additive functions on the set of divisors. *Acta Arith.* 30, 2, 9–12.

J.C. Kluyver (1903) An analytical expression for the greatest common divisor of two integers. *Proc. Roy. Acad. Amsterdam.* 5, 658–62.

J. Kubilius (1964) *Probabilistic methods in the theory of numbers.* Amer. Math. Soc. Monographs. No. 11 Providence R.I.

L. Kuipers & H. Niederreiter (1974) *Uniform distribution of sequences.* Wiley, New York.

E. Landau (1927) *Vorlesungen über Zahlentheorie.* S. Hirzel. Leipzig.

H. Maier (1985) Primes in short intervals. *Michigan Math. J.* 32, 221–25.

H. Maier & G. Tenenbaum (1984) On the set of divisors of an integer. *Invent. Math.* 76, 121–8.

R.J. Miech (1967) On a conjecture of Erdös and Rényi. *Illinois J. Math.* 11, 114–27.

K.K. Norton (1978) Estimates for partial sums of the exponential series. *J. Math. & Appl.* 63, 265–96.

A. Perelli & U. Zannier (1989) An extremal property of the Möbius function. *Arch. Math.* 53, 20–29.

S. Pillai (1939) On numbers which are not multiples of any other in the set. *Proc. Indian Acad. Sci.* A 10, 392–4.

K. Prachar (1957) *Primzahlverteilung.* Springer-Verlag.

H-E. Richert (1967) Zur Abschätzung der Riemannschen Zeta-funktion in der Nähe der Vertikalen $\sigma = 1$. *Math. Annalen.* 169, 97–101

H. Rohrbach (1937) Beweis einer zahlentheoretischen Ungleichung. *J. Reine Angew. Math.* 177, 193–6.

I.Z. Ruzsa (1976) Probabilistic generalization of a number theoretic inequality. *Amer. Math. Monthly.* 83, 723–5.

I.Z. Ruzsa (1988) On an additive property of squares and primes. *Acta Arith.* 49, 281–89.

I.Z. Ruzsa & G. Tenenbaum (199x) A note on Behrend sequences. *Acta Math. Hung.* (to appear).

I. Schoenberg (1928) Über die asymptotische Verteilung reeller Zahlen mod 1. *Math. Zeitschrift.* 28, 171–99.

P. Shiu (1980) A Brun–Titchmarsh theorem for multiplicative functions. *J. Reine Angew. Math.* 313, 161–70.

G. Tenenbaum (1979) Lois de répartition des diviseurs 4. *Ann. Inst. Fourier.* 29, 1–15.

G. Tenenbaum (1980) Lois de répartition des diviseurs 2. *Acta Arith.* 38, 1–36.

G. Tenenbaum (1982) Sur la densité divisorielle d'une suite d'entiers. *J. Number Theory.* 15, 331–46.

G. Tenenbaum (1984) Sur la probabilité qu'un entier possède un diviseur dans un intervalle donné. *Composito Math.* 51, 243–63.

G. Tenenbaum (1988) Un problème de probabilité conditionelle en Arithmétique, *Acta Arith.*49, 165–87.

G. Tenenbaum (1990) *Introduction à la theorie analytique et probabiliste des nombres*, Institut Elie Cartan **13**, Nancy. English translation: *Introduction to analytic and probabilistic number theory*, Cambridge University Press, 1995.

G. Tenenbaum (199x) On block Behrend sequences. *Math. Proc. Cam. Phil. Soc.* (to appear).

G. Tenenbaum (199y) Uniform distribution on divisors and Behrend sequences. *l'Enseignement Mathématique* (to appear).

E.C. Titchmarsh (1939) *The theory of functions.* Oxford University Press.

E.C. Titchmarsh (1951) *The theory of the Riemann zeta-function.* Oxford University Press.

J.H. van Lint & H-E. Richert (1965) On primes in arithmetic progressions. *Acta Arith.* 11, 209–216.

I.M. Vinogradov (1958) A new estimate for $\zeta(1+it)$ (in Russian). *Izv. Akad. Nauk. S.S.S.R. Ser. Math.* 22, 161–164.

J.C. Wilson (1994) On Franel–Kluyver integrals of order three. *Acta Arith.* 66, 71–87.

D. Wolke (1971) Multiplikative Funktionen auf schnell wachsenden Folgen. *J. Reine Angew. Math.* 251, 54–67.

Index